GW00706052

HVAC
Field Manual

HVAC
Field Manual

ROBERT O. PARMLEY, P.E.
Editor-in-Chief

McGRAW-HILL BOOK COMPANY
New York   St. Louis   San Francisco   Auckland
Bogotá   Hamburg   London   Madrid   Mexico
Milan   Montreal   New Delhi   Panama
Paris   São Paulo   Singapore
Sydney   Tokyo   Toronto

Library of Congress Cataloging-in-Publication Data

HVAC field manual/Robert O. Parmley, editor-in-chief.
    p. cm.
  ISBN 0-07-048524-0
  1. Heating—Handbooks, manuals, etc.   2. Ventilation—Handbooks, manuals, etc.   3. Air conditioning—Handbooks, manuals,
etc.   I. Parmley, Robert O.
TH7225.H85 1988   88-3532
697′.00212—dc19   CIP

1234567890   DOC/DOC   8921098

ISBN 0-07-048524-0

The editors for this book were Betty Sun and Lester Strong, and the
production supervisor was Suzanne Babeuf. It was set in Times Roman by Techna Type.

Printed and bound by R. R. Donnelley & Sons Company.

Dedicated to the Memory of Geo. H. Morgan, P.E.

CONTENTS

Section 4 Basic Theory and Fundamentals   4.1

Section 12   Air Duct Design

PREFACE

The concept of a series of small, pocket-sized field manuals, tailored to specific engineering disciplines, was initiated by McGraw-Hill several years ago. The first publication using this format was titled *Field Engineer's Manual*. Since that time, McGraw-Hill has published several additional pocket-sized manuals on a number of engineering topics.

In keeping with the innovative design, this manual has been developed using the same style and format as its companion forerunners. Basic heating, ventilating, and air conditioning (HVAC) data for field personnel are presented in a compact volume featuring quick reference to give assistance when an all-inclusive library is not available.

HVAC technology is a very broad and fast-moving field. No attempt has been made in this manual to cover the entire discipline, especially solar and electrical systems. For obvious reasons, nuclear systems and their data are not included. It is the desire of the Editor and the Board of Consultants to provide a general overview and organize basic technical information to give ready support to personnel in the field concerning conventional technology. Therefore, a blank page has been provided at the end of each section so that users can personalize their manuals by incorporating supplemental technical data geared to each individual's specific area of practice.

As in previous field manuals, we assume that every user has at hand a pocket-sized electronic calculator (slide-rule type). Trigonometric and logarithm tables plus similar mathematical listings are not included in this publication because they are stored in your hand-held calculator. This has provided space to allow inclusion of additional data.

At the beginning of this effort, we were overwhelmed by the voluminous technical data available on HVAC. It has been a lengthy and at times a frustrating task to condense this material and select information to include in this manual. Hopefully, we have been wise in our editorship and produced a practical collection of basic material. Each section is devoted to a particular category of HVAC. A complete Table of Contents, coupled with an indepth Index, will allow the user to easily locate specific technical information.

Material contained in this manual is so extensive and has been obtained from so many varied sources that it is impossible to list all of the publications, professional organizations, technical societies, manufacturers, and consultants who contributed. However, where possible, we have noted the source and given credit on the pages where their respective material appears. A special thanks to Robert A. Parsons and his staff at ASHRAE and the staff of The Trane Company for the assistance we received. This manual is liberally sprinkled with their technical data and we acknowledge their generosity in granting permission for the use of that information in this publication.

Credit is also due to those unknown individuals who have made significant contributions to this field since time immemorial. Their practical insight and inventive characteristics have largely gone unsung. From the first crude attempts to warm the air in a confined space by burning forest products until the eighteenth century, little "scientific" research into the nature of thermodynamics was conducted because of widely held superstitions. Most knowledge was gained by the "trial and error" of craft workers. Toward the end of the 1700s, two scientists took a deeper look into heat. They were the American-born Benjamin Thompson and the British chemist Sir Humphry Davy, who raised doubts about the age-old caloric theory. Their observations triggered research resulting in the idea that heat is a form of energy. During the mid-1800s, proof of this fact was developed by three men—James Joule, a British physicist, and Julius Robert von Mayer and Hermann von Helmholtz, both German physicists. In 1748, Dr. William Cullen at the University of Glasgow first demonstrated artificial refrigeration by evaporating ether in a partial vacuum. This work led to further research into chemical and mechanical methods for generating low temperatures. Closed-cycle compression refrigeration was proposed in the early 1800s. A chain of well-documented events followed at an ever increasing rate, resulting in practical air conditioning by the 1920s. Home air conditioning did not become cost-effective until the 1950s.

Parallel with the development of air conditioning, heating and ventilating technology experienced many innovations too. A wide range of heating fuels was harnessed into service for humanity. Fuel-efficient furnaces were developed to combat the energy crunch of the mid-1970s. Effluent restrictions on smoke stacks instituted by regulatory agencies and code requirements for ventilation refined in recent decades have made everyone more aware of how important HVAC technology has become in this modern world.

In summary, I want to thank all who have so generously participated in the preparation of this manual. A special thank you to John A. Spalding, A.I.A., for his help at the initial stage; to Wayne for his illustrating contributions; and, of course, to Ethne for her untiring efforts in typing correspondence and the final manuscript.

ROBERT O. PARMLEY, P.E.
Editor-in-Chief

GENERAL MATHEMATICAL DATA

TABLE 1.1   Functions of the Numbers 1 to 99

No	Square	Cube	Square Root	Cubic Root	Logarithm [1]	No = Diameter	
						Circum	Area
1	1	1	1 0000	1 0000	0 00000	3 142	0 7854
2	4	8	1 4142	1 2599	0 30103	6 283	3 1416
3	9	27	1 7321	1 4422	0 47712	9 425	7 0686
4	16	64	2 0000	1 5874	0 60206	12 566	12 5664
5	25	125	2 2361	1 7100	0 69897	15 708	19 6350
6	36	216	2 4495	1.8171	0 77815	18 850	28 2743
7	49	343	2 6458	1 9129	0 84510	21 991	38 4845
8	64	512	2 8284	2 0000	0 90309	25 133	50 2655
9	81	729	3 0000	2 0801	0 95424	28 274	63 6173
10	100	1000	3 1623	2 1544	1 00000	31 416	78 5398
11	121	1331	3 3166	2 2240	1 04139	34 558	95 0332
12	144	1728	3 4641	2 2894	1 07918	37 699	113 097
13	169	2197	3 6056	2 3513	1 11394	40 841	132 732
14	196	2744	3 7417	2 4101	1 14613	43 982	153 938
15	225	3375	3 8730	2 4662	1 17609	47 124	176 715
16	256	4096	4 0000	2 5198	1 20412	50 265	201 062
17	289	4913	4 1231	2 5713	1 23045	53 407	226 980
18	324	5832	4 2426	2 6207	1 25527	56 549	254 469
19	361	6859	4 3589	2 6684	1 27875	59 690	283 529
20	400	8000	4 4721	2 7144	1 30103	62 832	314 159
21	441	9261	4 5826	2 7589	1 32222	65 973	346 361
22	484	10648	4 6904	2 8020	1 34242	69 115	380 133
23	529	12167	4 7958	2 8439	1 36173	72 257	415 476
24	576	13824	4 8990	2 8845	1 38021	75 398	452 389
25	625	15625	5 0000	2 9240	1 39794	78 540	490 874
26	676	17576	5 0990	2 9625	1 41497	81 681	530 929
27	729	19683	5 1962	3 0000	1 43136	84 823	572 555
28	784	21952	5 2915	3 0366	1 44716	87 965	615 752
29	841	24389	5 3852	3 0723	1 46240	91 106	660 520
30	900	27000	5 4772	3 1072	1 47712	94 248	706 858
31	961	29791	5 5678	3 1414	1 49136	97 389	754 768
32	1024	32768	5 6569	3 1748	1 50515	100 531	804 248
33	1089	35937	5 7446	3 2075	1 51851	103 673	855 299
34	1156	39304	5 8310	3 2396	1 53148	106 814	907 920
35	1225	42875	5 9161	3 2711	1 54407	109 956	962 113
36	1296	46656	6 0000	3 3019	1 55630	113 097	1017 88
37	1369	50653	6 0828	3 3322	1 56820	116 239	1075 21
38	1444	54872	6 1644	3 3620	1 57978	119 381	1134 11
39	1521	59319	6 2450	3 3912	1 59106	122 522	1194 59
40	1600	64000	6 3246	3 4200	1 60206	125 66	1256 64
41	1681	68921	6 4031	3 4482	1 61278	128 81	1320 25
42	1764	74088	6 4807	3 4760	1 62325	131 95	1385 44
43	1849	79507	6 5574	3 5034	1 63347	135 09	1452 20
44	1936	85184	6 6332	3 5303	1 64345	138 23	1520 53
45	2025	91125	6 7082	3 5569	1 65321	141 37	1590 43
46	2116	97336	6 7823	3 5830	1 66276	144 51	1661 90
47	2209	103823	6 8557	3 6088	1 67210	147 65	1734 94
48	2304	110592	6 9282	3 6342	1 68124	150 80	1809 56
49	2401	117649	7 0000	3 6593	1 69020	153 94	1885 74

TABLE 1.1   Functions of the Numbers 1 to 99 (continued)

No	Square	Cube	Square Root	Cubic Root	Loga-rithm	Circum	Area
50	2500	125000	7 0711	3 6840	1 69897	157 08	1963 50
51	2601	132651	7 1414	3 7084	1 70757	160 22	2042 82
52	2704	140608	7 2111	3 7325	1 71600	163 36	2123 72
53	2809	148877	7 2801	3 7563	1 72428	166 50	2206 18
54	2916	157464	7 3485	3 7798	1 73239	169 65	2290 22
55	3025	166375	7 4162	3 8030	1 74036	172 79	2375 83
56	3136	175616	7 4838	3 8259	1 74819	175 93	2463 01
57	3249	185193	7 5498	3 8485	1 75587	179 07	2551 76
58	3364	195112	7 6158	3 8709	1 76343	182 21	2642 08
59	3481	205379	7 6811	3 8930	1 77085	185 35	2733 97
60	3600	216000	7 7460	3 9149	1 77815	188 50	2827 43
61	3721	226981	7 8102	3 9365	1 78533	191 64	2922 47
62	3844	238328	7 8740	3 9579	1 79239	194 78	3019 07
63	3969	250047	7 9373	3 9791	1 79934	197 92	3117 25
64	4096	262144	8 0000	4 0000	1 80618	201 06	3216 99
65	4225	274625	8 0623	4 0207	1 81291	204 20	3318 31
66	4356	287496	8 1240	4 0412	1 81954	207 35	3421 19
67	4489	300763	8 1854	4 0615	1 82607	210 49	3525 65
68	4624	314432	8 2462	4 0817	1 83251	213 63	3631 68
69	4761	328509	8 3066	4 1016	1 83885	216 77	3739 28
70	4900	343000	8 3666	4 1213	1 84510	219 91	3848 45
71	5041	357911	8 4261	4 1408	1 85126	223 05	3959 19
72	5184	373248	8 4853	4 1602	1 85733	226 19	4071 50
73	5329	389017	8 5440	4 1793	1 86332	229 34	4185 39
74	5476	405224	8 6023	4 1983	1 86923	232 48	4300 84
75	5625	421875	8 6603	4 2172	1 87506	235 62	4417 86
76	5776	438976	8 7178	4 2358	1 88081	238 76	4536 46
77	5929	456533	8 7750	4 2543	1 88649	241 90	4656 63
78	6084	474552	8 8318	4 2727	1 89209	245 04	4778 36
79	6241	493039	8 8882	4 2908	1 89763	248 19	4901 67
80	6400	512000	8 9443	4 3089	1 90309	251 33	5026 55
81	6561	531441	9 0000	4 3267	1 90849	254 47	5153 00
82	6724	551366	9 0554	4 3445	1 91381	257 61	5281 02
83	6889	571787	9 1104	4 3621	1 91908	260 75	5410 61
84	7056	592704	9 1652	4 3795	1 92428	263 89	5541 77
85	7225	614125	9 2195	4 3968	1 92942	267 04	5674 50
86	7396	636056	9 2736	4 4140	1 93450	270 18	5808 80
87	7569	658503	9 3274	4 4310	1 93952	273 32	5944 68
88	7744	681472	9 3808	4 4480	1 94448	276 46	6082 12
89	7921	704969	9 4340	4 4647	1 94939	279 60	6221 14
90	8100	729000	9 4868	4 4814	1 95424	282 74	6361 73
91	8281	753571	9 5394	4 4979	1 95904	285 88	6503 88
92	8464	778688	9 5917	4 5144	1 96379	289 03	6647 61
93	8649	804357	9 6437	4 5307	1 96848	292 17	6792 91
94	8836	830584	9 6954	4 5468	1 97313	295 31	6939 78
95	9025	857375	9 7468	4 5629	1 97772	298 45	7088 22
96	9216	884736	9 7980	4 5789	1 98227	301 59	7238 23
97	9409	912673	9 8489	4 5947	1 98677	304 73	7389 81
98	9604	941192	9 8995	4 6104	1 99123	307 88	7542 96
99	9801	970299	9 9499	4 6261	1 99564	311 02	7697 69

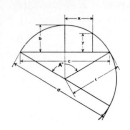

Circumference = $6.28318\ r = 3.14159\ d$
Diameter = 0.31831 circumference
Area = $3.14159\ r^2$

Arc $a = \dfrac{\pi r\,A°}{180°} = 0.017453\ r\,A°$

Angle $A° = \dfrac{180°\ a}{\pi r} = 57.29578\ \dfrac{a}{r}$

Radius $r = \dfrac{4\ b^2 + c^2}{8\ b}$

Chord $c = 2\sqrt{2\ br - b^2} = 2\ r\sin\dfrac{A}{2}$

Rise $b = r - \tfrac{1}{2}\sqrt{4\ r^2 - c^2} = \dfrac{c}{2}\tan\dfrac{A}{4}$

$\quad = 2\ r\sin^2\dfrac{A}{4} = r + y - \sqrt{r^2 - x^2}$

$y = b - r + \sqrt{r^2 - x^2}$

$x = \sqrt{r^2 - (r + y - b)^2}$

Diameter of circle of equal periphery as square = 1.27324 side of square
Side of square of equal periphery as circle = 0.78540 diameter of circle
Diameter of circle circumscribed about square = 1.41421 side of square
Side of square inscribed in circle = 0.70711 diameter of circle

---

CIRCULAR SECTOR

r = radius of circle y = angle ncp in degrees

Area of Sector ncpo = $\tfrac{1}{2}$ (length of arc nop $\times\ r$)

$\quad$ = Area of Circle $\times\ \dfrac{y}{360}$

$\quad$ = $0.0087266 \times r^2 \times y$

CIRCULAR SEGMENT

r = radius of circle x = chord b = rise

Area of Segment nop = Area of Sector ncpo − Area of triangle ncp

$\quad$ = $\dfrac{(\text{Length of arc nop} \times r) - x \times (r - b)}{2}$

Area of Segment nsp = Area of Circle − Area of Segment nop

VALUES FOR FUNCTIONS OF π

$\pi = 3.14159265359$,   log = 0.4971499

$\pi^2 = 9.8696044$, log = 0.9942997 $\dfrac{1}{\pi} = 0.3183099$, log = $\overline{1}.5028501$ $\sqrt{\dfrac{1}{\pi}} = 0.5641896$, log = $\overline{1}.7514251$

$\pi^3 = 31.0062767$, log = 1.4914496 $\dfrac{1}{\pi^2} = 0.1013212$, log = $\overline{1}.0057003$ $\dfrac{\pi}{180} = 0.0174533$, log = $\overline{2}.2418774$

$\sqrt{\pi} = 1.7724539$, log = 0.2485749 $\dfrac{1}{\pi^3} = 0.0322515$, log = $\overline{2}.5085504$ $\dfrac{180}{\pi} = 57.2957795$, log = 1.7581226

Note: Logs of fractions such as $\overline{1}.5028501$ and $\overline{2}.5085500$ may also be written 9.5028501 − 10 and 8.5085500 − 10 respectively.

FIG. 1.1   Properties of the circle. (*Courtesy American Institute of Steel Construction.*)

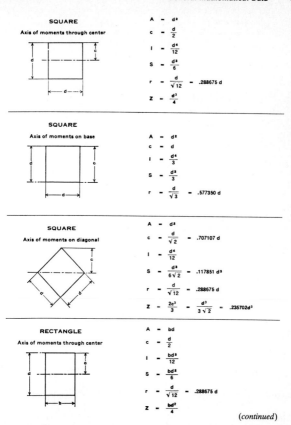

SQUARE

Axis of moments through center

$A = d^2$

$c = \dfrac{d}{2}$

$I = \dfrac{d^4}{12}$

$S = \dfrac{d^3}{6}$

$r = \dfrac{d}{\sqrt{12}} = .288675\, d$

$Z = \dfrac{d^3}{4}$

SQUARE

Axis of moments on base

$A = d^2$

$c = d$

$I = \dfrac{d^4}{3}$

$S = \dfrac{d^3}{3}$

$r = \dfrac{d}{\sqrt{3}} = .577350\, d$

SQUARE

Axis of moments on diagonal

$A = d^2$

$c = \dfrac{d}{\sqrt{2}} = .707107\, d$

$I = \dfrac{d^4}{12}$

$S = \dfrac{d^3}{6\sqrt{2}} = .117851\, d^3$

$r = \dfrac{d}{\sqrt{12}} = .288675\, d$

$Z = \dfrac{2c^3}{3} = \dfrac{d^3}{3\sqrt{2}} = .235702 d^3$

RECTANGLE

Axis of moments through center

$A = bd$

$c = \dfrac{d}{2}$

$I = \dfrac{bd^3}{12}$

$S = \dfrac{bd^2}{6}$

$r = \dfrac{d}{\sqrt{12}} = .288675\, d$

$Z = \dfrac{bd^2}{4}$

(*continued*)

FIG. 1.2 Properties of geometric sections. (*Courtesy American Institute of Steel Construction.*)

RECTANGLE

Axis of moments on base

$$A = bd$$

$$c = d$$

$$I = \frac{bd^3}{3}$$

$$S = \frac{bd^2}{3}$$

$$r = \frac{d}{\sqrt{3}} = .577350\,d$$

RECTANGLE

Axis of moments on diagonal

$$A = bd$$

$$c = \frac{bd}{\sqrt{b^2 + d^2}}$$

$$I = \frac{b^3 d^3}{6\,(b^2 + d^2)}$$

$$S = \frac{b^2 d^2}{6\sqrt{b^2 + d^2}}$$

$$r = \frac{bd}{\sqrt{6\,(b^2 + d^2)}}$$

RECTANGLE

Axis of moments any line
through center of gravity

$$A = bd$$

$$c = \frac{b \sin a + d \cos a}{2}$$

$$I = \frac{bd\,(b^2 \sin^2 a + d^2 \cos^2 a)}{12}$$

$$S = \frac{bd\,(b^2 \sin^2 a + d^2 \cos^2 a)}{6\,(b \sin a + d \cos a)}$$

$$r = \sqrt{\frac{b^2 \sin^2 a + d^2 \cos^2 a}{12}}$$

HOLLOW RECTANGLE

Axis of moments through center

$$A = bd - b_1 d_1$$

$$c = \frac{d}{2}$$

$$I = \frac{bd^3 - b_1 d_1^3}{12}$$

$$S = \frac{bd^3 - b_1 d_1^3}{6d}$$

$$r = \sqrt{\frac{bd^3 - b_1 d_1^3}{12\,A}}$$

$$Z = \frac{bd^2}{4} - \frac{b_1 d_1^2}{4}$$

FIG. 1.2—(*continued*)

EQUAL RECTANGLES

Axis of moments through center of gravity

$$A = b(d - d_1)$$

$$c = \frac{d}{2}$$

$$I = \frac{b(d^3 - d_1^3)}{12}$$

$$S = \frac{b(d^3 - d_1^3)}{6d}$$

$$r = \sqrt{\frac{d^3 - d_1^3}{12(d - d_1)}}$$

$$Z = \frac{b}{4}(d^2 - d_1^2)$$

UNEQUAL RECTANGLES

Axis of moments through center of gravity

$$A = bt + b_1 t_1$$

$$c = \frac{\frac{1}{2} bt^2 + b_1 t_1 (d - \frac{1}{2} t_1)}{A}$$

$$I = \frac{bt^3}{12} + bty^2 + \frac{b_1 t_1^3}{12} + b_1 t_1 y_1^2$$

$$S = \frac{I}{c} \qquad S_1 = \frac{I}{c_1}$$

$$r = \sqrt{\frac{I}{A}}$$

$$Z = \frac{A}{2}\left[d - \left(\frac{t + t_1}{2}\right)\right]$$

TRIANGLE

Axis of moments through center of gravity

$$A = \frac{bd}{2}$$

$$c = \frac{2d}{3}$$

$$I = \frac{bd^3}{36}$$

$$S = \frac{bd^2}{24}$$

$$r = \frac{d}{\sqrt{18}} = .235702\, d$$

TRIANGLE

Axis of moments on base

$$A = \frac{bd}{2}$$

$$c = d$$

$$I = \frac{bd^3}{12}$$

$$S = \frac{bd^2}{12}$$

$$r = \frac{d}{\sqrt{6}} = .408248\, d$$

(continued)

FIG. 1.2—*(continued)*

TRAPEZOID

Axis of moments through center of gravity

$$A = \frac{d(b + b_1)}{2}$$

$$c = \frac{d(2b + b_1)}{3(b + b_1)}$$

$$I = \frac{d^3 (b^2 + 4 bb_1 + b_1{}^2)}{36 (b + b_1)}$$

$$S = \frac{d^2 (b^2 + 4 bb_1 + b_1{}^2)}{12 (2b + b_1)}$$

$$r = \frac{d}{6(b + b_1)} \sqrt{2 (b^2 + 4 bb_1 + b_1{}^2)}$$

CIRCLE

Axis of moments through center

$$A = \frac{\pi d^2}{4} = \pi R^2 = .785398 \; d^2 = 3.141593 \; R^2$$

$$c = \frac{d}{2} = R$$

$$I = \frac{\pi d^4}{64} = \frac{\pi R^4}{4} = .049087 \; d^4 = .785398 \; R^4$$

$$S = \frac{\pi d^3}{32} = \frac{\pi R^3}{4} = .098175 \; d^3 = .785398 \; R^3$$

$$r = \frac{d}{4} = \frac{R}{2}$$

$$Z = \frac{d^3}{6}$$

HOLLOW CIRCLE

Axis of moments through center

$$A = \frac{\pi (d^2 - d_1{}^2)}{4} = .785398 \; (d^2 - d_1{}^2)$$

$$c = \frac{d}{2}$$

$$I = \frac{\pi (d^4 - d_1{}^4)}{64} = .049087 \; (d^4 - d_1{}^4)$$

$$S = \frac{\pi (d^4 - d_1{}^4)}{32d} = .098175 \; \frac{d^4 - d_1{}^4}{d}$$

$$r = \frac{\sqrt{d^2 + d_1{}^2}}{4}$$

$$Z = \frac{d^3}{6} - \frac{d_1{}^3}{6}$$

HALF CIRCLE

Axis of moments through center of gravity

$$A = \frac{\pi R^2}{2} \qquad\qquad = 1.570796 \; R^2$$

$$c = R \left(1 - \frac{4}{3\pi} \right) = .575587 \; R$$

$$I = R^4 \left(\frac{\pi}{8} - \frac{8}{9\pi} \right) = .109757 \; R^4$$

$$S = \frac{R^3}{24} \frac{(9\pi^2 - 64)}{(3\pi - 4)} = .190687 \; R^3$$

$$r = R \frac{\sqrt{9\pi^2 - 64}}{6\pi} = .264336 \; R$$

FIG. 1.2—*(continued)*

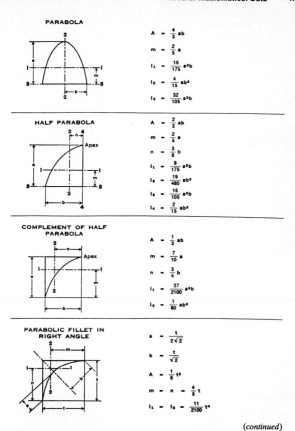

PARABOLA

$A = \frac{4}{3} ab$

$m = \frac{2}{5} a$

$I_1 = \frac{16}{175} a^3 b$

$I_2 = \frac{4}{15} ab^3$

$I_3 = \frac{32}{105} a^3 b$

HALF PARABOLA

$A = \frac{2}{3} ab$

$m = \frac{2}{5} a$

$n = \frac{3}{8} b$

$I_1 = \frac{8}{175} a^3 b$

$I_2 = \frac{19}{480} ab^3$

$I_3 = \frac{16}{105} a^3 b$

$I_4 = \frac{2}{15} ab^3$

COMPLEMENT OF HALF PARABOLA

$A = \frac{1}{3} ab$

$m = \frac{7}{10} a$

$n = \frac{3}{4} b$

$I_1 = \frac{37}{2100} a^3 b$

$I_2 = \frac{1}{80} ab^3$

PARABOLIC FILLET IN RIGHT ANGLE

$a = \frac{t}{2\sqrt{2}}$

$b = \frac{t}{\sqrt{2}}$

$A = \frac{1}{6} t^2$

$m = n = \frac{4}{5} t$

$I_1 = I_2 = \frac{11}{2100} t^4$

(*continued*)

FIG. 1.2—(*continued*)

* HALF ELLIPSE

$$A = \frac{1}{2}\pi ab$$

$$m = \frac{4a}{3\pi}$$

$$I_1 = a^3b\left(\frac{\pi}{8} - \frac{8}{9\pi}\right)$$

$$I_2 = \frac{1}{8}\pi ab^3$$

$$I_3 = \frac{1}{8}\pi a^3b$$

* QUARTER ELLIPSE

$$A = \frac{1}{4}\pi ab$$

$$m = \frac{4a}{3\pi}$$

$$n = \frac{4b}{3\pi}$$

$$I_1 = a^3b\left(\frac{\pi}{16} - \frac{4}{9\pi}\right)$$

$$I_2 = ab^3\left(\frac{\pi}{16} - \frac{4}{9\pi}\right)$$

$$I_3 = \frac{1}{16}\pi a^3b$$

$$I_4 = \frac{1}{16}\pi ab^3$$

* ELLIPTIC COMPLEMENT

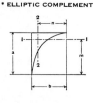

$$A = ab\left(1 - \frac{\pi}{4}\right)$$

$$m = \frac{a}{6\left(1 - \frac{\pi}{4}\right)}$$

$$n = \frac{b}{6\left(1 - \frac{\pi}{4}\right)}$$

$$I_1 = a^3b\left(\frac{1}{3} - \frac{\pi}{16} - \frac{1}{36\left(1 - \frac{\pi}{4}\right)}\right)$$

$$I_2 = ab^3\left(\frac{1}{3} - \frac{\pi}{16} - \frac{1}{36\left(1 - \frac{\pi}{4}\right)}\right)$$

* To obtain properties of half circle, quarter circle and circular complement substitute a = b = R.

FIG. 1.2—(continued)

REGULAR POLYGON

Axis of moments
through center

$$n = \text{Number of sides}$$

$$\phi = \frac{180°}{n}$$

$$a = 2\sqrt{R^2 - R_1^2}$$

$$R = \frac{a}{2\sin\phi}$$

$$R_1 = \frac{a}{2\tan\phi}$$

$$A = \frac{1}{4}na^2\cot\phi = \frac{1}{2}nR^2\sin 2\phi = nR_1^2\tan\phi$$

$$I_1 = I_2 = \frac{A(6R^2 - a^2)}{24} = \frac{A(12R_1^2 + a^2)}{48}$$

$$r_1 = r_2 = \sqrt{\frac{6R^2 - a^2}{24}} = \sqrt{\frac{12R_1^2 + a^2}{48}}$$

ANGLE

Axis of moments through
center of gravity

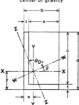

Z-Z is axis of minimum I

$$\tan 2\theta = \frac{2K}{I_y - I_x}$$

$$A = t(b + c) \quad x = \frac{b^2 + ct}{2(b + c)} \quad y = \frac{d^2 + at}{2(b + c)}$$

$$K = \text{Product of Inertia about X-X \& Y-Y}$$

$$= \mp \frac{abcdt}{4(b + c)}$$

$$I_x = \frac{1}{3}\left(t(d - y)^3 + by^3 - a(y - t)^3\right)$$

$$I_y = \frac{1}{3}\left(t(b - x)^3 + dx^3 - c(x - t)^3\right)$$

$$I_z = I_x\sin^2\theta + I_y\cos^2\theta + K\sin 2\theta$$

$$I_w = I_x\cos^2\theta + I_y\sin^2\theta - K\sin 2\theta$$

K is negative when heel of angle, with respect
to c. g., is in 1st or 3rd quadrant, positive
when in 2nd or 4th quadrant.

BEAMS AND CHANNELS

Transverse force oblique
through center of gravity

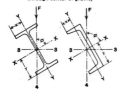

$$I_3 = I_x\sin^2\phi + I_y\cos^2\phi$$

$$I_4 = I_x\cos^2\phi + I_y\sin^2\phi$$

$$f_b = M\left(\frac{y}{I_x}\sin\phi + \frac{x}{I_y}\cos\phi\right)$$

where M is bending moment due to force F.

FIG. 1.3 Properties of geometric sections and structural shapes. (*Courtesy American Institute of Steel Construction.*)

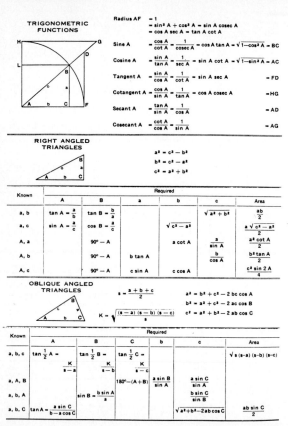

FIG. 1.4 Trigonometric formulas. (*Courtesy American Institute of Steel Construction.*)

TABLE 1.2   Decimals of a Foot for Each 32nd of an Inch

Inch	0	1	2	3	4	5
0	0	.0833	.1667	.2500	.3333	.4167
1/32	.0026	.0859	.1693	.2526	.3359	.4193
1/16	.0052	.0885	.1719	.2552	.3385	.4219
3/32	.0078	.0911	.1745	.2578	.3411	.4245
1/8	.0104	.0938	.1771	.2604	.3438	.4271
5/32	.0130	.0964	.1797	.2630	.3464	.4297
3/16	.0156	.0990	.1823	.2656	.3490	.4323
7/32	.0182	.1016	.1849	.2682	.3516	.4349
1/4	.0208	.1042	.1875	.2708	.3542	.4375
9/32	.0234	.1068	.1901	.2734	.3568	.4401
5/16	.0260	.1094	.1927	.2760	.3594	.4427
11/32	.0286	.1120	.1953	.2786	.3620	.4453
3/8	.0313	.1146	.1979	.2812	.3646	.4479
13/32	.0339	.1172	.2005	.2839	.3672	.4505
7/16	.0365	.1198	.2031	.2865	.3698	.4531
15/32	.0391	.1224	.2057	.2891	.3724	.4557
1/2	.0417	.1250	.2083	.2917	.3750	.4583
17/32	.0443	.1276	.2109	.2943	.3776	.4609
9/16	.0469	.1302	.2135	.2969	.3802	.4635
19/32	.0495	.1328	.2161	.2995	.3828	.4661
5/8	.0521	.1354	.2188	.3021	.3854	.4688
21/32	.0547	.1380	.2214	.3047	.3880	.4714
11/16	.0573	.1406	.2240	.3073	.3906	.4740
23/32	.0599	.1432	.2266	.3099	.3932	.4766
3/4	.0625	.1458	.2292	.3125	.3958	.4792
25/32	.0651	.1484	.2318	.3151	.3984	.4818
13/16	.0677	.1510	.2344	.3177	.4010	.4844
27/32	.0703	.1536	.2370	.3203	.4036	.4870
7/8	.0729	.1563	.2396	.3229	.4063	.4896
29/32	.0755	.1589	.2422	.3255	.4089	.4922
15/16	.0781	.1615	.2448	.3281	.4115	.4948
31/32	.0807	.1641	.2474	.3307	.4141	.4974

(continued)

SOURCE: American Institute of Steel Construction.

TABLE 1.2   Decimals of a Foot for Each 32nd of an Inch (*continued*)

Inch	6	7	8	9	10	11
0	.5000	.5833	.6667	.7500	.8333	.9167
1/32	.5026	.5859	.6693	.7526	.8359	.9193
1/16	.5052	.5885	.6719	.7552	.8385	.9219
3/32	.5078	.5911	.6745	.7578	.8411	.9245
1/8	.5104	.5938	.6771	.7604	.8438	.9271
5/32	.5130	.5964	.6797	.7630	.8464	.9297
3/16	.5156	.5990	.6823	.7656	.8490	.9323
7/32	.5182	.6016	.6849	.7682	.8516	.9349
1/4	.5208	.6042	.6875	.7708	.8542	.9375
9/32	.5234	.6068	.6901	.7734	.8568	.9401
5/16	.5260	.6094	.6927	.7760	.8594	.9427
11/32	.5286	.6120	.6953	.7786	.8620	.9453
3/8	.5313	.6146	.6979	.7813	.8646	.9479
13/32	.5339	.6172	.7005	.7839	.8672	.9505
7/16	.5365	.6198	.7031	.7865	.8698	.9531
15/32	.5391	.6224	.7057	.7891	.8724	.9557
1/2	.5417	.6250	.7083	.7917	.8750	.9583
17/32	.5443	.6276	.7109	.7943	.8776	.9609
9/16	.5469	.6302	.7135	.7969	.8802	.9635
19/32	.5495	.6328	.7161	.7995	.8828	.9661
5/8	.5521	.6354	.7188	.8021	.8854	.9688
21/32	.5547	.6380	.7214	.8047	.8880	.9714
11/16	.5573	.6406	.7240	.8073	.8906	.9740
23/32	.5599	.6432	.7266	.8099	.8932	.9766
3/4	.5625	.6458	.7292	.8125	.8958	.9792
25/32	.5651	.6484	.7318	.8151	.8984	.9818
13/16	.5677	.6510	.7344	.8177	.9010	.9844
27/32	.5703	.6536	.7370	.8203	.9036	.9870
7/8	.5729	.6563	.7396	.8229	.9063	.9896
29/32	.5755	.6589	.7422	.8255	.9089	.9922
15/16	.5781	.6615	.7448	.8281	.9115	.9948
31/32	.5807	.6641	.7474	.8307	.9141	.9974

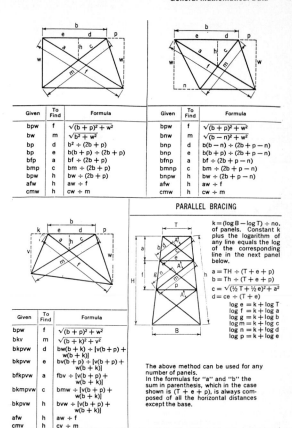

Given	To Find	Formula
bpw	f	$\sqrt{(b+p)^2 + w^2}$
bw	m	$\sqrt{b^2 + w^2}$
bp	d	$b^2 \div (2b + p)$
bp	e	$b(b+p) \div (2b + p)$
bfp	a	$bf \div (2b + p)$
bmp	c	$bm \div (2b + p)$
bpw	h	$bw \div (2b + p)$
afw	h	$aw \div f$
cmw	h	$cw \div m$

Given	To Find	Formula
bpw	f	$\sqrt{(b+p)^2 + w^2}$
bnw	m	$\sqrt{(b-n)^2 + w^2}$
bnp	d	$b(b-n) \div (2b + p - n)$
bnp	e	$b(b+p) \div (2b + p - n)$
bfnp	a	$bf \div (2b + p - n)$
bmnp	c	$bm \div (2b + p - n)$
bnpw	h	$bw \div (2b + p - n)$
afw	h	$aw \div f$
cmw	h	$cw \div m$

Given	To Find	Formula
bpw	f	$\sqrt{(b+p)^2 + w^2}$
bkv	m	$\sqrt{(b+k)^2 + v^2}$
bkpvw	d	$bw(b+k) \div [v(b+p) + w(b+k)]$
bkpvw	e	$bv(b+p) \div [v(b+p) + w(b+k)]$
bfkpvw	a	$fbv \div [v(b+p) + w(b+k)]$
bkmpvw	c	$bmw \div [v(b+p) + w(b+k)]$
bkpvw	h	$bvw \div [v(b+p) + w(b+k)]$
afw	h	$aw \div f$
cmv	h	$cv \div m$

PARALLEL BRACING

$k = (\log B - \log T) \div$ no. of panels. Constant k plus the logarithm of any line equals the log of the corresponding line in the next panel below.

$a = TH \div (T + e + p)$
$b = Th \div (T + e + p)$
$c = \sqrt{(\frac{1}{2}T + \frac{1}{2}e)^2 + a^2}$
$d = ce \div (T + e)$

$\log e = k + \log T$
$\log f = k + \log a$
$\log g = k + \log b$
$\log m = k + \log c$
$\log n = k + \log d$
$\log p = k + \log e$

The above method can be used for any number of panels.
In the formulas for "a" and "b" the sum in parenthesis, which in the case shown is $(T + e + p)$, is always composed of all the horizontal distances except the base.

FIG. 1.5 Bracing formulas. (*Courtesy American Institute of Steel Construction.*)

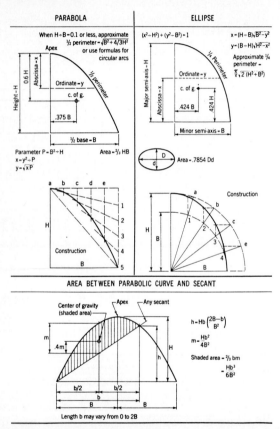

FIG. 1.6 Properties of the parabola and ellipse. (*Courtesy American Institute of Steel Construction.*)

TABLE 1.3 Weights and Measures—United States Customary System (USCS)

LINEAR MEASURE

Inches	Feet	Yards	Rods	Furlongs	Miles
1.0 =	.08333 =	.02778 =	.0050505 =	.00012626 =	.00001578
12.0 =	1.0 =	.33333 =	.0606061 =	.00151515 =	.00018939
36.0 =	3.0 =	1.0 =	.1818182 =	.00454545 =	.00056818
198.0 =	16.5 =	5.5 =	1.0 =	.025 =	.003125
7920.0 =	660.0 =	220.0 =	40.0 =	1.0 =	.125
63360.0 =	5280.0 =	1760.0 =	320.0 =	8.0 =	1.0

SQUARE AND LAND MEASURE

Sq. Inches	Square Feet	Square Yards	Sq. Rods	Acres	Sq. Miles
1.0 =	.006944 =	.000772			
144.0 =	1.0 =	.111111			
1296.0 =	9.0 =	1.0 =	.03306 =	.000207	
39204.0 =	272.25 =	30.25 =	1.0 =	.00625 =	.0000098
	43560.0 =	4840.0 =	160.0 =	1.0 =	.0015625
		3097600.0 =	102400.0 =	640.0 =	1.0

AVOIRDUPOIS WEIGHTS

Grains	Drams	Ounces	Pounds	Tons
1.0 =	.03657 =	.002286 =	.000143 =	.0000000714
27.34375 =	1.0 =	.0625 =	.003906 =	.00000195
437.5 =	16.0 =	1.0 =	.0625 =	.00003125
7000.0 =	256.0 =	16.0 =	1.0 =	.0005
14000000.0 =	512000.0 =	32000.0 =	2000.0 =	1.0

DRY MEASURE

Pints	Quarts	Pecks	Cubic Feet	Bushels
1.0 =	.5 =	.0625 =	.01945 =	.01563
2.0 =	1.0 =	.125 =	.03891 =	.03125
16.0 =	8.0 =	1.0 =	.31112 =	.25
51.42627 =	25.71314 =	3.21414 =	1.0 =	.80354
64.0 =	32.0 =	4.0 =	1.2445 =	1.0

LIQUID MEASURE

Gills	Pints	Quarts	U.S. Gallons	Cubic Feet
1.0 =	.25 =	.125 =	.03125 =	.00418
4.0 =	1.0 =	.5 =	.125 =	.01671
8.0 =	2.0 =	1.0 =	.250 =	.03342
32.0 =	8.0 =	4.0 =	1.0 =	.1337
			7.48052 =	1.0

SOURCE: American Institute of Steel Construction.

TABLE 1.4   Weights and Measures—International System of Units (SI)[a] (Metric Practice)

BASE UNITS

Quantity	Unit	Symbol
length	metre	m
mass	kilogram	kg
time	second	s
electric current	ampere	A
thermodynamic temperature	kelvin	K
amount of substance	mole	mol
luminous intensity	candela	cd

SUPPLEMENTARY UNITS

Quantity	Unit	Symbol
plane angle	radian	rad
solid angle	steradian	sr

DERIVED UNITS (WITH SPECIAL NAMES)

Quantity	Unit	Symbol	Formula
force	newton	N	$kg \cdot m/s^2$
pressure, stress	pascal	Pa	N/m^2
energy, work, quantity of heat	joule	J	$N \cdot m$
power	watt	W	J/s

DERIVED UNITS (WITHOUT SPECIAL NAMES)

Quantity	Unit	Formula
area	square metre	m^2
volume	cubic metre	m^3
velocity	metre per second	m/s
acceleration	metre per second squared	m/s^2
specific volume	cubic metre per kilogram	m^3/kg
density	kilogram per cubic metre	kg/m^3

SI PREFIXES

Multiplication Factor		Prefix	Symbol
1 000 000 000 000 000 000	$= 10^{18}$	exa	E
1 000 000 000 000 000	$= 10^{15}$	peta	P
1 000 000 000 000	$= 10^{12}$	tera	T
1 000 000 000	$= 10^{9}$	giga	G
1 000 000	$= 10^{6}$	mega	M
1 000	$= 10^{3}$	kilo	k
100	$= 10^{2}$	hecto[b]	h
10	$= 10^{1}$	deka[b]	da
0.1	$= 10^{-1}$	deci[b]	d
0.01	$= 10^{-2}$	centi[b]	c
0.001	$= 10^{-3}$	milli	m
0.000 001	$= 10^{-6}$	micro	μ
0.000 000 001	$= 10^{-9}$	nano	n
0.000 000 000 001	$= 10^{-12}$	pico	p
0.000 000 000 000 001	$= 10^{-15}$	femto	f
0.000 000 000 000 000 001	$= 10^{-18}$	atto	a

[a] Refer to ASTM E380-79 for more complete information on SI.
[b] Use is not recommended.

SOURCE: American Institute of Steel Construction.

TABLE 1.5 SI Conversion Factors[a]

Quantity	Multiply	by	to obtain	
Length	inch	[b]25.400	millimetre	mm
	foot	[b] 0.304 800	metre	m
	yard	[b] 0.914 400	metre	m
	mile (U.S. Statute)	1.609 347	kilometre	km
	millimetre	$39.370\ 079 \times 10^{-3}$	inch	in
	metre	3.280 840	foot	ft
	metre	1.093 613	yard	yd
	kilometre	0.621 370	mile	mi
Area	square inch	[b] $0.645\ 160 \times 10^3$	square millimetre	mm²
	square foot	[b] 0.092 903	square metre	m²
	square yard	0.836 127	square metre	m²
	square mile (U.S. Statute)	2.589 998	square kilometre	km²
	acre	$4.046\ 873 \times 10^3$	square metre	m²
	acre	0.404 687	hectare	
	square millimetre	$1.550\ 003 \times 10^{-3}$	square inch	in²
	square metre	10.763 910	square foot	ft²
	square metre	1.195 990	square yard	yd²
	square kilometre	0.386 101	square mile	mi²
	square metre	$0.247\ 104 \times 10^{-3}$	acre	
	hectare	2.471 044	acre	
Volume	cubic inch	[b]$16.387\ 06 \times 10^3$	cubic millimetre	mm³
	cubic foot	$28.316\ 85 \times 10^{-3}$	cubic metre	m³
	cubic yard	0.764 555	cubic metre	m³
	gallon (U.S. liquid)	3.785 412	litre	l
	quart (U.S. liquid)	0.946 353	litre	l
	cubic millimetre	$61.023\ 759 \times 10^{-6}$	cubic inch	in³
	cubic metre	35.314 662	cubic foot	ft³
	cubic metre	1.307 951	cubic yard	yd³
	litre	0.264 172	gallon (U.S. liquid)	gal
	litre	1.056 688	quart (U.S. liquid)	qt
Mass	ounce (avoirdupois)	28.349 52	gram	g
	pound (avoirdupois)	0.453 592	kilogram	kg
	short ton	$0.907\ 185 \times 10^3$.	kilogram	kg
	gram	$35.273\ 966 \times 10^{-3}$	ounce (avoirdupois)	oz av
	kilogram	2.204 622	pound (avoirdupois)	lb av
	kilogram	$1.102\ 311 \times 10^{-3}$	short ton	

[a] Refer to ASTM E380-79 for more complete information on SI.

[b] Indicates exact value.

SOURCE: American Institute of Steel Construction.

TABLE 1.5 SI Conversion Factors (continued)

Quantity	Multiply	by	to obtain	
Force	ounce-force	0.278 014	newton	N
	pound-force	4.448 222	newton	N
	newton	3.596 942	ounce-force	
	newton	0.224 809	pound-force	lbf
Bending Moment	pound-force-inch	0.112 985	newton-metre	N·m
	pound-force-foot	1.355 818	newton-metre	N·m
	newton-metre	8.850 748	pound-force-inch	lbf·in
	newton-metre	0.737 562	pound-force-foot	lbf·ft
Pressure, Stress	pound-force per square inch	6.894 757	kilopascal	kPa
	foot of water (39.2 F)	2.988 98	kilopascal	kPa
	inch of mercury (32 F)	3.386 38	kilopascal	kPa
	kilopascal	0.145 038	pound-force per square inch	lbf/in²
	kilopascal	0.334 562	foot of water (39.2 F)	
	kilopascal	0.295 301	inch of mercury (32 F)	
Energy, Work, Heat	foot-pound-force	1.355 818	joule	J
	^{c}British thermal unit	$1.055\ 056 \times 10^3$	joule	J
	^{c}calorie	$^b\ 4.186\ 800$	joule	J
	kilowatt hour	$^b\ 3.600\ 000 \times 10^6$	joule	J
	joule	0.737 562	foot-pound-force	ft·lbf
	joule	$0.947\ 817 \times 10^{-3}$	^{c}British thermal unit	Btu
	joule	0.238 846	^{c}calorie	
	joule	$0.277\ 778 \times 10^{-6}$	kilowatt hour	kW·h
Power	foot-pound-force/second	1.355 818	watt	W
	^{c}British thermal unit per hour	0.293 071	watt	W
	horsepower (550 ft. lbf/s)	0.745 700	kilowatt	kW
	watt	0.737 562	foot-pound-force/second	ft·lbf/s
	watt	3.412 141	^{c}British thermal unit per hour	Btu/h
	kilowatt	1.341 022	horsepower (550 ft.·lbf/s)	hp
Angle	degree	$17.453\ 29 \times 10^{-3}$	radian	rad
	radian	57.295 788	degree	
Temperature	degree Fahrenheit	$t°C = (t°F - 32)/1.8$	degree Celsius	
	degree Celsius	$t°F = 1.8 \times t°C + 32$	degree Fahrenheit	

^{a} Refer to ASTM E380-79 for more complete information on SI.

^{b} Indicates exact value.

^{c} International Table.

TABLE 1.6 Decimals of an Inch for Each 64th of an Inch, with Millimeter Equivalents

Fraction	1/64ths	Decimal	Millimeters (Approx.)	Fraction	1/64ths	Decimal	Millimeters (Approx.)
...	1	.015625	0.397	...	33	.515625	13.097
1/32	2	.03125	0.794	17/32	34	.53125	13.494
...	3	.046875	1.191	...	35	.546875	13.891
1/16	4	.0625	1.588	9/16	36	.5625	14.288
...	5	.078125	1.984	...	37	.578125	14.684
3/32	6	.09375	2.381	19/32	38	.59375	15.081
...	7	.109375	2.778	...	39	.609375	15.478
1/8	8	.125	3.175	5/8	40	.625	15.875
...	9	.140625	3.572	...	41	.640625	16.272
5/32	10	.15625	3.969	...	42	.65625	16.669
...	11	.171875	4.366	21/32	43	.671875	17.066
3/16	12	.1875	4.763	11/16	44	.6875	17.463
...	13	.203125	5.159	...	45	.703125	17.859
7/32	14	.21875	5.556	23/32	46	.71875	18.256
...	15	.234375	5.953	...	47	.734375	18.653
1/4	16	.250	6.350	3/4	48	.750	19.050
...	17	.265625	6.747	...	49	.765625	19.447
9/32	18	.28125	7.144	25/32	50	.78125	19.844
...	19	.296875	7.541	...	51	.796875	20.241
5/16	20	.3125	7.938	13/16	52	.8125	20.638
...	21	.328125	8.334	...	53	.828125	21.034
11/32	22	.34375	8.731	27/32	54	.84375	21.431
...	23	.359375	9.128	...	55	.859375	21.828
3/8	24	.375	9.525	7/8	56	.875	22.225
...	25	.390625	9.922	...	57	.890625	22.622
13/32	26	.40625	10.319	29/32	58	.90625	23.019
...	27	.421875	10.716	...	59	.921875	23.416
7/16	28	.4375	11.113	15/16	60	.9375	23.813
...	29	.453125	11.509	...	61	.953125	24.209
15/32	30	.46875	11.906	31/32	62	.96875	24.606
...	31	.484375	12.303	...	63	.984375	25.003
1/2	32	.500	12.700	1	64	1.000	25.400

SOURCE: American Institute of Steel Construction.

TABLE 1.7   Equivalent Listing

Fraction–Decimal–Millimeter

Fract. inch.	Inch decimal equiv.	Millimeter equiv.	Fract. inch.	Inch decimal equiv.	Millimeter equiv.
1/64	.0156	.397	33/64	.5156	13.097
1/32	.0312	.794	17/32	.5312	13.494
3/64	.0469	1.191	35/64	.5469	13.891
1/16	.0625	1.588	9/16	.5625	14.288
5/64	.0781	1.984	37/64	.5781	14.684
3/32	.0937	2.381	19/32	.5937	15.081
7/64	.1094	2.778	39/64	.6094	15.478
1/8	.1250	3.175	5/8	.6250	15.875
9/64	.1406	3.572	41/64	.6406	16.272
5/32	.1562	3.969	21/32	.6562	16.669
11/64	.1719	4.366	43/64	.6719	17.066
3/16	.1875	4.763	11/16	.6875	17.463
13/64	.2031	5.159	45/64	.7031	17.859
7/32	.2187	5.556	23/32	.7187	18.256
15/64	.2344	5.953	47/64	.7344	18.653
1/4	.2500	6.350	3/4	.7500	19.050
17/64	.2656	6.747	49/64	.7656	19.447
9/32	.2812	7.144	25/32	.7812	19.844
19/64	.2969	7.541	51/64	.7969	20.241
5/16	.3125	7.938	13/16	.8125	20.638
21/64	.3281	8.334	53/64	.8281	21.034
11/32	.3437	8.731	27/32	.8437	21.431
23/64	.3594	9.128	55/64	.8594	21.828
3/8	.3750	9.525	7/8	.8750	22.225
25/64	.3906	9.922	57/64	.8906	22.622
13/32	.4062	10.319	29/32	.9062	23.019
27/64	.4219	10.716	59/64	.9219	23.416
7/16	.4375	11.113	15/16	.9375	23.813
29/64	.4531	11.509	61/64	.9531	24.209
15/32	.4687	11.906	31/32	.9687	24.606
31/64	.4844	12.303	63/64	.9844	25.003
1/2	.5000	12.700	1	1.0000	25.400

Millimeters–Inches

Millimeters	Inches	Millimeters	Inches
1	.0394	16	.6299
2	.0787	17	.6693
3	.1181	18	.7087
4	.1575	19	.7480
5	.1968	20	.7874
6	.2362	21	.8268
7	.2756	22	.8661
8	.3150	23	.9055
9	.3543	24	.9449
10	.3937	25	.9842
11	.4331	26	1.0236
12	.4724	27	1.0630
13	.5118	28	1.1024
14	.5512	29	1.1417
15	.5905	30	1.1811

SOURCE: Boston Gear, Quincy, MA.

TABLE 1.8   General Conversions, USCS and Metric

Multiply	by	to obtain
acres	43,560	square feet
acres	4047	square meters
acres	1.562×10^{-3}	square miles
acres	4840	square yards
amperes	1/10	abamperes
atmospheres	76.0	cms. of mercury
atmospheres	29.92	inches of mercury
atmospheres	33.90	feet of water
atmospheres	14.70	lbs. per sq. inch
British thermal units	0.2520	kilogram-calories
British thermal units	777.5	foot-pounds
British thermal units	3.927×10^{-4}	horse-power-hours
British thermal units	1054	joules
British thermal units	107.5	kilogram-meters
British thermal units	2.928×10^{-4}	kilowatt-hours
B.t.u. per min.	12.96	foot-pounds per sec.
B.t.u. per min.	0.02356	horse-power
B.t.u. per min.	0.01757	kilowatts
B.t.u. per min.	17.57	watts
B.t.u. per sq. ft. per min.	0.1220	watts per sq. inch
bushels	1.244	cubic feet
bushels	2150	cubic inches
centimeters	0.3937	inches
centimeters	0.01	meters
centimeter-grams	980.7	centimeter-dynes
centimeters of mercury	0.01316	atmospheres
centimeters of mercury	0.4461	feet of water
centimeters per second	1.969	feet per minute
centimeters per second	0.03281	feet per second
cubic centimeters	3.531×10^{-5}	cubic feet
cubic centimeters	6.102×10^{-2}	cubic inches
cubic feet	2.832×10^{4}	cubic cms.
cubic feet	1728	cubic inches
cubic feet	0.02832	cubic meters
cubic feet	0.03704	cubic yards
cubic feet	7.481	gallons
cubic feet	62.43	pounds of water

SOURCE: Dietzgen Company.

(continued)

TABLE 1.8   General Conversions, USCS and Metric (continued)

Multiply	by	to obtain
cubic feet per minute...	472.0	cubic cms. per sec.
cubic feet per minute...	0.1247	gallons per sec.
cubic feet per minute...	0.4720	liters per second
cubic feet per minute...	62.4	lbs. of water per min.
cubic inches............	16.39	cubic centimeters
cubic inches............	5.787x10⁻⁴	cubic feet
cubic yards.............	27	cubic feet
cubic yards.............	0.7646	cubic meters
cubic yards per minute..	0.45	cubic feet per sec.
degrees (angle).........	60	minutes
degrees (angle).........	0.01745	radians
degrees (angle).........	3600	seconds
dynes..................	7.233x10⁻⁵	poundals
ergs...................	2.390x10⁻¹¹	kilogram-calories
feet...................	30.48	centimeters
feet...................	12	inches
feet...................	0.3048	meters
feet...................	.36	varas
feet...................	1/3	yards
feet of water..........	0.02950	atmospheres
feet of water..........	0.8826	inches of mercury
feet of water..........	304.8	kgs. per sq. meter
feet of water..........	62.43	pounds per sq. ft.
feet of water..........	0.4335	pounds per sq. inch.
foot-pounds............	1.286x10⁻³	British thermal units
foot-pounds............	1.356x10⁷	ergs
foot-pounds............	5.050x10⁻⁷	horse-power-hours
foot-pounds............	1.356	joules
foot-pounds per min....	1.286x10⁻³	B.t. units per minute
foot-pounds per min....	0.01667	foot-pounds per sec.
foot-pounds per min....	3.030x10⁻⁵	horse-power
foot-pounds per min....	2.260x10⁻⁵	kilowatts
foot-pounds per sec.....	7.717x10⁻²	B.t.units per minute
gallons................	8.345	pounds of water
gallons................	231	cubic inches
gallons................	3.785	liters
gallons per minute......	2.228x10⁻³	cubic feet per sec.
grains (troy)..........	0.06480	grams

TABLE 1.8   General Conversions, USCS and Metric (*continued*)

Multiply	by	to obtain
grams	980.7	dynes
grams	0.03527	ounces
horse-power	42.44	B.t.units per minute
horse-power	33,000	foot-pounds per min.
horse-power	550	foot-pounds per sec.
horse-power	1.014	horse-power(metric)
horse-power	10.70	kg.-calories per min.
horse-power	0.7457	kilowatts
horse-power	745.7	watts
horse-power (boiler)	33,520	B.t.u. per hour
horse-power (boiler)	9.804	kilowatts
horse-power-hours	2547	British thermal units
horse-power-hours	1.98×10^6	foot-pounds
inches	2.540	centimeters
inches of mercury	0.03342	atmospheres
inches of mercury	1.133	feet of water
inches of water	0.002458	atmospheres
inches of water	0.03613	pounds per sq. inch
kilograms	980,665	dynes
kilograms	2.2046	pounds
kilogram-calories	3.968	British thermal units
kilogram-calories	3086	foot-pounds
k.g.-calories per min.	51.43	foot-pounds per sec.
k.g.-calories per min.	0.06972	kilowatts
kilometers	10^5	centimeters
kilometers	3281	feet
kilometers	10^3	meters
kilometers	0.6214	miles
kilometers	1093.6	yards
kilowatts	56.92	B.t.units per min.
kilowatts	4.425×10^4	foot-pounds per min.
kilowatts	737.6	foot-pounds per sec.
kilowatts	1.341	horse-power
kilowatts	14.34	kg.-calories per min.
kilowatts	10^3	watts
kilowatt-hours	3415	British thermal units
kilowatt-hours	2.655×10^6	foot-pounds

(*continued*)

TABLE 1.8   General Conversions, USCS and Metric (*continued*)

Multiply	by	to obtain
kilowatt-hours	1.341	horse-power-hours
kilowatt-hours	3.6×10^6	joules
kilowatt-hours	860.5	kilogram-calories
kilowatt-hours	3.671×10^5	kilogram-meters
$\log^{10} N$	2.303	$\log_\epsilon N$ or $\ln N$
$\log_\epsilon N$ or $\ln N$	0.4343	$\log_{10} N$
meters	100	centimeters
meters	3.2808	feet
meters	39.37	inches
miles	5280	feet
miles	1.6093	kilometers
miles per hour	88	feet per minute
ounces	28.35	grams
ounces per sq. inch	0.0625	pounds per sq. inch
pints (dry)	33.60	cubic inches
pints (liq.)	28.87	cubic inches
pounds	453.6	grams
pounds	16	ounces
pounds of water	0.01602	cubic feet
pounds of water	27.68	cubic inches
pounds of water	0.1198	gallons
pounds of water per min.	2.669×10^{-4}	cubic feet per sec.
pounds per cubic foot	5.787×10^{-4}	pounds per cubic in.
pounds per sq. foot	0.01602	feet of water
pounds per sq. inch	0.06804	atmospheres
pounds per sq. inch	2.307	feet of water
quarts	32	fluid ounces
quarts (dry)	67.20	cubic inches
quarts (liq.)	57.75	cubic inches
rods	16.5	feet
square centimeters	0.1550	square inches
square inches	6.452	square centimeters
square miles	640	acres
square miles	27.88×10^6	square feet
square yards	0.8361	square meters
temp. (degs. C.)+17.8	1.8	temp. (degs. Fahr.)
temp. (degs. F.)−32	5/9	temp. (degs. Cent.)
yards	.9144	meters

TABLE 1.9   Length Equivalents, USCS and Metric

One	Is Equal to	One	Is Equal to
Foot	12 inches	Inch	0.0254 meter
Yard	3 feet	Foot	0.3048 meter
Mile	5280 feet	Yard	0.9144 meter
Kilometer	1000 meters	Mile	1.6090 kilometers

SOURCE: A. M. Khashab, *Heating, Ventilating, and Air Conditioning Systems Estimating Manual,* 2d ed., McGraw-Hill, New York, © 1984. Used with permission of the publisher.

TABLE 1.10   Area Equivalents, USCS and Metric

One	Is Equal to	One	Is Equal to
Square foot	144 square inches	Square inch	6.541×10^{-4} square meter
Square yard	9 square feet	Square foot	0.0929 square meter
Acre	43,560 square feet	Square yard	0.8361 square meter
Hectare	2.471 acres	Acre	4047 square meter
		Hectare	10,000 square meters

SOURCE: A. M. Khashab, *Heating, Ventilating, and Air Conditioning Systems Estimating Manual,* 2d ed., McGraw-Hill, New York, © 1984. Used with permission of the publisher.

TABLE 1.11 Volume Equivalents, USCS and Metric

One	Is Equal to	One	Is Equal to
Cubic foot	1728 cubic inches	Cubic foot	0.02832 cubic meter
Cubic yard	27 cubic feet	Cubic yard	0.7646 cubic meter

SOURCE: A. M. Khashab, *Heating, Ventilating, and Air Conditioning Systems Estimating Manual*, 2d ed., McGraw-Hill, New York, © 1984. Used with permission of the publisher.

TABLE 1.12 Capacity Equivalents, USCS and Metric

One	Is Equal to	One	Is Equal to
U.S. gallon	0.1337 cubic foot	U.S. gallon	3.785 liters
Liter	0.03531 cubic foot	Liter	0.001 cubic meter

SOURCE: A. M. Khashab, *Heating, Ventilating, and Air Conditioning Systems Estimating Manual*, 2d ed., McGraw-Hill, New York, © 1984. Used with permission of the publisher.

TABLE 1.13 Flow Equivalents, USCS and Metric

One	Is Equal to	One	Is Equal to
Cubic foot per minute	4.72×10^{-4} cubic meter per second (4.72×10^{-1} liter per second)	Gallon per minute	6.309×10^{-5} cubic meter per second (6.309×10^{-2} liter per second)

SOURCE: A. M. Khashab, *Heating, Ventilating, and Air Conditioning Systems Estimating Manual*, 2d ed., McGraw-Hill, New York, © 1984. Used with permission of the publisher.

TABLE 1.14   Temperature Conversion Table—Celsius (Centigrade) to Fahrenheit

C	F	C	F	C	F	C	F	C	F
-30	-22.0	20	68.0	70	158.0	120	248.0	170	338.0
-29	-20.2	21	69.8	71	159.8	121	249.8	171	339.8
-28	-18.4	22	71.6	72	161.6	122	251.6	172	341.6
-27	-16.6	23	73.4	73	163.4	123	253.4	173	343.4
-26	-14.8	24	75.2	74	165.2	124	255.2	174	345.2
-25	-13.0	25	77.0	75	167.0	125	257.0	175	347.0
-24	-11.2	26	78.8	76	168.8	126	258.8	176	348.8
-23	-9.4	27	80.6	77	170.6	127	260.6	177	350.6
-22	-7.6	28	82.4	78	172.4	128	262.4	178	352.4
-21	-5.8	29	84.2	79	174.2	129	264.2	179	354.2
-20	-4.0	30	86.0	80	176.0	130	266.0	180	356.0
-19	-2.2	31	87.8	81	177.8	131	267.8	181	357.8
-18	-0.4	32	89.6	82	179.6	132	269.6	182	359.6
-17	+1.4	33	91.4	83	181.4	133	271.4	183	361.4
-16	3.2	34	93.2	84	183.2	134	273.2	184	363.2
-15	5.0	35	95.0	85	185.0	135	275.0	185	365.0
-14	6.8	36	96.8	86	186.8	136	276.8	186	366.8
-13	8.6	37	98.6	87	188.6	137	278.6	187	368.6
-12	10.4	38	100.4	88	190.4	138	280.4	188	370.4
-11	12.2	39	102.2	89	192.2	139	282.2	189	372.2
-10	14.0	40	104.0	90	194.0	140	284.0	190	374.0
-9	15.8	41	105.8	91	195.8	141	285.8	191	375.8
-8	17.6	42	107.6	92	197.6	142	287.6	192	377.6
-7	19.4	43	109.4	93	199.4	143	289.4	193	379.4
-6	21.2	44	111.2	94	201.2	144	291.2	194	381.2
-5	23.0	45	113.0	95	203.0	145	293.0	195	383.0
-4	24.8	46	114.8	96	204.8	146	294.8	196	384.8
-3	26.6	47	116.6	97	206.6	147	296.6	197	386.6
-2	28.4	48	118.4	98	208.4	148	298.4	198	388.4
-1	30.2	49	120.2	99	210.2	149	300.2	199	390.2

(continued)

SOURCE: The Trane Company, LaCrosse, WI. Reproduced by permission.

TABLE 1.14 Temperature Conversion Table—Celsius (Centigrade) to Fahrenheit (*continued*)

C	F	C	F	C	F	C	F	C	F
0	32.0	50	122.0	100	212.0	150	302.0	200	392.0
1	33.8	51	123.8	101	213.8	151	303.8	201	393.8
2	35.6	52	125.6	102	215.6	152	305.6	202	395.6
3	37.4	53	127.4	103	217.4	153	307.4	203	397.4
4	39.2	54	129.2	104	219.2	154	309.2	204	399.2
5	41.0	55	131.0	105	221.0	155	311.0	205	401.0
6	42.8	56	132.8	106	222.8	156	312.8	206	402.8
7	44.6	57	134.6	107	224.6	157	314.6	207	404.6
8	46.4	58	136.4	108	226.4	158	316.4	208	406.4
9	48.2	59	138.2	109	228.2	159	318.2	209	408.2
10	50.0	60	140.0	110	230.0	160	320.0	210	410.0
11	51.8	61	141.8	111	231.8	161	321.8	211	411.8
12	53.6	62	143.6	112	233.6	162	323.6	212	413.6
13	55.4	63	145.4	113	235.4	163	325.4	213	415.4
14	57.2	64	147.2	114	237.2	164	327.2	214	417.2
15	59.0	65	149.0	115	239.0	165	329.0	215	419.0
16	60.8	66	150.8	116	240.8	166	330.8	216	420.8
17	62.6	67	152.6	117	242.6	167	332.6	217	422.6
18	64.4	68	154.4	118	244.4	168	334.4	218	424.4
19	66.2	69	156.2	119	246.2	169	336.2	219	426.2

NOTES

NOTES

METRIC SYSTEM
AND CONVERSION[1]

[1] This section adapted with permission from R. O. Parmley, *Mechanical Components Handbook,* McGraw-Hill, © 1985.

2.1 CONVERSION FACTORS

Units derived from SI base units are shown in Table 2.1.

Table 2.2 gives the definitions of various units of measure that are exact numerical multiples of coherent SI units, and provides multiplication factors for converting numbers and miscellaneous units to corresponding new numbers and SI units.

The first two digits of each numerical entry represent a power of 10. An asterisk follows each number that expresses an exact definition. For example, the entry " − 02 2.54*" expresses the fact that 1 inch = 2.54 × 10⁻² meter, exactly, by definition. Most of the definitions are extracted from National Bureau of Standards documents. Numbers not followed by an asterisk are only approximate or are the results of physical measurements. The conversion factors are listed alphabetically and by physical quantity.

The listing by physical quantity includes only relationships which are frequently encountered and deliberately omits the many combinations of units which are used for more specialized purposes. Conversion factors for combinations of units are easily generated from numbers given in the alphabetical listing by the technique of direct substitution or by other well-known rules for manipulating units. These units are adequately discussed in many science and engineering textbooks and are not repeated here.

TABLE 2.1   Derived Units of the International System

Quantity	Name of unit	Unit symbol or abbreviation, where differing from base form	Unit expressed in terms of base or supplementary units†
Area	square meter		m²
Volume	cubic meter		m³
Frequency	hertz, cycle per second‡	Hz	s⁻¹
Density	kilogram per cubic meter		kg/m³
Velocity	meter per second		m/s
Angular velocity	radian per second		rad/s
Acceleration	meter per second squared		m/s²
Angular acceleration	radian per second squared		rad/s²
Volumetric flow rate	cubic meter per second		m³/s
Force	newton	N	kg·m/s²
Surface tension	newton per meter, joule per square meter	N/m, J/m²	kg/s²
Pressure	newton per square meter, pascal	N/m², Pa	kg/m·s²
Viscosity, dynamic	newton-second per square meter, pascal-second	N·s/m², Pa·s	kg/m·s
Viscosity, kinematic	meter squared per second		m²/s
Work, torque, energy, quantity of heat	joule, newton-meter, watt-second	J, N·m, W·s	kg·m²/s²
Power, heat flux	watt, joule per second	W, J/s	kg·m²/s³
Heat flux density	watt per square meter	W/m²	kg/s³
Volumetric heat release rate	watt per cubic meter	W/m³	kg/m·s³
Heat transfer coefficient	watt per square meter-kelvin	W/m²·K	kg/s³·K
Heat capacity (specific)	joule per kilogram-kelvin	J/kg·K	m²/s²·K

(continued)

† Supplementary units are plane angle, radian (rad), solid angle, steradian (sr).

TABLE 2.1   Derived Units of the International System (*continued*)

Quantity	Name of unit	Unit symbol or abbreviation, where differing from base form	Unit expressed in terms of base or supplementary units†
Capacity rate	watt per kelvin	W/K	$kg \cdot m^2/s^3 \cdot K$
Thermal conductivity	watt per meter-kelvin	W/m-deg, $J/m \cdot s \cdot m^2 \cdot K$	$kg \cdot m/s^3 \cdot K$
Quantity of electricity	coulomb	C	$A \cdot s$
Electromotive force	volt	V, W/A	$kg \cdot m^2/A \cdot s^3$
Electric field strength	volt per meter	V/m	V/m
Electric resistance	ohm	Ω, V/A	$kg \cdot m^2/A^2 \cdot s^3$
Electric conductivity	ampere per volt-meter	A/V·m	$A^2 s^3/kg \cdot m^3$
Electric capacitance	farad	F, A·s/V	$A^3 s^3/kg \cdot m^2$
Magnetic flux	weber	Wb, V·s	$kg \cdot m^2/A \cdot s^2$
Inductance	henry	H, V·s/A	$kg \cdot m^2/A^2 \cdot s^2$
Magnetic permeability	henry per meter	H/m	$kg \cdot m/A^2 \cdot s^2$
Magnetic field density	tesla, weber per square meter	T, Wb/m^2	$kg/A \cdot s^2$
Magnetic field strength	ampere per meter	A/m	A/m
Magnetomotive force	ampere		A
Luminous flux	lumen	lm	cd sr
Luminance	candela per square meter		cd/m^2
Illumination	lux, lumen per square meter	lx, lm/m^2	$cd \cdot sr/m^2$

SOURCE: Adapted from Tyler G. Hicks, *Metrication Manual*, McGraw-Hill, New York, © 1972. Used with permission of the publisher.

TABLE 2.2 Conversion Factors as Extracted Multiples of SI Units

To convert from	To	Multiply by
abampere	ampere	+01 1.00*
abcoulomb	coulomb	+01 1.00*
abfarad	farad	+09 1.00*
abhenry	henry	−09 1.00*
abmho	siemens	+09 1.00*
abohm	ohm	−09 1.00*
abvolt	volt	−08 1.00*
acre	meter^2	+03 4.046 873
ampere (international of 1948)	ampere	−01 9.998 35
angstrom	meter	−10 1.00*
are	meter^2	+02 1.00*
astronomical unit	meter	+11 1.495 979
atmosphere (standard)	pascal (newton/meter^2)	+05 1.013 250*
bar	pascal (newton/meter^2)	+05 1.00*
barn	meter^2	−28 1.00*
barrel (petroleum, 42 gallons)	meter^3	−01 1.589 873
barye	newton/meter^2	−01 1.00*
British thermal unit (ISO/TC 12)	joule	+03 1.055 06
British thermal unit (International Steam Table)	joule	+03 1.055 04
British thermal unit (mean)	joule	+03 1.055 87
British thermal unit (thermochemical)	joule	+03 1.054 350 264 488
British thermal unit (39°F)	joule	+03 1.059 67
British thermal unit (60°F)	joule	+03 1.054 68
bushel (U.S.)	meter^3	−02 3.523 907 016 688*
cable	meter	+02 2.194 56*
caliber	meter	−04 2.54*
calorie (International Steam Table)	joule	+00 4.1868
calorie (mean)	joule	+00 4.190 02
calorie (thermochemical)	joule	+00 4.184*
calorie (15°C)	joule	+00 4.185 80
calorie (20°C)	joule	+00 4.181 90
calorie (kilogram, International Steam Table)	joule	+03 4.1868
calorie (kilogram, mean)	joule	+03 4.190 02
calorie (kilogram, thermochemical)	joule	+03 4.184*
carat (metric)	kilogram	−04 2.00*
Celsius (temperature)	kelvin	$t_K = t_C = 273.15$
centimeter of mercury (0°C)	newton/meter^2	+03 1.333 22
centimeter of water (4°C)	newton/meter^2	+01 9.806 38
chain (engineer or ramden)	meter	+01 3.048*
chain (surveyor or gunter)	meter	+01 2.011 68*
circular mil	meter^2	−10 5.067 074 8
cord	meter^3	+00 3.624 556 3
coulomb (international of 1948)	coulomb	−01 9.998 35
cubit	meter	−01 4.572*

(continued)

SOURCE: Adapted from Tyler G. Hicks, *Metrication Manual*, McGraw-Hill, New York, © 1972. Used with permission of the publisher.

TABLE 2.2   *(continued)*

To convert from	To	Multiply by
cup	meter^3	−04 2.365 882 365*
curie	disintegration/second	+10 3.70*
day (mean solar)	second (mean solar)	+04 8.64*
day (sidereal)	second (mean solar)	+04 8.616 409 0
degree (angle)	radian	−02 1.745 329 251 994 3
denier (international)	kilogram/meter	−07 1.00*
dram (avoirdupois)	kilogram	−03 1.771 845 195 312 5*
dram (troy or apothecary)	kilogram	−03 3.887 934 6*
dram (U.S. fluid)	meter^3	−06 3.696 691 195 312 5*
dyne	newton	−05 1.00*
electron-volt	joule	−19 1.602 10
erg	joule	−07 1.00*
Fahrenheit (temperature)	kelvin	$t_K = (5/9) (t_F + 459.67)$
Fahrenheit (temperature)	Celsius	$t_C = (5/9) (t_F − 32)$
farad (international of 1948)	farad	−01 9.995 05
faraday (based on carbon 12)	coulomb	+04 9.648 70
faraday (chemical)	coulomb	+04 9.649 57
faraday (physical)	coulomb	+04 9.652 19
fathom	meter	+00 1.828 8*
fermi (femtometer)	meter	−15 1.00*
fluid ounce (U.S.)	meter^3	−05 2.957 352 956 25*
foot	meter	−01 3.048*
foot (U.S. survey)	meter	+00 1200/3937*
foot (U.S. survey)	meter	−01 3.048 006 096
foot of water (39.2°F)	newton/meter^2	+03 2.988 98
foot-candle	lumen/meter^2	+01 1.076 391 0
foot-lambert	candela/meter^2	+00 3.426 259
furlong	meter	+02 2.011 68*
gal (galileo)	meter/second^2	−02 1.00*
gallon (U.K. liquid)	meter^3	−03 4.546 087
gallon (U.S. dry)	meter^3	−03 4.404 883 770 86*
gallon (U.S. liquid)	meter^3	−03 3.785 411 784*
gamma	tesla	−09 1.00*
gauss	tesla	−04 1.00*
gilbert	ampere-turn	−01 7.957 747 2
gill (U.K.)	meter^3	−04 1.420 652
gill (U.S.)	meter^3	−04 1.182 941 2
grad	degree (angular)	−01 9.00*
grad	radian	−02 1.570 796 3
grain	kilogram	−05 6.479 891*
gram	kilogram	−03 1.00*
hand	meter	−01 1.016*
hectare	meter^2	+04 1.00*
henry (international of 1948)	henry	+00 1.000 495
hogshead (U.S.)	meter^3	−01 2.384 809 423 92*
horsepower (550 foot lbf/second)	watt	+02 7.456 998 7

TABLE 2.2 *(continued)*

To convert from	To	Multiply by
horsepower (boiler)	watt	+03 9.809 50
horsepower (electric)	watt	+02 7.46*
horsepower (metric)	watt	+02 7.354 99
horsepower (U.K.)	watt	+02 7.457
horsepower (water)	watt	+02 7.460 43
hour (mean solar)	second (mean solar)	+03 3.60*
hour (sidereal)	second (mean solar)	+03 3.590 170 4
hundredweight (long)	kilogram	+01 5.080 234 544*
hundredweight (short)	kilogram	+01 4.535 923 7*
inch	meter	−02 2.54*
inch of mercury (32°F)	pascal (newton/meter²)	+03 3.386 38
inch of mercury (60°F)	pascal (newton/meter²)	+03 3.376 85
inch of water (39.2°F)	pascal (newton/meter²)	+02 2.490 82
inch of water (60°F)	pascal (newton/meter²)	+02 2.488 4
joule (international of 1948)	joule	+00 1.000 165
kayser	1/meter	+02 1.00*
kilocalorie (International Steam Table)	joule	+03 4.186 74
kilocalorie (mean)	joule	+03 4.190 02
kilocalorie (thermochemical)	joule	+03 4.184*
kilogram mass	kilogram	+00 1.00*
kilogram force (kgf)	newton	+00 9.806 65*
kilopond force	newton	+00 9.806 65*
kip	newton	+03 4.448 221 615 260 5*
knot (international)	meter/second	−01 5.144 444 444
lambert	candela/meter²	+04 1/π*
lambert	candela/meter²	+03 3.183 098 8
langley	joule/meter²	+04 4.184*
lbf (pound force, avoirdupois)	newton	+00 4.448 221 615 260 5*
lbm (pound mass, avoirdupois)	kilogram	−01 4.535 923 7*
league (British nautical)	meter	+03 5.559 552*
league (international nautical)	meter	+03 5.556*
league (statute)	meter	+03 4.828 032*
light year	meter	+15 9.460 55
link (engineer or ramden)	meter	−01 3.048*
link (surveyor or gunter)	meter	−01 2.011 68*
liter	meter³	−03 1.00*
lux	lumen/meter²	+00 1.00*
maxwell	weber	−08 1.00*
meter	wavelengths Kr 86	+06 1.650 763 73*
micron	meter	−06 1.00*
mil	meter	−05 2.54*
mile (U.S. statute)	meter	+03 1.609 344*

TABLE 2.2   (*continued*)

To convert from	To	Multiply by
mile (U.K. nautical)	meter	+03 1.853 184*
mile (international nautical)	meter	+03 1.852*
mile (U.S. nautical)	meter	+03 1.852*
millibar	newton/meter²	+02 1.00*
millimeter of mercury (0°C)	newton/meter²	+02 1.333 224
minute (angle)	radian	−04 2.908 882 086 66
minute (mean solar)	second (mean solar)	+01 6.00*
minute (sidereal)	second (mean solar)	+01 5.983 617 4
month (mean calendar)	second (mean solar)	+06 2.628*
nautical mile (international)	meter	+03 1.852*
nautical mile (U.S.)	meter	+03 1.852*
nautical mile (U.K.)	meter	+03 1.853 184*
oersted	ampere/meter	+01 7.957 747 2
ohm (international of 1948)	ohm	+00 1.000 495
ounce force (avoirdupois)	newton	−01 2.780 138 5
ounce mass (avoirdupois)	kilogram	−02 2.834 952 312 5*
ounce mass (troy or apothecary)	kilogram	−02 3.110 347 68*
ounce (U.S. fluid)	meter³	−05 2.957 352 956 25*
pace	meter	−01 7.62*
parsec	meter	+16 3.083 74
pascal	newton/meter²	+00 1.00*
peck (U.S.)	meter³	−03 8.809 767 541 72*
pennyweight	kilogram	−03 1.555 173 84*
perch	meter	+00 5.0292*
phot	lumen/meter²	+04 1.00*
pica (printers)	meter	−03 4.217 517 6*
pint (U.S. dry)	meter³	−04 5.506 104 713 575*
pint (U.S. liquid)	meter³	−04 4.731 764 73*
point (printers)	meter	−04 3.514 598*
poise	newton-second/meter²	−01 1.00*
pole	meter	+00 5.0292*
pound force (lbf avoirdupois)	newton	+00 4.448 221 615 260 5*
pound mass (lbm avoirdupois)	kilogram	−01 4.535 923 7*
pound mass (troy or apothecary)	kilogram	−01 3.732 417 216*
poundal	newton	−01 1.382 549 543 76*
quart (U.S. dry)	meter³	−03 1.101 220 942 715*
quart (U.S. liquid)	meter³	−04 9.463 529 5
rad (radiation dose absorbed)	joule/kilogram	−02 1.00*
Rankine (temperature)	kelvin	$t_K = (5/9)t_R$
rayleigh (rate of photon emission)	1/second-meter²	+10 1.00*
rhe	meter²/newton-second	+01 1.00*
rod	meter	+00 5.0292*
roentgen	coulomb/kilogram	−04 2.579 76*
rutherford	disintegration/second	+06 1.00*

TABLE 2.2   *(continued)*

To convert from	To	Multiply by
second (angle)	radian	−06 4.848 136 811
second (ephemeris)	second	+00 1.000 000 000
second (mean solar)	second (ephemeris)	Consult American Ephemeris and Nautical Almanac
second (sidereal)	second (mean solar)	−01 9.972 695 7
section	meter²	+06 2.589 988 110 336*
scruple (apothecary)	kilogram	−03 1.295 978 2*
shake	second	−08 1.00
skein	meter	+02 1.097 28*
slug	kilogram	+01 1.459 390 29
span	meter	−01 2.286*
statampere	ampere	−10 3.335 640
statcoulomb	coulomb	−10 3.335 640
statfarad	farad	−12 1.112 650
stathenry	henry	+11 8.987 554
statmho	mho	−12 1.112 650
statohm	ohm	+11 8.987 554
statute mile (U.S.)	meter	+03 1.609 344*
statvolt	volt	+02 2.997 925
stere	meter³	+00 1.00*
stilb	candela/meter²	+04 1.00
stoke	meter²/second	−04 1.00*
tablespoon	meter³	−05 1.478 676 478 125*
teaspoon	meter³	−06 4.928 921 593 75*
ton (assay)	kilogram	−02 2.916 666 6
ton (long)	kilogram	+03 1.016 046 908 8*
ton (metric)	kilogram	+03 1.00*
ton (explosive energy of one ton of TNT)	joule	+09 4.184
ton (register)	meter³	+00 2.831 684 659 2*
ton (short, 2000 pound)	kilogram	+02 9.071 847 4*
tonne	kilogram	+03 1.00*
torr (0°C)	newton/meter²	+02 1.333 22
township	meter²	+07 9.323 957 2
unit pole	weber	−07 1.256 637
volt (international of 1948)	volt	+00 1.000 330
watt (international of 1948)	watt	+00 1.000 165
yard	meter	−01 9.144*
year (calendar)	second (mean solar)	+07 3.1536*
year (sidereal)	second (mean solar)	+07 3.155 815 0
year (tropical)	second (mean solar)	+07 3.155 692 6
year 1900, tropical, Jan., day 0, hour 12	second (ephemeris)	+07 3.155 692 597 47*
year 1900, tropical, Jan., day 0, hour 12	second	+07 3.155 692 597 47

TABLE 2.2   *(continued)*

To convert from	To	Multiply by
LISTING BY PHYSICAL QUANTITY		

LISTING BY PHYSICAL QUANTITY

Acceleration

To convert from	To	Multiply by
foot/second^2	meter/second^2	-01 3.048*
free fall, standard	meter/second^2	$+00$ 9.806 65*
gal (galileo)	meter/second^2	-02 1.00*
inch/second^2	meter/second^2	-02 2.54*

Area

To convert from	To	Multiply by
acre	meter^2	$+03$ 4.046 856 422 4*
are	meter^2	$+02$ 1.00*
barn	meter^2	-28 1.00*
circular mil	meter^2	-10 5.067 074 8
foot^2	meter^2	-02 9.290 304*
hectare	meter^2	$+04$ 1.00*
inch^2	meter^2	-04 6.4516*
mile^2 (U.S. statute)	meter^2	$+06$ 2.589 988 110 336*
section	meter^2	$+06$ 2.589 988 110 336*
township	meter^2	$+07$ 9.323 957 2
yard^2	meter^2	-01 8.361 273 6*

Density

To convert from	To	Multiply by
gram/centimeter^3	kilogram/meter^3	$+03$ 1.00*
lbm/inch^3	kilogram/meter^3	$+04$ 2.767 990 5
lbm/foot^3	kilogram/meter^3	$+01$ 1.601 846 3
slug/foot^3	kilogram/meter^3	$+02$ 5.153 79

Energy

To convert from	To	Multiply by
British thermal unit (ISO/TC 12)	joule	$+03$ 1.055 06
British thermal unit (International Steam Table)	joule	$+03$ 1.055 04
British thermal unit (mean)	joule	$+03$ 1.055 87
British thermal unit (thermochemical)	joule	$+03$ 1.054 350 264 488
British thermal unit (39°F)	joule	$+03$ 1.059 67
British thermal unit (60°F)	joule	$+03$ 1.054 68
calorie (International Steam Table)	joule	$+00$ 4.1868
calorie (mean)	joule	$+00$ 4.190 02
calorie (thermochemical)	joule	$+00$ 4.184*
calorie (15°C)	joule	$+00$ 4.185 80
calorie (20°C)	joule	$+00$ 4.181 90
calorie (kilogram, International Steam Table)	joule	$+03$ 4.1868
calorie (kilogram, mean)	joule	$+03$ 4.190 02
calorie (kilogram, thermochemical)	joule	$+03$ 4.184*

TABLE 2.2   (*continued*)

To convert from	To	Multiply by
electron-volt	joule	−19 1.602 10
erg	joule	−07 1.00*
foot-lbf	joule	+00 1.355 817 9
foot-poundal	joule	−02 4.214 011 0
joule (international of 1948)	joule	+00 1.000 165
kilocalorie (International Steam Table)	joule	+03 4.1868
kilocalorie (mean)	joule	+03 4.190 02
kilocalorie (thermochemical)	joule	+03 4.184*
kilowatt-hour	joule	+06 3.60*
kilowatt-hour (international of 1948)	joule	+06 3.600 59
ton (nuclear equivalent of TNT)	joule	+09 4.20
watt-hour	joule	+03 3.60*

	Energy/area time	
Btu (thermochemical)/foot²-second	watt/meter²	+04 1.134 893 1
Btu (thermochemical)/foot²-minute	watt/meter²	+02 1.891 488 5
Btu (thermochemical)/foot²-hour	watt/meter²	+00 3.152 480 8
Btu (thermochemical)/inch²-second	watt/meter²	+06 1.634 246 2
calorie (thermochemical)/ centimeter²-minute	watt/meter²	+02 6.973 333 3
erg/centimeter²-second	watt/meter²	−03 1.00*
watt/centimeter²	watt/meter²	+04 1.00*

	Force	
dyne	newton	−05 1.00*
kilogram force (kgf)	newton	+00 9.806 65*
kilopond force	newton	+00 9.806 65*
kip	newton	+03 4.448 221 615 260 5*
lbf (pound force, avoirdupois)	newton	+00 4.448 221 615 260 5*
ounce force (avoirdupois)	newton	−01 2.780 138 5
pound force, lbf (avoirdupois)	newton	+00 4.448 221 615 260 5*
poundal	newton	−01 1.382 549 543 76*

	Length	
angstrom	meter	−10 1.00*
astronomical unit	meter	+11 1.495 978 9
cable	meter	+02 2.194 56*
caliber	meter	−04 2.54*
chain (surveyor or gunter)	meter	+01 2.011 68*
chain (engineer or ramden)	meter	+01 3.048*
cubit	meter	−01 4.572*
fathom	meter	+00 1.8288*
fermi (femtometer)	meter	−15 1.00*

TABLE 2.2   (continued)

To convert from	To	Multiply by
foot	meter	−01 3.048*
foot (U.S. survey)	meter	+00 1200/3937*
foot (U.S. survey)	meter	−01 3.048 006 096
furlong	meter	+02 2.011 68*
hand	meter	−01 1.016*
inch	meter	−02 2.54*
league (U.K. nautical)	meter	+03 5.559 552*
league (international nautical)	meter	+03 5.556*
league (statute)	meter	+03 4.828 032*
light year	meter	+15 9.460 55*
link (engineer or ramden)	meter	−01 3.048*
link (surveyor or gunter)	meter	−01 2.011 68*
meter	wavelengths Kr 86	+06 1.650 763 73*
micron	meter	−06 1.00*
mil	meter	−05 2.54*
mile (U.S. statute)	meter	+03 1.609 344*
mile (U.K. nautical)	meter	+03 1.853 184*
mile (international nautical)	meter	+03 1.852*
mile (U.S. nautical)	meter	+03 1.852*
nautical mile (U.K.)	meter	+03 1.853 184*
nautical mile (international)	meter	+03 1.852*
nautical mile (U.S.)	meter	+03 1.852*
pace	meter	−01 7.62*
parsec	meter	+16 3.083 74
perch	meter	+00 5.0292*
pica (printers)	meter	−03 4.217 517 6*
point (printers)	meter	−04 3.514 598*
pole	meter	+00 5.0292*
rod	meter	+00 5.0292*
skein	meter	+02 1.097 28*
span	meter	−01 2.286*
statute mile (U.S.)	meter	+03 1.609 344*
yard	meter	−01 9.144*

	Mass	
carat (metric)	kilogram	−04 2.00*
dram (avoirdupois)	kilogram	−03 1.771 845 195 312 5*
dram (troy or apothecary)	kilogram	−03 3.887 934 6*
grain	kilogram	−05 6.479 891*
gram	kilogram	−03 1.00*
hundredweight (long)	kilogram	+01 5.080 234 544*
hundredweight (short)	kilogram	+01 4.535 923 7*
kgf-second²-meter (mass)	kilogram	+00 9.806 65*
kilogram mass	kilogram	+00 1.00*
lbm (pound mass, avoirdupois)	kilogram	−01 4.535 923 7*
ounce mass (avoirdupois)	kilogram	−02 2.834 952 312 5*
ounce mass (troy or apothecary)	kilogram	−02 3.110 347 68*

TABLE 2.2   (*continued*)

To convert from	To	Multiply by
pennyweight	kilogram	−03 1.555 173 84*
pound mass, lbm (avoirdupois)	kilogram	−01 4.535 923 7*
pound mass (troy or apothecary)	kilogram	−01 3.732 417 216*
scruple (apothecary)	kilogram	−03 1.295 978 2*
slug	kilogram	+01 1.459 390 29
ton (assay)	kilogram	−02 2.916 666 6
ton (long)	kilogram	+03 1.016 046 908 8*
ton (metric)	kilogram	+03 1.00*
ton (short, 2000 pound)	kilogram	+02 9.071 847 4*
tonne	kilogram	+03 1.00*

	Power	
Btu (thermochemical)/second	watt	+03 1.054 350 264 488
Btu (thermochemical)/minute	watt	+01 1.757 250 4
calorie (thermochemical)/second	watt	+00 4.184*
calorie (thermochemical)/minute	watt	−02 6.973 333 3
foot-lbf/hour	watt	−04 3.766 161 0
foot-lbf/minute	watt	−02 2.259 696 6
foot-lbf/second	watt	+00 1.355 817 9
horsepower (550 foot lbf/second)	watt	+02 7.456 998 7
horsepower (boiler)	watt	+03 9.809 50
horsepower (electric)	watt	+02 7.46*
horsepower (metric)	watt	+02 7.354 99
horsepower (U.K.)	watt	+02 7.457
horsepower (water)	watt	+02 7.460 43
kilocalorie (thermochemical)/ minute	watt	+01 6.973 333 3
kilocalorie (thermochemical)/ second	watt	+03 4.184*
watt (international of 1948)	watt	+00 1.000 165

	Pressure	
atmosphere	newton/meter²	+05 1.013 25*
bar	newton/meter²	+05 1.00*
barye	newton/meter²	−01 1.00*
centimeter of mercury (0°C)	newton/meter²	+03 1.333 22
centimeter of water (4°C)	newton/meter²	+01 9.806 38
dyne/centimeter²	newton/meter²	−01 1.00*
foot of water (39.2°F)	newton/meter²	+03 2.988 98
inch of mercury (32°F)	newton/meter²	+03 3.386 389
inch of mercury (60°F)	newton/meter²	+03 3.376 85
inch of water (39.2°F)	newton/meter²	+02 2.490 82
inch of water (60°F)	newton/meter²	+02 2.4884
kgf centimeter²	newton/meter²	+04 9.806 65*
kgf/meter²	newton/meter²	+00 9.806 65*
lbf/foot²	newton/meter²	+01 4.788 025 8
lbf/inch²(psi)	newton/meter²	+03 6.894 757 2
millibar	newton/meter²	+02 1.00*

TABLE 2.2   (*continued*)

To convert from	To	Multiply by
millimeter of mercury (0°C)	newton/meter²	+02 1.333 224
pascal	newton/meter²	+00 1.00*
psi (lbf/inch²)	newton/meter²	+03 6.894 757 2
torr (0°C)	newton/meter²	+02 1.333 22
Speed		
foot/hour	meter/second	−05 8.466 666 6
foot/minute	meter/second	−03 5.08*
foot/second	meter/second	−01 3.048*
inch/second	meter/second	−02 2.54*
kilometer/hour	meter/second	−01 2.777 777 8
knot (international)	meter/second	−01 5.144 444 444
mile hour (U.S. statute)	meter/second	−01 4.4704*
mile/minute (U.S. statute)	meter/second	+01 2.682 24*
mile/second (U.S. statute)	meter/second	+03 1.609 344*
Temperature		
Celsius	kelvin	$t_K = t_C + 273.15$
Fahrenheit	kelvin	$t_K = (5/9)(t_F + 459.67)$
Fahrenheit	Celsius	$t_C = (5/9)(t_F - 32)$
Rankine	kelvin	$t_K = (5/9)t_R$
Time		
day (mean solar)	second (mean solar)	+04 8.64*
day (sidereal)	second (mean solar)	+04 8.616 409 0
hour (mean solar)	second (mean solar)	+03 3.60*
hour (sidereal)	second (mean solar)	+03 3.590 170 4
minute (mean solar)	second (mean solar)	+01 6.00*
minute (sidereal)	second (mean solar)	+01 5.983 617 4
month (mean calendar)	second (mean solar)	+06 2.628*
second (ephemeris)	second	+00 1.000 000 000
second (mean solar)	second (ephemeris)	Consult American Ephemeris and Nautical Almanac
second (sidereal)	second (mean solar)	−01 9.972 695 7
year (calendar)	second (mean solar)	+07 3.1536*
year (sidereal)	second (mean solar)	+07 3.155 815 0
year (tropical)	second (mean solar)	+07 3.155 692 6
year 1900, tropical, Jan., day 0, hour 12	second (ephemeris)	+07 3.155 692 597 47*
year 1900, tropical, Jan., day 0, hour 12	second	+07 3.155 692 597 47
Viscosity		
centistoke	meter²/second	−06 1.00*
stoke	meter²/second	−04 1.00*

TABLE 2.2   (continued)

To convert from	To	Multiply by
foot²/second	meter²/second	−02 9.290 304*
centipoise	newton-second/meter²	−03 1.00*
lbm/foot-second	newton-second/meter²	+00 1.488 163 9
lbf-second/foot²	newton-second/meter²	+01 4.788 025 8
poise	newton-second/meter²	−01 1.00*
poundal-second/foot²	newton-second/meter²	+00 1.488 163 9
slug/foot-second	newton-second/meter²	+01 4.788 025 8
rhe	meter²/newton-second	+01 1.00*

	Volume	
acre-foot	meter³	+03 1.233 481 9
barrel (petroleum, 42 gallons)	meter³	−01 1.589 873
board foot	meter³	−03 2.359 737 216*
bushel (U.S.)	meter³	−02 3.523 907 016 688*
cord	meter³	+00 3.624 556 3
cup	meter³	−04 2.365 882 365*
dram (U.S. fluid)	meter³	−06 3.696 691 195 312 5*
fluid ounce (U.S.)	meter³	−05 2.957 352 956 25*
foot³	meter³	−02 2.831 684 659 2*
gallon (U.K. liquid)	meter³	−03 4.546 087
gallon (U.S. dry)	meter³	−03 4.404 883 770 86*
gallon (U.S. liquid)	meter³	−03 3.785 411 784*
gill (U.K.)	meter³	−04 1.420 652
gill (U.S.)	meter³	−04 1.182 941 2
hogshead (U.S.)	meter³	−01 2.384 809 423 92*
inch³	meter³	−05 1.638 706 4*
liter	meter³	−03 1.00*
ounce (U.S. fluid)	meter³	−05 2.957 352 956 25*
peck (U.S.)	meter³	−03 8.809 767 541 72*
pint (U.S. dry)	meter³	−04 5.506 104 713 575*
pint (U.S. liquid)	meter³	−04 4.731 764 73*
quart (U.S. dry)	meter³	−03 1.101 220 942 715*
quart (U.S. liquid)	meter³	−04 9.463 529 5
stere	meter³	+00 1.00*
tablespoon	meter³	−05 1.478 676 478 125*
teaspoon	meter³	−06 4.928 921 593 75*
ton (register)	meter³	+00 2.831 684 659 2*
yard³	meter³	−01 7.645 548 579 84*

NOTES

CONSTRUCTION MATERIAL PROPERTIES

TABLE 3.1 Weights of Building Materials

Materials	Weight Lb. per Sq. Ft.	Materials	Weight Lb. per Sq. Ft.
CEILINGS		**PARTITIONS**	
Channel suspended system	1	Clay Tile	
Lathing and plastering	See Partitions	3 in.	17
Acoustical fiber tile	1	4 in.	18
		6 in.	28
FLOORS		8 in.	34
Steel Deck	See Manufacturer	10 in.	40
		Gypsum Block	
Concrete-Reinforced 1 in.		2 in.	9½
Stone	12½	3 in.	10½
Slag	11½	4 in.	12½
Lightweight	6 to 10	5 in.	14
		6 in.	18½
Concrete-Plain 1 in.		Wood Studs 2 × 4	
Stone	12	12–16 in. o.c.	2
Slag	11	Steel partitions	4
Lightweight	3 to 9	Plaster 1 inch	
		Cement	10
Fills 1 inch		Gypsum	5
Gypsum	6	Lathing	
Sand	8	Metal	½
Cinders	4	Gypsum Board ½ in.	2
Finishes			
Terrazzo 1 in.	13	**WALLS**	
Ceramic or Quarry Tile ¾ in.	10	Brick	
Linoleum ¼ in.	1	4 in.	40
Mastic ¾ in.	9	8 in.	80
Hardwood ⅞ in.	4	12 in.	120
Softwood ¾ in.	2½	Hollow Concrete Block (Heavy Aggregate)	
		4 in.	30
ROOFS		6 in.	43
Copper or tin	1	8 in.	55
Corrugated steel	See p. 6-5, *Steel Construction Manual*, 8th ed.	12½ in.	80
		Hollow Concrete Block (Light Aggregate)	
3-ply ready roofing	1	4 in.	21
3-ply felt and gravel	5½	6 in.	30
5-ply felt and gravel	6	8 in.	38
		12 in.	55
Shingles		Clay tile (Load Bearing)	
Wood	2	4 in.	25
Asphalt	3	6 in.	30
Clay tile	9 to 14	8 in.	33
Slate ¼	10	12 in.	45
		Stone 4 in.	55
Sheathing		Glass Block 4 in.	18
Wood ¾ in.	3	Windows, Glass, Frame & Sash	8
Gypsum 1 in.	4	Curtain Walls	See Manufacturer
Insulation 1 in.		Structural Glass 1 in.	15
Loose	½	Corrugated Cement Asbestos ¼ in.	3
Poured in place	2		
Rigid	1½		

SOURCE: American Institute of Steel Construction.

TABLE 3.2   Weights and Specific Gravities

Substance	Weight Lb. per Cu. Ft.	Specific Gravity	Substance	Weight Lb. per Cu. Ft.	Specific Gravity
ASHLAR MASONRY			**MINERALS**		
Granite, syenite, gneiss	165	2.3-3.0	Asbestos	153	2.1-2.8
Limestone, marble	160	2.3-2.8	Barytes	281	4.50
Sandstone, bluestone	140	2.1-2.4	Basalt	184	2.7-3.2
			Bauxite	159	2.55
MORTAR RUBBLE			Borax	109	1.7-1.8
MASONRY			Chalk	137	1.8-2.6
Granite, syenite, gneiss	155	2.2-2.8	Clay, marl	137	1.8-2.6
Limestone, marble	150	2.2-2.6	Dolomite	181	2.9
Sandstone, bluestone	130	2.0-2.2	Feldspar, orthoclase	159	2.5-2.6
			Gneiss, serpentine	159	2.4-2.7
DRY RUBBLE MASONRY			Granite, syenite	175	2.5-3.1
Granite, syenite, gneiss	130	1.9-2.3	Greenstone, trap	187	2.8-3.2
Limestone, marble	125	1.9-2.1	Gypsum, alabaster	159	2.3-2.8
Sandstone, bluestone	110	1.8-1.9	Hornblende	187	3.0
			Limestone, marble	165	2.5-2.8
BRICK MASONRY			Magnesite	187	3.0
Pressed brick	140	2.2-2.3	Phosphate rock, apatite	200	3.2
Common brick	120	1.8-2.0	Porphyry	172	2.6-2.9
Soft brick	100	1.5-1.7	Pumice, natural	40	0.37-0.90
			Quartz, flint	165	2.5-2.8
CONCRETE MASONRY			Sandstone, bluestone	147	2.2-2.5
Cement, stone, sand	144	2.2-2.4	Shale, slate	175	2.7-2.9
Cement, slag, etc.	130	1.9-2.3	Soapstone, talc	169	2.6-2.8
Cement, cinder, etc.	100	1.5-1.7			
VARIOUS BUILDING					
MATERIALS			**STONE, QUARRIED, PILED**		
Ashes, cinders	40-45	———	Basalt, granite, gneiss	96	———
Cement, portland, loose	90	———	Limestone, marble, quartz	95	———
Cement, portland, set	183	2.7-3.2	Sandstone	82	———
Lime, gypsum, loose	53-64	———	Shale	92	———
Mortar, set	103	1.4-1.9	Greenstone, hornblende	107	———
Slags, bank slag	67-72	———			
Slags, bank screenings	98-117	———			
Slags, machine slag	96	———			
Slags, slag sand	49-55	———	**BITUMINOUS SUBSTANCES**		
			Asphaltum	81	1.1-1.5
EARTH, ETC., EXCAVATED			Coal, anthracite	97	1.4-1.7
Clay, dry	63	———	Coal, bituminous	84	1.2-1.5
Clay, damp, plastic	110	———	Coal, lignite	78	1.1-1.4
Clay and gravel, dry	100	———	Coal, peat, turf, dry	47	0.65-0.85
Earth, dry, loose	76	———	Coal, charcoal, pine	23	0.28-0.44
Earth, dry, packed	95	———	Coal, charcoal, oak	33	0.47-0.57
Earth, moist, loose	78	———	Coal, coke	75	1.0-1.4
Earth, moist, packed	96	———	Graphite	131	1.9-2.3
Earth, mud, flowing	108	———	Paraffine	56	0.87-0.91
Earth, mud, packed	115	———	Petroleum	54	0.87
Riprap, limestone	80-85	———	Petroleum, refined	50	0.79-0.82
Riprap, sandstone	90	———	Petroleum, benzine	46	0.73-0.75
Riprap, shale	105	———	Petroleum, gasoline	42	0.66-0.69
Sand, gravel, dry, loose	90-105	———	Pitch	69	1.07-1.15
Sand, gravel, dry, packed	100-120	———	Tar, bituminous	75	1.20
Sand, gravel, wet	118-120	———			
EXCAVATIONS IN WATER					
Sand or gravel	60	———	**COAL AND COKE, PILED**		
Sand or gravel and clay	65	———	Coal, anthracite	47-58	———
Clay	80	———	Coal, bituminous, lignite	40-54	———
River mud	90	———	Coal, peat, turf	20-26	———
Soil	70	———	Coal, charcoal	10-14	———
Stone riprap	65	———	Coal, coke	23-32	———

The specific gravities of solids and liquids refer to water at 4°C., those of gases to air at 0°C. and 760 mm. pressure. The weights per cubic foot are derived from average specific gravities, except where stated that weights are for bulk, heaped or loose material, etc.

(continued)

TABLE 3.2 Weights and Specific Gravities (*continued*)

Substance	Weight Lb. per Cu. Ft.	Specific Gravity	Substance	Weight Lb. per Cu. Ft.	Specific Gravity
METALS, ALLOYS, ORES			**TIMBER, U. S. SEASONED**		
Aluminum, cast,			Moisture Content by		
hammered	165	2.55-2.75	Weight:		
Brass, cast, rolled	534	8.4-8.7	Seasoned timber 15 to 20%		
Bronze, 7.9 to 14% Sn	509	7.4-8.9	Green timber up to 50%		
Bronze, aluminum	481	7.7	Ash, white, red	40	0.62-0.65
Copper, cast, rolled	556	8.8-9.0	Cedar, white, red	22	0.32-0.38
Copper ore, pyrites	262	4.1-4.3	Chestnut	41	0.66
Gold, cast, hammered	1205	19.25-19.3	Cypress	30	0.48
Iron, cast, pig	450	7.2	Fir, Douglas spruce	32	0.51
Iron, wrought	485	7.6-7.9	Fir, eastern	25	0.40
Iron, spiegel-eisen	468	7.5	Elm, white	45	0.72
Iron, ferro-silicon	437	6.7-7.3	Hemlock	29	0.42-0.52
Iron ore, hematite	325	5.2	Hickory	49	0.74-0.84
Iron ore, hematite in bank	160-180	....	Locust	46	0.73
Iron ore, hematite loose	130-160	....	Maple, hard	43	0.68
Iron ore, limonite	237	3.6-4.0	Maple, white	33	0.53
Iron ore, magnetite	315	4.9-5.2	Oak, chestnut	54	0.86
Iron slag	172	2.5-3.0	Oak, live	59	0.95
Lead	710	11.37	Oak, red, black	41	0.65
Lead ore, galena	465	7.3-7.6	Oak, white	46	0.74
Magnesium, alloys	112	1.74-1.83	Pine, Oregon	32	0.51
Manganese	475	7.2-8.0	Pine, red	30	0.48
Manganese ore, pyrolusite	259	3.7-4.6	Pine, white	26	0.41
Mercury	849	13.6	Pine, yellow, long-leaf	44	0.70
Monel Metal	556	8.8-9.0	Pine, yellow, short-leaf	38	0.61
Nickel	565	8.9-9.2	Poplar	30	0.48
Platinum, cast, hammered	1330	21.1-21.5	Redwood, California	26	0.42
Silver, cast, hammered	656	10.4-10.6	Spruce, white, black	27	0.40-0.46
Steel, rolled	490	7.85	Walnut, black	38	0.61
Tin, cast, hammered	459	7.2-7.5	Walnut, white	26	0.41
Tin ore, cassiterite	418	6.4-7.0			
Zinc, cast, rolled	440	6.9-7.2			
Zinc ore, blende	253	3.9-4.2	**VARIOUS LIQUIDS**		
			Alcohol, 100%	49	0.79
			Acids, muriatic 40%	75	1.20
			Acids, nitric 91%	94	1.50
VARIOUS SOLIDS			Acids, sulphuric 87%	112	1.80
Cereals, oats .... bulk	32	....	Lye, soda 66%	106	1.70
Cereals, barley .... bulk	39	....	Oils, vegetable	58	0.91-0.94
Cereals, corn, rye .... bulk	48	....	Oils, mineral, lubricants	57	0.90-0.93
Cereals, wheat .... bulk	48	....	Water, 4°C. max. density	62.428	1.0
Hay and Straw .... bales	20	....	Water, 100°C.	59.830	0.9584
Cotton, Flax, Hemp .... bales	93	1.47-1.50	Water, ice	56	0.88-0.92
Fats	58	0.90-0.97	Water, snow, fresh fallen	8	.125
Flour, loose	28	0.40-0.50	Water, sea water	64	1.02-1.03
Flour, pressed	47	0.70-0.80			
Glass, common	156	2.40-2.60			
Glass, plate or crown	161	2.45-2.72			
Glass, crystal	184	2.90-3.00	**GASES**		
Leather	59	0.86-1.02	Air, 0°C. 760 mm.	.08071	1.0
Paper	58	0.70-1.15	Ammonia	.0478	0.5920
Potatoes, piled	42	....	Carbon dioxide	.1234	1.5291
Rubber, caoutchouc	59	0.92-0.96	Carbon monoxide	.0781	0.9673
Rubber goods	94	1.0-2.0	Gas, illuminating	.028-.036	0.35-0.45
Salt, granulated, piled	48	....	Gas, natural	.038-.039	0.47-0.48
Saltpeter	67	....	Hydrogen	.00559	0.0693
Starch	96	1.53	Nitrogen	.0784	0.9714
Sulphur	125	1.93-2.07	Oxygen	.0892	1.1056
Wool	82	1.32			

The specific gravities of solids and liquids refer to water at 4°C., those of gases to air at 0°C. and 760 mm. pressure. The weights per cubic foot are derived from average specific gravities, except where stated that weights are for bulk, heaped or loose material, etc.

SOURCE: American Institute of Steel Construction.

TABLE 3.3   Thermal Properties of Typical Building and Insulation Materials—Design Values[a]

Description	Density (lb/ft³)	Conductivity (k)	Conductance (C)	Customary Unit Resistance[b] (R)		Specific Heat, Btu/(lb) (deg F)	SI Unit Resistance[b] (R)	
				Per inch thickness (1/k)	For thickness listed (1/C)		(m·K) / W	(m²·K) / W
BUILDING BOARD								
Boards, Panels, Subflooring, Sheathing Woodboard Panel Products								
Asbestos-cement board	120	4.0	—	0.25	—	0.24	1.73	
Asbestos-cement board .... 0.125 in.	120	—	33.00	—	0.03			0.005
Asbestos-cement board .... 0.25 in.	120	—	16.50	—	0.06			0.01
Gypsum or plaster board ...... 0.375 in.	50	—	3.10	—	0.32	0.26		0.06
Gypsum or plaster board .... 0.5 in.	50	—	2.22	—	0.45			0.08
Gypsum or plaster board ...... 0.625 in.	50	—	1.78	—	0.56			0.10
Plywood (Douglas Fir)[c]	34	0.80	—	1.25	—	0.29	8.66	
Plywood (Douglas Fir) .... 0.25 in.	34	—	3.20	—	0.31			0.05
Plywood (Douglas Fir) .... 0.375 in.	34	—	2.13	—	0.47			0.08
Plywood (Douglas Fir) .... 0.5 in.	34	—	1.60	—	0.62			0.11
Plywood (Douglas Fir) .... 0.625 in.	34	—	1.29	—	0.77			0.14
Plywood or wood panels .... 0.75 in.	34	—	1.07	—	0.93	0.29		0.16
Vegetable Fiber Board								
Sheathing, regular density ..... 0.5 in.	18	—	0.76	—	1.32	0.31		0.23
..........................0.78125 in.	18	—	0.49	—	2.06			0.36
Sheathing intermediate density .. 0.5 in.	22	—	0.82	—	1.22	0.31		0.21
Nail-base sheathing .... 0.5 in.	25	—	0.88	—	1.14	0.31		0.20
Shingle backer .... 0.375 in.	18	—	1.06	—	0.94	0.31		0.17
Shingle backer .... 0.3125 in.	18	—	1.28	—	0.78			0.14
Sound deadening board .... 0.5 in.	15	—	0.74	—	1.35	0.30		0.24
Tile and lay-in panels, plain or acoustic	18	0.40	—	2.50	—	0.14	17.33	
..............................0.5 in.	18	—	0.80	—	1.25			0.22
.............................0.75 in.	18	—	0.53	—	1.89			0.33
Laminated paperboard	18	0.50	—	2.00	—	0.33	13.86	0.22
Homogeneous board from repulped paper	30	0.50	—	2.00	—	0.28	13.86	0.33

(continued)

TABLE 3.3   (continued)

Description	Density (lb/ft³)	Conductivity (k)	Conductance (C)	Resistance[b] (R) Per inch thickness (1/k)	Resistance[b] (R) For thickness listed (1/C)	Specific Heat, Btu/(lb)(deg F)	SI Unit Resistance[b] (R) (m·K)/W	SI Unit Resistance[b] (R) (m²·K)/W
Hardboard								
Medium density	50	0.73	—	1.37	—	0.31	9.49	
High density, service temp. service underlay	55	0.82	—	1.22	—	0.32	8.46	
High density, std. tempered	63	1.00	—	1.00	—	0.32	6.93	
Particleboard								
Low density	37	0.54	—	1.85	—	0.31	12.82	
Medium density	50	0.94	—	1.06	—	0.31	7.35	
High density	62.5	1.18	—	0.85	—	0.31	5.89	
Underlayment ....0.625 in.	40	—	1.22	—	0.82	0.29		0.14
Wood subfloor ....0.75 in.		—	1.06	—	0.94	0.33		0.17
BUILDING MEMBRANE								
Vapor—permeable felt.	—		16.70	—	0.06			0.01
Vapor—seal, 2 layers of mopped 15-lb felt	—		8.35	—	0.12			0.02
Vapor—seal, plastic film	—			—	Negl.			
FINISH FLOORING MATERIALS								
Carpet and fibrous pad	—		0.48	—	2.08	0.34		0.37
Carpet and rubber pad	—		0.81	—	1.23	0.33		0.22
Cork tile ....0.125 in.	—		3.60	—	0.28	0.48		0.05
Terrazzo ....1 in.	—		12.50	—	0.08	0.19		0.01
Tile—asphalt, linoleum, vinyl, rubber	—		20.00	—	0.05	0.30		0.01
vinyl asbestos						0.24		
ceramic.						0.19		
Wood, hardwood finish ....0.75 in.			1.47	—	0.68	0.19		0.12
INSULATING MATERIALS BLANKET AND BATT[d]								
Mineral Fiber, fibrous form processed from rock, slag, or glass								
approx.[e] 3–3.5 in.	0.3–2.0	—	0.091	—	11[d]			1.94
approx.[e] 5.50–6.5	0.3–2.0	—	0.053	—	19[d]			3.35
approx.[e] 6–7 in.	0.3–2.0	—	0.045	—	22[d]			3.87
approx.[e] 8.5–9 in.	0.3–2.0	—	0.033	—	30[d]			5.28
approx.[e] 12 in.	0.3–2.0	—	0.026	—	38[d]			6.69

BOARD AND SLABS

Material								
Cellular glass	8.5	0.35	—	2.86	—	0.18	19.81	
Glass fiber, organic bonded	4–9	0.25	—	4.00	—	0.23	27.72	
Expanded perlite, organic bonded	1.0	0.36	—	2.78	—	0.30	19.26	
Expanded rubber (rigid)	4.5	0.22	—	4.55	—	0.40	31.53	
Expanded polystyrene extruded								
Cut cell surface	1.8	0.25	—	4.00	—	0.29	27.72	
Smooth skin surface	1.8–3.5	0.20	—	5.00	—	0.29	34.65	
Expanded polystyrene, molded beads	1.0	0.26	—	—	—	—	26.3	3.8
	1.25	0.25	—	—	—	—	27.8	4.0
	1.5	0.24	—	—	—	—	29.1	4.2
	1.75	0.24	—	—	—	—	29.1	4.2
	2.0	0.23	—	—	—	—	29.8	4.3
Cellular polyurethane[a] (R-11 exp.)(unfaced)	1.5	0.16	—	6.25	—	0.38	43.82	
(Thickness 1 in. or greater)	2.5							
(Thickness 1 in. or greater—high resistance to gas permeation facing)	1.5	0.14						
Foil-faced, glass fiber-reinforced cellular								
Polyisocyanurate (R-11 exp.)[b]	2	0.14	—	7.04	—	0.22	48.79	
Nominal 0.5 in.		0.278	—	—	3.6	—		0.63
Nominal 1.0 in.		0.139	—	—	7.2	—		1.27
Nominal 2.0 in.		0.069	—	—	14.4	—		2.53
Mineral fiber with resin binder	15	0.29	—	3.45	—	0.17	23.91	
Mineral fiberboard, wet felted								
Core or roof insulation	16–17	0.34	—	2.94	—		20.38	
Acoustical tile	18	0.35	—	2.86	—	0.19	19.82	
Acoustical tile	21	0.37	—	2.70	—		18.71	
Mineral fiberboard, wet molded								
Acoustical tile[c]	23	0.42	—	2.38	—	0.14	16.49	
Wood or cane fiberboard								
Acoustical tile[c] ... 0.5 in.	—	—	0.80	—	1.25	—	0.31	20.38
Acoustical tile[c] ... 0.75 in.	—	—	0.53	—	1.89	—		19.82
Interior finish (plank, tile)	15	0.35	—	2.86	—	0.32	19.82	
Cement fiber slabs (shredded wood with Portland cement binder	25–27	0.50–0.53	—	2.0–1.89	—	—	13.87	
Cement fiber slabs (shredded wood with magnesia oxysulfide binder)	22	0.57	—	1.75	—	0.31	12.16	

(continued)

TABLE 3.3 (continued)

Description	Density (lb/ft³)	Conductivity (k)	Conductance (C)	Customary Unit Resistance[b] (R) Per inch thickness (1/k)	Customary Unit Resistance[b] (R) For thickness listed (1/C)	Specific Heat, Btu/(lb)(deg F)	SI Unit Resistance[b] (R) (m·K)/W	SI Unit Resistance[b] (R) (m²·K)/W
LOOSE FILL								
Cellulosic insulation (milled paper or wood pulp)	2.3–3.2	0.27–0.32	—	3.13–3.70	—	0.33	21.69–25.64	
Sawdust or shavings	8.0–15.0	0.45	—	2.22	—	0.33	15.39	
Wood fiber, softwoods	2.0–3.5	0.30	—	3.33	—	0.33	23.08	
Perlite, expanded			—	2.70	—	0.26	18.71	
Mineral fiber (rock, slag or glass)								
approx.[c] 3.75–5 in.	0.6–2.0	0.27–0.31	3.7–3.3		11	0.17		1.94
approx.[c] 6.5–8.75 in.	0.6–2.0	0.31–0.36	3.3–2.8		19			3.35
approx.[c] 7.5–10 in.	0.6–2.0	0.36–0.42	2.8–2.4		22			3.87
approx.[c] 10.25–13.75 in.					30			5.28
Vermiculite, exfoliated	0.6–8.0	0.47		2.13		3.20	14.76	
	4.0–6.0	0.44		2.27			15.73	
ROOF INSULATION[h]								
Preformed, for use above deck								
Different roof insulations are available in different thicknesses to provide the design C values listed.[h] Consult individual manufacturers for actual thickness of their material.		0.36 to 0.05			2.7 to 20			0.49 to 3.52
MASONRY MATERIALS								
CONCRETES								
Cement mortar	116	5.0	—	0.20	—		1.39	
Gypsum-fiber concrete 87.5% gypsum, 12.5% wood chips	51	1.66	—	0.60	—	0.21	4.16	
Lightweight aggregates including expanded shale, clay or slate; expanded slags; cinders; pumice; vermiculite; also cellular concretes	120	5.2	—	0.19	—		1.32	
	100	3.6	—	0.28	—		1.94	
	80	2.5	—	0.40	—		2.77	
	60	1.7	—	0.59	—		4.09	
	40	1.15	—	0.86	—		5.96	
	30	0.90	—	1.11	—		7.69	
	20	0.70	—	1.43	—		9.91	
Perlite, expanded	40	0.93	—	1.08	—		7.48	
	30	0.71	—	1.41	—		9.77	
	20	0.50	—	2.00	—	0.32	13.86	

Sand and gravel or stone aggregate (oven dried)	140	9.0		—			0.22	0.76
Sand and gravel or stone aggregate (not dried)	140	12.0	5.0	0.11	0.08			0.55 / 1.39
Stucco	116	5.0	9.0	0.20	0.20			1.39 / 0.76
MASONRY UNITS								
Brick, common	120	5.0		0.20	0.11		0.19	
Brick, face	130	9.0						
Clay tile, hollow:								
1 cell deep ... 3 in.				1.25		0.80	0.21	0.14
1 cell deep ... 4 in.				0.90		1.11		0.20
2 cells deep ... 6 in.				0.66		1.52		0.27
2 cells deep ... 8 in.				0.54		1.85		0.33
2 cells deep ... 10 in.				0.45		2.22		0.39
3 cells deep ... 12 in.				0.40		2.50		0.44
Concrete blocks, three oval core:								
Sand and gravel aggregate ... 4 in.				1.40		0.71	0.22	0.13
... 8 in.				0.90		1.11		0.20
... 12 in.				0.78		1.28		0.23
Cinder aggregate ... 3 in.				1.16		0.86	0.21	0.15
... 4 in.				0.90		1.11		0.20
... 8 in.				0.58		1.72		0.30
... 12 in.				0.53		1.89		0.33
Lightweight aggregate (expanded shale, clay, slate or slag; pumice) ... 3 in.				0.79		1.27	0.21	0.22
... 4 in.				0.67		1.50		0.26
... 8 in.				0.50		2.00		0.35
... 12 in.				0.44		2.27		0.40
Concrete blocks, rectangular core. *)								
Sand and gravel aggregate								
2 core, 8 in. 36 lb.**				0.96		1.04	0.22	0.18
Same with filled cores**				0.52		1.93	0.22	0.34
Lightweight aggregate (expanded shale, clay, slate or slag; pumice):								
3 core, 6 in. 19 lb.**				0.61		1.65	0.21	0.29
Same with filled cores**				0.33		2.99		0.53
2 core, 8 in. 24 lb.**				0.46		2.18		0.38
Same with filled cores**				0.20		5.03		0.89
3 core, 12 in. 38 lb.**				0.40		2.48		0.44
Same with filled cores**				0.17		5.82		1.02
Stone, lime or sand.			12.50		0.08		0.19	0.55
Gypsum partition tile:								
3 x 12 x 30 in. solid				0.79		1.26		0.22
3 x 12 x 30 in. 4-cell				0.74		1.35		0.24
4 x 12 x 30 in. 3-cell				0.60		1.67		0.29

(continued)

METALS   See Table 3.4.

TABLE 3.3 (continued)

Description	Density (lb/ft³)	Conductivity (k)	Conductance (C)	Resistance (R) Per inch thickness (1/k)	Resistance (R) For thickness listed (1/C)	Specific Heat, Btu/lb (deg F)	SI Unit Resistance (R) (m·K)/W	SI Unit Resistance (R) (m²·K)/W
PLASTERING MATERIALS								
Cement plaster, sand aggregate	116	5.0	—	0.20	—	0.20	1.39	
.........0.375 in.	—		13.3		0.08	0.20		0.01
.........0.75 in.	—		6.66		0.15	0.20		0.03
Gypsum plaster:								
Lightweigh aggregate.........0.5 in.	45	—	3.12		0.32			0.06
Lightweigh aggregate.........0.625 in.	45	—	2.67		0.39			0.07
Lightweigh agg. on metal lath.........0.75 in.	—	—	2.13		0.47			0.08
Perlite aggregate	45	1.5	—	0.67	—	0.32	4.64	
Sand aggregate	105	5.6	—	0.18	—	0.20	1.25	
Sand aggregate.........0.5 in.	105	—	11.10		0.09			0.02
Sand aggregate.........0.625 in.	105	—	9.10		0.11			0.02
Sand aggregate on metal lath.........0.75 in.	—	—	7.70		0.13			0.02
Vermiculite aggregate	45	1.7	—	0.59	—		4.09	
ROOFING								
Asbestos-cement shingles	120		4.76		0.21	0.24		0.04
Asphalt roll roofing	70		6.50		0.15	0.36		0.03
Asphalt shingles	70		2.27		0.44	0.30		0.08
Built-up roofing.........0.375 in.	70		3.00		0.33	0.35		0.06
Slate.........0.5 in.	—		20.00		0.05	0.30		0.01
Wood shingles, plain and plastic film faced	—		1.06		0.94	0.31		0.17
SIDING MATERIALS (ON FLAT SURFACE)								
Shingles								
Asbestos-cement	120		4.75		0.21	0.24		0.04
Wood, 16 in., 7.5 exposure	—		1.15		0.87	0.31		0.15
Wood, double, 16-in., 12-in. exposure	—		0.84		1.19	0.28		0.21
Wood, plus insul. backer board, 0.3125 in.	—		0.71		1.40	0.31		0.25
Siding								
Asbestos-cement, 0.25 in., lapped	—		4.76		0.21	0.24		0.04
Asphalt roll siding	—		6.50		0.15	0.35		0.03
Asphalt insulating siding (0.5 in. bed.)	—		0.69		1.46	0.35		0.26
Hardboard siding, 0.4375 in.	40	1.49	—	0.67	—	0.28	4.65	
Wood, drop, 1 × 8 in.	—		1.27		0.79	0.28		0.14
Wood, bevel, 0.5 × 8 in., lapped	—		1.23		0.81	0.28		0.14
Wood, bevel, 0.75 × 10 in., lapped	—		0.95		1.05	0.28		0.18
Wood, plywood, 0.375 in., lapped	—		1.59		0.59	0.29		0.10

Material	Density	k	C	1/k	1/C		Sp. ht.	
Aluminum or Steel,[m] over sheathing								
Hollow-backed			1.61		0.61			0.11
Insulating-board backed nominal								
0.375 in.			0.55		1.82		0.29	0.32
Insulating-board backed nominal								
0.375 in., foil backed			0.34		2.96		0.32	0.52
Architectural glass			10.00		0.10		0.20	0.02
WOODS[a,p]								
Maple, oak, and similar hardwoods	45	1.10		0.91		6.31	0.30	
Fir, pine, etc.	32	0.80		1.25		8.66	0.33	
0.75 in.			1.06		0.94			0.17
1.5 in.			0.53		1.88			0.33
2.5 in.			0.32		3.12			0.55
3.5 in.			0.23		4.38			0.77
5.5 in.			0.14		7.14			1.26
7.25 in.			0.11		9.09			1.60
9.25 in.			0.09		11.11			1.96
11.25 in.			0.07		14.28			2.15

[a] Representative values for dry materials were selected by ASHRAE TC 4.4, Thermal Insulation and Moisture Retarders (Total Thermal Performance Design Criteria). They are intended as design (not specification) values for materials in normal use. Insulation materials in actual service may have thermal values which vary from design values depending on their in-situ properties such as density and moisture content. For properties of a particular product, use the value supplied by the manufacturer or by unbiased tests.

[b] Resistance values are the reciprocals of C before rounding of C to two decimal places.

[c] Also see Insulating Materials, Board.

[d] Does not include paper backing and facing, if any. Where insulation forms a boundary (reflective or otherwise) of an air space, see Tables 1 and 2 for the insulating value of an air space for the appropriate effective emittance and temperature conditions of the space.

[e] Conductivity varies with fiber diameter. (See Chapter 21, Thermal Conductivity section, and Fig. 1) Insulation is produced by different densities; therefore, there is wide variation in thickness for the same R-value among manufacturers. No effort should be made to relate any specific R-value to any specific thickness. Commercial thicknesses generally available range from 2 to 8.5.

[f] Values are for aged, unfaced, board stock. For change in conductivity with age of expanded urethane, see Chapter 20, Factors Affecting Thermal Conductivity.

[g] Insulating value of acoustical tile varies, depending on density of the board and on type, size, and depth of perforations.

[h] ASTM C-855-77 recognizes the specification of roof insulation on the basis of the C-values shown. Roof insulation is made in thicknesses to meet these values.

[i] Face brick and common brick do not always have these specific densities. When density is different from that shown, there will be a change in thermal conductivity.

[j] Data on rectangular core concrete blocks differ from the above data on oval core blocks, due to core configuration, different mean temperatures, and possibly differences in unit weights. Weight data on the oval core blocks tested are not available.

[k] Weights of units approximately 7.625 in. high and 15.75 in. long. These weights are given as a means of describing the blocks tested, but conductance values are all for 1 ft² of area.

[l] Vermiculite, perlite, or mineral wool insulation. Where insulation is used, vapor barriers or other precautions must be considered to keep insulation dry.

[m] Values for metal siding applied over flat surfaces vary widely, depending on amount of ventilation of air space beneath the siding; whether air space is reflective or nonreflective; and on thickness, type, and application of insulating backing-board used. Values given are averages for use as design guides, and were obtained from several guarded hotbox tests (ASTM C236) or calibrated hotbox (BSS 77) on hollow-backed types and types made using backing-boards of wood fiber, foamed plastic, and glass fiber. Departures of ±50% or more from the values given may occur.

[n] Time-aged values for board stock with gas-barrier quality (0.001 in. thickness or greater) aluminum foil facers on two major surfaces.

[p] Forest Products Laboratory Wood Handbook, U.S. Dept. of Agriculture #72, 1974, Tables 3 and 4.

P.L. Adams: Supporting cryogenic equipment with wood (*Chemical Engineering*, May 17, 1971).

SOURCE: Reprinted by permission from *ASHRAE Handbook—1981 Fundamentals.*

TABLE 3.4 Properties of Solids

Material Description	Specific Heat J/kg · K	Density kg/m³	Thermal Conductivity W/m · K	Emissivity Ratio	Emissivity Surface Condition
Aluminum (alloy 1100)	896[u]	2 740[b]	221[u]	0.09[b] 0.20[u]	commercial sheet heavily oxidized
Aluminum Bronze (76%Cu, 22%Zn, 2% Al)	400[u]	8 280[u]			
Alundum (aluminum oxide)	779[b]		100[u]		
Asbestos: fiber		2 400[u]			
insulation	1050[b]		0.170[u]	0.93[b]	"paper"
Ashes, wood	800[t]	580[b]	0.16[b]		
Asphalt	920[b]	640[u]	0.071[b] [323]		
Bakelite	1500[b]	2 110[u]	0.74[b]		
Bell Metal	360[t] [323]	1 300[u]	17[u]		
Bismuth Tin	170[u]		65.0*		
Brick, building	800[b]	1 970[u]	0.7[b]	0.93*	about 394 K
Brass:					
red (85% Cu, 15% Zn)	400[u]	8 780[b]	150[u]	0.030[b]	highly polished
yellow (65% Cu, 35% Zn)	400[t]	8 310[u]	120[u] [273]	0.033[b]	highly polished
Bronze	435[t]	8 490[u]	29[u] [273]	0.02[d]	
Cadmium	230[d]	8 650[t]	92.9[b] [256]	0.81[a]	
Carbon (gas retort)	710[a]		0.07[b]		
Cardboard	1300[b]		0.057[t]		
Cellulose	670[t]	54[t]	0.029[t]		
Cement (Portland clinker)	900[t]	1 920[t]	0.83*	0.34*	
Chalk	840[t]	2 290[t]			
Charcoal (wood)	710[b]		0.05[a] [473]		
Chrome Brick	920[b]	3 200[b]	1.2[b]		
Clay	1000[t]	3 000[t]			
Coal	1500[t] [313]	1 400[t]	0.17[t] [273]		
Coal Tars	1500[b] [673]	1 200[b]	0.1[b]		
Coke (petroleum, powdered)	653[t] [473]	990[b]	0.95[b] [673]		
Concrete (stone)	390[t]	2 300[u]	0.93[b]		
Copper (electrolytic)	2030[u]	8 910[u]	393	0.072[n]	commercial, shiny
Cork (granulated)	1340[u]	86[t]	0.048[t] [268]		
Cotton (fiber)	1060[b]	1 500[b]	0.042[u]		
Cryolite (AlF₃ · 3NaF)	616[t]	2 000[b]			
Diamond		2 420[t]	47[t]		
Earth (dry and packed)		1 330[b]	0.064[e]	0.41*	
Felt			0.05[b]		
Fireclay Brick	829[b] [373]	1 790[b]		0.75[u]	at 1273 K
Fluorspar (CaF₂)	880[b]	3 190[u]	1.1[v] [473]		
German Silver (nickel silver)	400[u]	8 730[u]	33[u]	0.135[n]	polished

Material					
Glass:					
crown (soda-lime)	750[b]	2 470[b]	1.0[i] [366]	0.94[n]	smooth
flint (lead)	490[b]	4 280[u]	1.4[i]		
pyrex	840[b]	2 230[i]	1.0[i] [366]		
"wool"	657[b]	52.0[i]	0.038[i]		
Gold	131[i]	19 350[u]	297[i]	0.02[n]	highly polished
Graphite:					
powder	691*	1 870[u]	0.183*	0.75[b]	on a smooth plate
"Karbate" (impervious)	670[u]	1 200[b]	130[u]	0.903[b]	on a smooth plate
Gypsum	1080[i]	1 500[u]	0.43[b]		
Hemp (fiber)	1352.3[u]	921[u]		0.95*	
Ice: [0°C]	2040[i]		2.24[b]		
[−20°C]	1950[i]		2.44[b]		
Iron:					
cast	500[i] [373]	7 210[i]	47.5[b] [327]	0.435[b]	freshly turned
wrought		7 770[u]	60.4[b]	0.94[b]	dull, oxidized
Lead	129[i]	11 300[u]	34.8[b]	0.28[b]	gray, oxidized
Leather (sole)		1 000[u]	0.16[b]		
Limestone	909[i]	1 650[u]	0.93[b]	0.36* to 0.90	at 336 to 467 K
Linen			0.09[b]		
Litharge (lead monoxide)	230[b]	7 850[b]			
Magnesia:					
powdered	980[b] [373]	796[b]	0.61[b] [320]		
light carbonate	930[b] [373]	210[b]	0.059[b] [373]		
Magnesite Brick		2 530[u]	3.8[b] [478]		
Magnesium	1000[b]	1 730[u]	160[u]	0.55[n]	oxidized
Marble	880[b]	2 600[b]	2.6[b]		
Nickel	440[b]	8 890[u]	59.5[u]	0.931[b]	light gray, polished electroplated,
				0.045[u]	polished
Paints:					
White lacquer				0.80[n]	on rough plate
White enamel				0.91[n]	
Black lacquer				0.80[n]	"matte" finish
Black shellac		1 000[u]	0.26[u]	0.91[n]	
Flat black lacquer				0.96[n]	on rough plate pasted on tinned plate
Aluminum lacquer				0.39[b]	
Paper	1300*	930[b]	0.13[b]	0.92[b]	
Paraffin	2900[i]	900[i]	0.24[b] [273]		
Plaster		2 110[u]	0.74[b] [348]	0.91[b]	rough
Platinum	130[b]	21 470[u]	69.0[u]	0.054[b]	polished
Porcelain	750	260[u]	2.2[b]	0.92[b]	glazed
Pyrites (Copper)	549[b]	4 200[b]			
Pyrites (Iron)	569[b] [342]	4 970[v]			

*Data source not determined.

(continued)

TABLE 3.4 Properties of Solids (continued)

Material Description	Specific Heat J/kg·K	Density kg/m³	Thermal Conductivity W/m·K	Emissivity Ratio	Emissivity Surface Condition
Rock Salt	917[u]	2 180[u]			
Rubber:					
Vulcanized (soft)	2000[*]	1 100[t]	0.1[t]	0.86[b]	rough
(hard)		1 190[t]	0.16[t]	0.95[b]	glossy
Sand	800[b]	1 520[b]	0.33[b]		
Sawdust		190[b]	0.05[b]		
Silica	1320[b]	2 240[v]	1.4[t] [366]	0.02[u]	polished and at 500 K
Silver	235[u]	10 500[u]	424[u] [366]		
Snow (freshly fallen)		100[t]	0.598[t]		
(at 32 F)		500[t]	2.2[t]		
Steel (mild)	500[b]	7 830[b]	45.3[b]	0.12[n]	cleaned
Stone (quarried)	800[b]	1 500[t]			
Tar:					
pitch	2500[v]	1 100[u]	0.88[v]		
bituminous		1 200[t]	0.71[u]		
Tin	233[u]	7 290[u]	64.9[u]	0.06[h]	bright and at 323 K
Tungsten	130[u]	19 400[u]	201[u]	0.032[n]	filament at 300 K
Wood:					
Hardwoods:	1900/2700[b]	370/1 100[z]	0.11/0.255[z]		
Ash, white		690[z]	0.172[z]		
Elm, American		580[z]	0.153[z]		
Hickory		800[z]			
Mahogany		550[z]	0.13[u]		
Maple, sugar	2390[u]	720[z]	0.187[z]	0.90[u]	planed
Oak, white		750[z]	0.176[z]		
Walnut, black		630[z]			
Softwoods:	See Table 3A, Chap. 23, source text	350/740[z]	0.11/0.16[z]		
Fir, white		430[z]	0.12[z]		
Pine, white		430[z]	0.11[z]		
Spruce		420[z]			
Wool:					
Fiber	1360[u]	1 300[u]	0.036/0.063[u]		
Fabric		110/330[u]			
Zinc:					
Cast	390[u]	7 130[u]	110[u]	0.05[u]	polished
Hot-rolled	390[b]	7 130[b]	110[b]		
Galvanizing				0.23[u]	fairly bright

[a] Source unknown.

[a] *Handbook of Chemistry & Physics* (Chemical Rubber Publishing Co., Cleveland, OH, 47th ed., 1966, 49th ed., 1969).

[b] J. H. Perry: *Chemical Engineers' Handbook* (McGraw-Hill Book Co., Inc., New York, NY, 2nd ed., 1941, 4th ed., 1963).

[c] *Tables of Thermodynamic and Transport Properties of Air, Argon, Carbon Dioxide, Carbon Monoxide, Hydrogen, Nitrogen, Oxygen and Steam* (Pergamon Press, Elmsford, NY, 1960).

[d] *American Institute of Physics Handbook* (McGraw-Hill Book Co., Inc., New York, NY, 2nd ed., 1963).

[e] Organick and Studhalter: Thermodynamic properties of benzene (*Chemical Engineering Progress*, November 1948, p. 847).

[f] Lange: *Handbook of Chemistry* (McGraw-Hill Book Co., Inc., New York, NY, revised 10th ed., 1967).

[g] *ASHRAE Thermodynamic Properties of Refrigerants* (ASHRAE, 1969).

[h] Reid and Sherwood: *The Properties of Gases and Liquids* (McGraw-Hill Book Co., Inc., New York, NY, 2nd ed., 1966).

[i] *ASHRAE Handbook of Fundamentals* (ASHRAE, 1967).

[j] *T.P.R.C. Data Book* (Thermophysical Properties Research Center, W. Lafayette, IN, 1966).

[k] Estimated.

[l] L. N. Canjar, Max Goldman, and Henry Marchman: Thermodynamic properties of propylene (*Industrial and Engineering Chemistry*, May 1951, p. 1183).

[m] *ASME Steam Tables* (American Society of Mechanical Engineers, New York, NY, 1967).

[n] W. H. McAdams: *Heat Transmission* (McGraw-Hill Book Co.,

Inc., New York, NY, 3rd ed., 1954).

[o] D. R. Stull: Vapor pressure of pure substances (organic compounds) (*Industrial and Engineering Chemistry*, April 1947, p. 517).

[p] *JANAF Thermochemical Tables* (PB 168 370, National Technical Information Service, Springfield, VA, 1965).

[q] *Physical Properties of Chemical Compounds* (American Chemical Society, Washington, DC, 1955-1961).

[r] *International Critical Tables of Numerical Data* (National Research Council of U. S. A., published by McGraw-Hill Book Co., Inc., New York, NY, 1928).

[s] *Matheson Gas Data Book* (Matheson Company, Inc., East Rutherford, NJ, 4th ed., 1966).

[t] Baumeister and Marks: *Standard Handbook for Mechanical Engineers* (McGraw-Hill Book Co., Inc., New York, NY, 1967).

[u] Miner and Seastone: *Handbook of Engineering Materials* (John Wiley and Sons, New York, NY, 1955).

[v] Kirk and Othmer: *Encyclopedia of Chemical Technology*, Interscience Division of John Wiley and Sons, New York, NY, 1966).

[w] Gouse and Stevens: *Chemical Technology of Petroleum* (McGraw-Hill Book Co., Inc., New York, NY, 3rd ed., 1960).

[x] *Saline Water Conversion Engineering Data Book* (M. W. Kellogg Co. for U. S., Dept. of Interior, 1955).

[y] J. Timmermans: *Physicochemical Constants of Pure Organic Compounds* (American Elsevier, New York, NY, Vol. 2, 2nd ed., 1965).

[z] *Wood Handbook* (Handbook No. 72, Forest Products Laboratory, U.S. Dept. of Agriculture, 1955).

[aa] Thermophysical Properties, of Refrigerants (ASHRAE 1976).

SOURCE: Reprinted by permission from *ASHRAE Handbook—1981 Fundamentals.*

TABLE 3.5 Coefficients of Transmission (U) of Frame Walls[a]

These coefficients are expressed in Btu per (hour) (square foot) (degree Fahrenheit difference in temperature between the air on the two sides), and are based on an outside wind velocity of 15 mph

Construction	Replace Air Space with 3.5-in. R-11 Blanket Insulation (New Item 4)			
	Resistance (R)			
	1		2	
	Between Framing	At Framing	Between Framing	At Framing
1. Outside surface (15 mph wind)	0.17	0.17	0.17	0.17
2. Siding, wood, 0.5 in. × 8 in. lapped (average)	0.81	0.81	0.81	0.81
3. Sheathing, 0.5-in. vegetable fiber board	1.32	1.32	1.32	1.32
4. Nonreflective air space, 3.5 in. (50 F mean; 10 deg F temperature difference)	1.01	—	11.00	—
5. Nominal 2-in. × 4-in. wood stud	—	4.35	—	4.35
6. Gypsum wallboard, 0.5 in.	0.45	0.45	0.45	0.45
7. Inside surface (still air)	0.68	0.68	0.68	0.68
Total Thermal Resistance (R)	R_i =4.44	R_s =7.78	R_I =14.43	R_s =7.78

Construction No. 1: U_i = 1/4.44 =0.225; U_s =1/7.81 =0.128. With 20% framing (typical of 2-in. × 4-in. studs @ 16-in. o.c.), U_{av} = 0.8(0.225) + 0.20(0.128) = 0.199 (See Eq 9)

Construction No. 2: U_i = 1/14.43 = 0.069; U_s = 0.128. With framing unchanged, U_{av} = 0.8(0.069) + 0.20(0.128) = 0.081

[a]See section Calculating Overall Coefficients in Chapter 23 of source text for basis of calculations.

SOURCE: Reprinted by permission from *ASHRAE Handbook—1981 Fundamentals*.

TABLE 3.6 Coefficients of Transmission (U) of Frame Partitions of Interior Walls[a]

Coefficients are expressed in Btu per (hour) (square foot) (degree Fahrenheit difference in temperature between the air on the two sides), and are based on still air (no wind) conditions on both sides

Replace Air Space with 3.5-in. R-11 Blanket Insulation (New Item 3)

| | Resistance (R) | | | |
| | 1 | | 2 | |
Construction	Between Framing	At Framing	Between Framing	At Framing
1. Inside surface (still air)	0.68	0.68	0.68	0.68
2. Gypsum wallboard, 0.5 in.	0.45	0.45	0.45	0.45
3. Nonreflective air space, 3.5 in. (50 F mean; 10 deg F temperature difference)	1.01	—	11.00	—
4. Nominal 2-in. × 4-in. wood stud	—	4.38	—	4.38
5. Gypsum wallboard, 0.5 in.	0.45	0.45	0.45	0.45
6. Inside surface (still air)	0.68	0.68	0.68	0.68
Total Thermal Resistance (R)	$R_1 = 3.27$	$R_2 = 6.64$	$R_1 = 13.26$	$R_2 = 6.64$

Construction No. 1: U_s = 1/3.27 = 0.306; U_i = 1/6.64 = 0.151. With 12% framing (typical of 2-in. × 4-in. studs @ 24-in. o.c.), U_{av} = 0.9 (0.306) + 0.12(0.151) = 0.293.

Construction No. 2: U_i = 1/13.26 = 0.075$_6$ U_s = 1/6.64 = 0.151. With framing unchanged, U_{av} = 0.9(0.075) + 0.1(0.151) = 0.083.

[a]See section Calculating Overall Coefficients in Chapter 23 of source text for basis of calculations.

SOURCE: Reprinted by permission from *ASHRAE Handbook—1981 Fundamentals*.

TABLE 3.7 Coefficients of Transmission (U) of Solid Masonry Walls[a]

Coefficients are expressed in Btu per (hour) (square foot) (degree Fahrenheit difference in temperature between the air on the two sides), and are based on an outside wind velocity of 15 mph

Construction	Replace Furring Strips and Air Space with 1-in. Expanded Polystyrene Extruded, Smooth Skin Surface, 2.2 lb/ft³ (New Item 4)		
	Resistance (R)		
	1		2
	Between Furring	At Furring	
1. Outside surface (15 mph wind)	0.17	0.17	0.17
2. Common brick, 8 in.	1.60	1.60	1.60
3. Nominal 1-in. ×3-in. vertical furring	—	0.94	—
4. Nonreflective air space, 0.75 in. (50 F mean; 10 deg F temperature difference)	1.01	—	5.00
5. Gypsum wallboard, 0.5 in.	0.45	0.45	0.45
6. Inside surface (still air)	0.68	0.68	0.68
Total Thermal Resistance (R)	$R_1 = 3.91$	$R_s = 3.84$	$R_1 = 7.90 = R_s$

Construction No. 1: $U_i = 1/3.91 = 0.256$; $U_s = 1/3.84 = 0.260$. With 20% framing (typical of 1-in. × 3-in. vertical furring on masonry @ 16-in. o.c.)
$U_{av} = 0.8 (0.256) + 0.2 (0.260) = 0.257$
Construction No. 2: $U_i = U_s = U_{av} = 1/7.90 = 0.127$

[a]See section Calculating Overall Coefficients in Chapter 23 of source text for basis of calculations.
SOURCE: Reprinted by permission from ASHRAE Handbook—1981 Fundamentals.

TABLE 3.8 Coefficients of Transmission (U) of Masonry Walls[a]

Coefficients are expressed in Btu per (hour) (square foot) (degree Fahrenheit difference in temperature between the air on the two sides), and are based on an outside wind velocity of 15 mph

	Replace Cinder Aggregate Block with 6-in. Light-weight Aggregate Block with Cores Filled (New Item 4)			
			Resistance (R)	
	1		2	
Construction	Between Furring	At Furring	Between Furring	At Furring
1. Outside surface (15 mph wind)	0.17	0.17	0.17	0.17
2. Face brick, 4 in.	0.44	0.44	0.44	0.44
3. Cement mortar, 0.5 in.	0.10	0.10	0.10	0.10
4. Concrete block, cinder aggregate, 8 in.	1.72	1.72	2.99	2.99
5. Reflective air space, 0.75 in. (50 F mean; 30 deg F temperature difference) E = 0.05[b,c]	2.77	—	2.77	—
6. Nominal 1-in. × 3-in. vertical furring	—	0.94	—	0.94
7. Gypsum wallboard, 0.5 in.	0.45	0.45	0.45	0.45
8. Inside surface (still air)	0.68	0.68	0.68	0.68
Total Thermal Resistance (R)	$R_s = 6.33$	$R_i = 4.50$	$R_i = 7.60$	$R_s = 5.77$

Construction No. 1: $U_s = 1/6.33 = 0.158$; $U_i = 1/4.50 = 0.222$. With 20% framing (typical of 1-in. × 3-in. vertical furring on masonry @ 16-in. o.c.), $U_{av} = 0.8(0.158) + 0.2(0.222) = 0.171$.

Construction No. 2: $U_i = 1/7.60 = 0.132$, $U_s = 1/5.77 = 0.173$. With framing unchanged, $U_{av} = 0.8(0.132) + 0.2(0.173) = 0.140$.

[a] See section Calculating Overall Coefficients in Chapter 23 of source text for basis of calculations.

[b] Based on E =0.05 for bright, polished surface. Due to surface oxidation, and other factors, actual value of the effective emissivity as it ages is expected to be E = 0.10 to 20.

[c] See "Caution" p. 23.5 para. 2—re: air leakage and adjusted U-values.

SOURCE: Reprinted by permission from *ASHRAE Handbook—1981 Fundamentals*.

TABLE 3.9   Coefficients of Transmission (U) of Masonry Cavity Walls[a]

Coefficients are expressed in Btu per (hour) (square foot) (degree Fahrenheit difference in temperature between the air on the two sides), and are based on an outside wind velocity of 15 mph

	Resistance (R)		
	Replace Furring Strips and Gypsum Wallboard with 0.625-in. Plaster (Sand Aggregate) Applied Directly to Concrete Block-Fill 2.5-in. Air Space with Vermiculite Insulation, 7–8.2 lb/ft³ (New Items 3 and 7)		
	1		2
Construction	Between Furring	At Furring	
1. Outside surface (15 mph wind)	0.17	0.17	0.17
2. Common brick, 4 in.	0.80	0.80	0.80
3. Nonreflective air space, 2.5 in. (30 F mean; 10 deg F temperature difference)	1.10*	1.10*	5.32**
4. Concrete block, three-oval core, stone and gravel aggregate, 4 in.	0.71	0.71	0.71
5. Nonreflective air space 0.75 in. (50 F mean; 10 deg F temperature difference)	1.01	—	—
6. Nominal 1-in. × 3-in. vertical furring	—	0.94	—
7. Gypsum wallboard, 0.5 in.	0.45	0.45	0.11
8. Inside surface (still air)	0.68	0.68	0.68
Total Thermal Resistance (R)	$R_i = 4.92$	$R_s = 4.85$	$R_i = R_s = 7.79$

Construction No. 1: $U_i = 1/4.92 = 0.203$; $U_s = 1/4.85 = 0.206$. With 20% framing (typical of 1-in. × 3-in. vertical furring on masonry @16-in. o.c.) $U_{av} = 0.8(0.203) + 0.2(0.206) = 0.204$

Construction No. 2: $U_s = U_2 = 1/7.79 = 0.128$

[a] See section Calculating Overall Coefficients in Chapter 23 of source text for basis of calculations.

*Interpolated value from Table 2 in Chapter 23 of source text.

**Calculated value from Table 3 in Chapter 23 of source text.

SOURCE: Reprinted by permission from *ASHRAE Handbook—1981 Fundamentals.*

TABLE 3.10 Coefficients of Transmission (U) of Masonry Partitions[a]

Coefficients are expressed in Btu per (hour) (square foot) (degree Fahrenheit difference in temperature between the air on the two sides), and are based on still air (no wind) conditions on both sides

Replace Concrete Block with 4-in. Gypsum Tile (New Item 3) Construction	1	2
1. Inside surface (still air)	0.68	0.68
2. Plaster, lightweight aggregate, 0.625 in.	0.39	0.39
3. Concrete block, cinder aggregate, 4 in.	1.11	1.67
4. Plaster, lightweight aggregate, 0.625 in.	0.39	0.39
5. Inside surface (still air)	0.68	0.68
Total Thermal Resistance(R)	3.25	3.81

Construction No. 1: $U = 1/3.25 = 0.308$
Construction No. 2: $U = 1/3.81 = 0.262$

[a]See section Calculating Overall Coefficients in Chapter 23 of source text for basis of calculations.
SOURCE: Reprinted by permission from *ASHRAE Handbook—1981 Fundamentals.*

TABLE 3.11 Coefficients of Transmission (U) of Frame Construction Ceilings and Floors[a]

Coefficients are expressed in Btu per (hour) (square foot) (degree Fahrenheit difference between the air on the two sides), and are based on still air (no wind) on both sides

Assume Unheated Attic Space above Heated Room with Heat Flow Up—Remove Tile, Felt, Plywood, Subfloor and Air Space—Replace with R-19 Blanket Insulation (New Item 4)
Heated Room Below Unheated Space

Construction (Heat Flow Up)	Resistance (R) 1		Resistance (R) 2	
	Between Floor Joists	At Floor Joists	Between Floor Joists	At Floor Joists
1. Bottom surface (still air)	0.61	0.61	0.61	0.61
2. Metal lath and lightweight aggregate, plaster, 0.75 in.	0.47	0.47	0.47	0.47
3. Nominal 2-in. × 8-in. floor joist	—	9.06	—	9.06
4. Nonreflective airspace, 7.25-in. (50 F mean; 10 deg F temperature difference)	0.93*	—	19.00	—
5. Wood subfloor, 0.75 in.	0.94	0.94	—	—
6. Plywood, 0.625 in.	0.77	0.77	—	—
7. Felt building membrane	0.06	0.06	—	—
8. Tile	0.05	0.05	—	—
9. Top surface (still air)	0.61	0.61	0.61	0.61
Total Thermal Resistance (R)	$R_s = 4.44$	$R_s = 12.57$	$R_s = 20.69$	$R_s = 10.75$

Construction No. 1: $U_i = 1/4.45 = 0.225$; $U_s = 1/12.58 = 0.079$. With 10% framing (typical of 2-in. joists @ 16-in. o.c.), $U_{av} = 0.9 (0.225) + 0.1 (0.079) = 0.210$

Construction No. 2: $U_i = 1/20.69 = 0.048$; $U_s = 1/10.75 = 0.093$. With framing unchanged, $U_{av} = 0.9 (0.048) + 0.1 (0.093) = 0.053$

[a]See section Calculating Overall Coefficients in Chapter 23 of source text for basis of calculations.

*Use largest air space (3.5-in.) value shown in Table 2, Chapter 23, of source text.

SOURCE: Reprinted by permission from *ASHRAE Handbook—1981 Fundamentals*.

TABLE 3.12 Coefficients of Transmission (U) of Flat Masonry Roofs with Built-Up Roofing, with and without Suspended Ceilings[a,b] (Winter Conditions, Upward Flow)

These Coefficients are expressed in Btu per (hour) (square foot) (degree Fahrenheit difference in temperature between the air on the two sides), and are based upon an outside wind velocity of 15 mph.

Add Rigid Roof Deck Insulation, $C = 0.24$ ($R = 1/C = 4.17$) (New Item 7) Construction (Heat Flow Up)	1	2
1. Inside surface (still air)	0.61	0.61
1. Metal lath and lightweight aggregate plaster, 0.75 in.	0.47	0.47
3. Nonreflective air space, greater than 3.5 in. (50 F mean; 10 deg F temperature difference)	0.93*	0.93*
4. Metal ceiling suspension system with metal hanger rods	0**	0**
5. Corrugated metal deck	0	0
6. Concrete slab, lightweight aggregate, 2 in. (30 lb/ft³)	2.22	2.22
7. Rigid roof deck insulation (none)	—	4.17
8. Built-up roofing, 0.375 in.	0.33	0.33
9. Outside surface (15 mph wind)	0.17	0.17
Total Thermal Resistance (R).	4.73	8.90

Construction No. 1: $U_{av} = 1/4.73 = 0.211$
Construction No. 2: $U_{av} = 1/8.90 = 0.112$

[a]See section Calculating Overall Coefficients in Chapter 23 of source text for basis of calculations.
[b]To adjust U values for the effect of added insulation between framing members, see Table 5 or 6 of source text.
*Use largest air space (3.5-in.) value shown in Table 2, Chapter 23, of source text.
** Area of hanger rods is negligible in relation to ceiling area.
SOURCE: Reprinted by permission from *ASHRAE Handbook—1981 Fundamentals.*

TABLE 3.13 Coefficients of Transmission (U) of Wood Construction Flat Roofs and Ceilings[a] (Winter Conditions, Upward Flow)

Coefficients are expressed in Btu per (hour) (square foot) (degree Fahrenheit difference in temperature between the air on the two sides), and are based upon an outside wind velocity of 15 mph.

	1		2 — Replace Roof Deck Insulation and Partially Fill the 7.25-in. Air Space with 6-in. R-19 Blanket Insulation and 1.25-in. Air Space (New Items 5 and 7)	
	Resistance (R)			
Construction (Heat Flow Up)	Between Joists	At Joists	Between Joists	At Joists
1. Inside surface (still air)	0.61	0.61	0.61	0.61
2. Acoustical tile, fiberboard, 0.5 in.	1.25	1.25	1.25	1.25
3. Gypsum wallboard, 0.5 in.	0.45	0.45	0.45	0.45
4. Nominal 2-in. × 8-in. ceiling joists	—	9.06	—	9.06
5. Nonreflective air space, 7.25 in. (50 F mean; 10 deg F temperature difference)	0.93*	—	1.05**	0.78
6. Plywood deck, 0.625 in.	0.78	0.78	0.78	0.78
7. Rigid roof deck insulation, c = 0.72, (R = 1/C)	1.39	1.39	19.00	—
8. Built-up roof	0.33	0.33	0.33	0.33
9. Outside surface (15 mph wind)	0.17	0.17	0.17	0.17
Total Thermal Resistance (R)	$R_i = 5.91$	$R_s = 14.04$	$R_i = 23.64$	$R_s = 12.65$

Construction No. 1: $U_i = 1/5.91 = 0.169$; $U_s = 1/14.04 = 0.071$. With 10% framing (typical of 2-in. joists @ 16-in. o.c.), $U_{av} = 0.9$ $(0.169) + 0.1 (0.071) = 0.159$.

Construction No. 2: $U_i = 1/23.64 = 0.042$; $U_s = 1/12.65 = 0.079$. With framing unchanged, $U_{av} = 0.9 (0.042) + 0.1 (0.079) = 0.046$.

[a]See section Calculating Overall Coefficients in Chapter 23 of source text for basis of calculations.
*Use largest air space (3.5-in.) value shown in Table 2, Chapter 23, of source text.
**Interpolated value (0°F mean; 10°F temperature difference).
SOURCE: Reprinted by permission from *ASHRAE Handbook—1981 Fundamentals.*

TABLE 3.14   Coefficients of Transmission (U) of Metal Construction Flat Roofs and Ceilings[a] (Winter Conditions, Upward Flow)

Coefficients are expressed in Btu per (hour) (square foot) (degree Fahrenheit difference in temperature between the air on the two sides), and are based on upon outside wind velocity of 15 mph.

Construction (Heat Flow Up)	Replace Rigid Roof Deck Insulation (C = 0.24) and Sand Aggregate Plaster with Rigid Roof Deck Insulation, C = 0.36 and Lightweight Aggregate Plaster 0.75 in. on Metal Lath (New Items 2 and 6)	
	1	2
1. Inside surface (still air)	0.61	0.61
2. Metal lath and sand aggregate plaster, 0.75 in	0.13	0.47
3. Structural beam	0.00*	0.00*
4. Nonreflective air space (50 F mean; 10 deg F temperature difference) thickness	0.93**	0.93**
5. Metal deck	0.00*	0.00*
6. Rigid roof deck insulation, C = 0.24 (R = 1/c)	4.17	2.78
7. Built-up roofing, 0.375 in.	0.33	0.33
8. Outside surface (15 mph wind)	0.17	0.17
Total Thermal Resistance (R)	6.34	5.29

Construction No. 1: $U = 1/6.34 = 0.158$
Construction No. 2: $U = 1/5.29 = 0.189$

[a]See section Calculating Overall Coefficients in Chapter 23 of source text for basis of calculations.

*If structural beams and metal deck are to be considered, the technique shown in Examples 1 and 2 and Fig. 3 of Chapter 23 of the source text may be used to estimate total R. Full scale testing of a suitable portion of the construction is, however, preferable.

**Use largest air space (3.5-in.) value shown in Table 2, Chapter 23, of source text.

SOURCE: Reprinted by permission from *ASHRAE Handbook—1981 Fundamentals.*

TABLE 3.15   Coefficients of Transmission (U) of Pitched Roofs[a,b]

Coefficients are expressed in Btu per (hour) (square foot) (degree Fahrenheit difference in temperature between the air on the two sides), and are based on an outside wind velocity of 15 mph for heat flow upward and 7.5 mph for heat flow downward

Find U_{av} for same Construction 2 with Heat Flow Down (Summer Conditions)

| | 1 | | | 2 | |
Construction 1 (Heat Flow Up) (Reflective Air Space)	Between Rafters	At Rafters		Between Rafters	At Rafters
1. Inside surface (still air)	0.62	0.62		0.76	0.76
2. Gypsum wallboard 0.5 in., foil backed	0.45	0.45		0.45	0.45
3. Nominal 2-in. × 4-in. ceiling rafter	—	4.35		—	4.35
4. 45 deg slope reflective air space, 3.5 in. (50 F mean, 30 deg F temperature difference) E = 0.05	2.17	—		4.33	—
5. Plywood sheathing, 0.625 in.	0.77	0.77		0.77	0.77
6. Permeable felt building membrane	0.06	0.06		0.06	0.06
7. Asphalt shingle roofing	0.44	0.44		0.44	0.44
8. Outside surface (15 mph wind)	0.17	0.17		0.25••	0.25••
Total Thermal Resistance (R)	$R_1 = 4.68$	$R_3 = 6.86$		$R_1 = 7.06$	$R_3 = 7.08$

Construction No. 1: $U_1 = 1/4.69 = 0.213$; $U_3 = 1/6.90 = 0.145$. With 10% framing (typical of 2-in. rafters @16-in. o.c.), $U_{av} = 0.9(0.213) + 0.1 (0.145) = 0.206$

Construction No. 2: $U_1 = 1/7.07 = 0.141$; $U_3 = 1/7.12 = 0.140$. With framing unchanged, $U_{av} = 0.9 (0.141) + 0.1 (0.140) = 0.141$

Find U_{av} for same Construction 2 with Heat Flow Down (Summer Conditions)

| | 3 | | | 4 | |
Construction 1 (Heat Flow Up) (Non-Reflective Air Space)	Between Rafters	At Rafters		Between Rafters	At Rafters
1. Inside surface (still air)	0.62	0.62		0.76	0.76
2. Gypsum wallboard, 0.5 in.	0.45	0.45		0.45	0.45
3. Nominal 2-in. × 4-in. ceiling rafter	—	4.35		—	4.35
4. 45 deg slope, nonreflective air space, 3.5 in. (50 F mean, 10 deg F temperature difference)	0.96	—		0.90*	—
5. Plywood sheathing, 0.625 in.	0.77	0.77		0.77	0.77
6. Permeable felt building membrane	0.06	0.06		0.06	0.06
7. Asphalt shingle roofing	0.44	0.44		0.44	0.44
8. Outside surface (15-mph wind)	0.17	0.17		0.25••	0.25••
Total Thermal Resistance (R)	$R_1 = 3.47$	$R_3 = 6.86$		$R_1 = 3.63$	$R_3 = 7.08$

Construction No. 3: $U = 1/3.48 = 0.287$; $U_s = 1/6.90 = 0.145$. With 10% framing (typical of 2-in. rafters @ 16-in. o.c.), $U_{av} = 0.9 (0.287)+ 0.1 (0.145) = 0.273$

Construction No. 4: $U_I = 1/3.64 = 0.275$; $U_s = 1/7.12 = 0.140$. With framing unchanged, $U_{av} = 0.9 (0.275) + 0.1 (0.140) = 0.262$

[a]See section Calculating Overall Coefficients in Chapter 23 of source text for basis of calculation.

[b]Pitch of roof—45 deg.

* Air space value at 90°F, 10°F temperature difference.

**7.5-mi/h wind.

Reprinted by permission from *ASHRAE Handbook—1981 Fundamentals*.

TABLE 3.16 Coefficients of Transmission (U) of Windows, Skylights, and Light Transmitting Partitions

These values are for heat transfer from air to air, Btu/(hr · ft² · °F). To calculate total heat gain including solar transmission, see Chapter 28.

PART A—VERTICAL PANELS (EXTERIOR WINDOWS, SLIDING PATIO DOORS, AND PARTITIONS)—FLAT GLASS, GLASS BLOCK, AND PLASTIC SHEET

Description	Exterior[a]		Interior
	Winter	Summer	Interior
Flat Glass[b]			
single glass	1.10	1.04	0.73
insulating glass—double[c]			
0.1875-in. air space[d]	0.62	0.65	0.51
0.25-in. air space[d]	0.58	0.61	0.49
0.5-in. air space[c]	0.49	0.56	0.46
0.5-in. air space, low emittance coating[f]			
e = 0.20	0.32	0.38	0.32
e = 0.40	0.38	0.45	0.38
e = 0.60	0.43	0.51	0.42
insulating glass—triple[c]			
0.25-in. air spaces[d]	0.39	0.44	0.38
0.5-in. air spaces[d]	0.31	0.39	0.30
storm windows			
1-in. to 4-in. air space[d]	0.50	0.50	0.44
Plastic Sheet			
single glazed			

PART B—HORIZONTAL PANELS (SKYLIGHTS)—FLAT GLASS, GLASS BLOCK, AND PLASTIC DOMES

Description	Exterior[a]		Interior[f]
	Winter[i]	Summer[i]	Interior[f]
Flat Glass[c]			
single glass	1.23	0.83	0.96
insulating glass—double[c]			
0.1875-in. air space[d]	0.70	0.57	0.62
0.25-in. air space[d]	0.65	0.54	0.59
0.5-in. air space[c]	0.59	0.49	0.56
0.5-in. air space, low emittance coating[f]			
e = 0.20	0.48	0.36	0.39
e = 0.40	0.52	0.42	0.45
e = 0.60	0.56	0.46	0.50
Glass Block[h]			
11 × 11 × 3 in. thick with cavity divider	0.53	0.35	0.44
12 × 12 × 4 in. thick with cavity divider	0.51	0.34	0.42
Plastic Domes[k]			
single-walled	1.15	0.80	—

	Single Glass	Double or Triple Glass	Storm Windows
0.5-in. thick	0.81	0.76	—
insulating unit—double^c			
0.25-in. air space^d	0.55	0.56	—
0.5-in. air space^e	0.43	0.45	—
Glass Block^h			
6 × 6 × 4 in. thick	0.60	0.57	0.46
8 × 8 × 4 in. thick	0.56	0.54	0.44
—with cavity divider	0.48	0.46	0.38
12 × 12 × 4 in. thick	0.52	0.50	0.41
—with cavity divider	0.44	0.42	0.36
12 × 12 × 2 in. thick	0.60	0.57	0.46

PART C—ADJUSTMENT FACTORS FOR VARIOUS WINDOW AND SLIDING PATIO DOOR TYPES (MULTIPLY U VALUES IN PARTS A AND B BY THESE FACTORS)

Description	Single Glass	Double or Triple Glass	Storm Windows
Windows			
All Glass^l	1.00	1.00	1.00
Wood Sash—80% Glass	0.90	0.95	0.90
Wood Sash—60% Glass	0.80	0.85	0.80
Metal Sash—80% Glass	1.00	1.20^m	1.20^m
Sliding^k Patio Doors			
Wood Frame	0.95	1.00	—
Metal Frame	1.00	1.10^m	—

a See Part C of this table for adjustment for various window and sliding patio door types.
b Emittance of uncooled glass surface = 0.84.
c Double and triple refer to the number of lights of glass.
d 0.125-in. glass.
e 0.25-in. glass.
f Coating on either glass surface facing air space; all other glass surfaces uncoated.
g Window design: 0.25-in. glass—0.125-in. glass—0.25-in. glass.
h Dimensions are nominal.
i For heat flow up.
j For heat flow down.
k Based on area of opening, not total surface area.
l Refers to windows with negligible opaque area.
m Values will be less than these when metal sash and frame incorporate thermal breaks. In some thermal break designs, U-values will be equal to or less than those for the glass. Window manufacturers should be consulted for specific data.

SOURCE: Reprinted by permission from *ASHRAE Handbook and Product Directory—1977 Fundamentals*.

TABLE 3.17 Coefficients of Transmission (*U*) for Slab Doors

	Btu per (hr · ft² · F)			
	Winter			Summer
	Solid Wood, No Storm Door	Storm Door[b]		No Storm Door
Thickness[a]		Wood	Metal	
1-in.	0.64	0.30	0.39	0.61
1.25-in.	0.55	0.28	0.34	0.53
1.5-in.	0.49	0.27	0.33	0.47
2-in.	0.43	0.24	0.29	0.42
	Steel Door			
1.75 in.				
A[c]	0.59	—	—	0.58
B[d]	0.19	—	—	0.18
C[e]	0.47	—	—	0.46

[a]Nominal thickness.

[b]Values for wood storm doors are for approximately 50% glass; for metal storm door values apply for any percent of glass.

[c]A = Mineral fiber core (2 lb/ft³).

[d]B = Solid urethane foam core with thermal break.

[e]C = Solid polystyrene core with thermal break.

SOURCE: Reprinted by permission from *ASHRAE Handbook and Product Directory—1977 Fundamentals.*

NOTES

NOTES

BASIC THEORY
AND
FUNDAMENTALS

TABLE 4.1   Basic Heating and Ventilating Formulas

Formula	Legend
$\text{Btu} = TC/55$ $\text{Btu/h} = 60TC/55$ $\text{DS} = 144C/60v$	$\text{Btu} = $ British thermal unit $T = $ temperature rise, °F $C = $ cubic feet of air delivery $\text{DS} = $ cross section of duct, in $v = $ flow velocity, ft/min $\Delta T = $ final temperature of air— temperature of entering air
$\text{Btu}/C = 1.085 \times C \times \Delta T$ $\text{SHL} = 24hd\,(t_i - t_a)/t_i - t_e$	$\text{Btu}/C = $ heating capacity of coil, Btu $\text{SHL} = $ seasonal heat loss, Btu $h = $ hourly building heat loss, Btu $d = $ total days of heating season $t_i = $ interior design temperature, °F $t_e = $ exterior design temperature, °F $t_a = $ average outside temperature for heating season, °F $24 = $ h/day
$U = 1/R$	$U = $ overall conductance (equal to number of Btu that will flow through 1 ft^2 of structure from air to air owing to a temperature difference of 1°F, in 1 h)
$R = 1/U$	$R = $ thermal resistance (reciprocal of conductance)
$R = 1/c$	$c = $ thermal conductance (heat flow through a given thickness of 1 ft^2 of material with 1°F temperature differential)
$Q = VN/60$	$Q = $ air changes, ft^3/min $V = $ volume of room, ft^3 $N = $ number of hourly air changes
$q = 1.08QwT$	$q = $ heat required to warm up cold air, Btu/h $Qw = $ air to be warmed, expressed in cubic feet per minute

TABLE 4.2   Minimum Ventilation Air for Various Activities

Type of occupancy	Ventilation air, ft³/min per person
Inactive, theaters	5
Light activity, offices	10
Light activity with some odor generations, restaurant	15
Light activity with moderate odor generations, bars	20
Active work, shipping rooms	30
Very active work, gymnasiums	50

TABLE 4.3   Basic Heating Data

---

1 Btu (British thermal unit) = heat required to raise 1 lb of water 1°F
Btu ÷ 3.413 = W
Specific heat of water = 1 Btu
Specific heat of air = 0.24 Btu
1 ft^3 of air @ 32°F = 0.0807 lb
1 Btu will heat 1 ft^3 of air 55°F
1 gal (U.S.) of water = 8.33 lb
1 gal (U.S.) of water = 231 in^3 @ 39.2°F
1 ft^3 of water = 7.48 gal
1 ft^3 of water = 62.418 lb @ 39.2°F
1 ft^2 of steam radiation = 240 Btu
1 ft^2 of hot water radiation = 150 Btu
No. 2 fuel oil = 20,571 Btu/lb or 144,000 Btu/gal
Natural gas = 18,000 Btu/lb or 1,030 Btu/ft^3
Propane = 21,564 Btu/lb or 2,572 Btu/ft^3
Butane = 21,440 Btu/lb or 3,200 Btu/ft^3
1 degree day = 1°F, in mean outdoor temperature, below 65°F (24-h avg.) per day
1 watt (W) = 1 joule (J) per second (s)
1 kWh = 1.341 hph
1 Btu = 778 ft·lb = 1055 J = 252 cal
1 hph = 2545 Btu = 0.7457 kWh
1 gal of LP gas = 36.6 ft^3

TABLE 4.4   Application Formulas

TO OBTAIN	HAVING	FORMULA
Velocity (V) Feet Per Minute	Pitch Diameter (D) of Gear or Sprocket – Inches & Rev. Per Min. (RPM)	$V = .2618 \times D \times RPM$
Rev. Per Min. (RPM)	Velocity (V) Ft. Per Min. & Pitch Diameter (D) of Gear or Sprocket – Inches	$RPM = \dfrac{V}{.2618 \times D}$
Pitch Diameter (D) of Gear or Sprocket – Inches	Velocity (V) Ft. Per Min. & Rev. Per Min. (RPM)	$D = \dfrac{V}{.2618 \times RPM}$
Torque (T) In. Lbs.	Force (W) Lbs. & Radius (R) Inches	$T = W \times R$
Horsepower (HP)	Force (W) Lbs. & Velocity (V) Ft. Per Min.	$HP = \dfrac{W \times V}{33000}$
Horsepower (HP)	Torque (T) In. Lbs. & Rev. Per Min. (RPM)	$HP = \dfrac{T \times RPM}{63025}$
Torque (T)	Horsepower (HP) & Rev. Per Min. (RPM)	$T = \dfrac{63025 \times HP}{RPM}$
Force (W) Lbs.	Horsepower (HP) & Velocity (V) Ft. Per Min.	$W = \dfrac{33000 \times HP}{V}$
Rev. Per Min. (RPM)	Horsepower (HP) & Torque (T) In. Lbs.	$RPM = \dfrac{63025 \times HP}{T}$

SOURCE: Boston Gear, Quincy, MA.

TABLE 4.5 Pressure Facts

	Vapor pressure, psig														
	Outside temperature, °F														
	−30	−20	−10	0	10	20	30	40	50	60	70	80	90	100	110
100% Propane	6.8	11.5	17.5	24.5	34	42	53	65	78	93	110	128	150	172	204
70% Propane 30% Butane	…	4.7	9	15	20.5	28	36.5	46	56	68	82	96	114	134	158
50% Propane 50% Butane	…	…	3.5	7.6	12.3	17.8	24.5	32.4	41	50	61	74	88	104	122
70% Butane 30% Propane	…	…	…	2.3	5.9	10.2	15.4	21.5	28.5	36.5	45	54	66	79	93
100% Butane	…	…	…	…	…	…	…	3.1	6.9	11.5	17	23	30	38	47

TABLE 4.6   Power, Mass, Volume, and Weight of Water Formulas

	Power
1 Btu per h	= 0.293 W
	= 12.96 ft·lb/min
	= 0.00039 hp
1 ton refrigeration (U.S.)	= 288,000 Btu per 24 h
	= 12,000 Btu/h
	= 200 Btu/min
	= 83.33 lb ice melted per h from and at 32°F
	= 2000 lb ice melted per 24 h from and at 32°F
1 hp	= 550 ft·lb/s
	= 746 W
	= 2545 Btu/h
1 boiler hp	= 33,480 Btu/h
	= 34.5 lb water evap. per h from and at 212°F
	= 9.8 kW
1 kW	= 3413 Btu/h
	Mass
1 lb (avoir.)	= 16 oz (avoir.)
	= 7000 grain
1 ton (short)	= 2000 lb
1 ton (long)	= 2240 lb
	Volume
1 gal (U.S.)	= 128 fl oz (U.S.)
	= 231 in^3
	= 0.833 gal (Brit.)
1 ft^3	= 7.48 gal (U.S.)
	Weight of Water

1 ft^3 at 50°F weighs 62.41 lb
1 gal at 50°F weighs 8.34 lb
1 ft^3 of ice weighs 57.2 lb
Water is at its greatest density at 39.2°F
1 ft^3 at 39.2°F weighs 62.43 lb

TABLE 4.7   Water Pressure Conversion: Pounds per Square Inch to Feet Head and Feet Head to Pounds per Square Inch

		Water Pressure to Feet Head	
Pounds per square inch	Feet head	Pounds per square inch	Feet head
1	2.31	100	230.90
2	4.62	110	253.98
3	6.93	120	277.07
4	9.24	130	300.16
5	11.54	140	323.25
6	13.85	150	346.34
7	16.16	160	369.43
8	18.47	170	392.52
9	20.78	180	415.61
10	23.09	200	461.78
15	34.63	250	577.24
20	46.18	300	692.69
25	57.72	350	808.13
30	69.27	400	922.58
40	92.36	500	1154.48
50	115.45	600	1385.39
60	138.54	700	1616.30
70	161.63	800	1847.20
80	184.72	900	2078.10
90	207.81	1000	2309.00

Note: One pound of pressure per square inch of water equals 2.309 ft of water at 62°F. Therefore, to find the feet head of water for any pressure not given in the table above, multiply the pressure pounds per square inch by 2.309.

TABLE 4.7   Water Pressure Conversion: Pounds per Square Inch to Feet Head and Feet Head to Pounds per Square Inch (*continued*)

	Feet Head of Water to PSI		
Feet head	Pounds per square inch	Feet head	Pounds per square inch
1	0.43	100	43.31
2	0.87	110	47.64
3	1.30	120	51.97
4	1.73	130	56.30
5	2.17	140	60.63
6	2.60	150	64.96
7	3.03	160	69.29
8	3.46	170	73.63
9	3.90	180	77.96
10	4.33	200	86.62
15	6.50	250	108.27
20	8.66	300	129.93
25	10.83	350	151.58
30	12.99	400	173.24
40	17.32	500	216.55
50	21.65	600	259.85
60	25.99	700	303.16
70	30.32	800	346.47
80	34.65	900	389.78
90	38.98	1000	433.00

Note: One foot of water at 62°F equals 0.433 lb pressure per square inch. To find the pressure per square inch for any feet head not given in the table above, multiply the feet head by 0.433.

TABLE 4.8 Conversion Factors for Pressure, Temperature, Weight of Liquid, Flow, and Work, USCS

Pressure	
1 lb/in²	= 2.31 ft water at 60°F
	= 2.04 inHg at 60°F
1 ft water at 60°F	= 0.433 lb/in²
	= 0.884 inHg at 60°F
1 inHg at 60°F	= 0.49 lb/in²
	= 1.13 ft water at 60°F
lb/in² absolute (psia)	= lb/in² gauge (psig) + 14.7
Temperature	
°C	= (°F − 32) × 5/9
Weight of Liquid	
1 gal (U.S.)	= 8.34 lb × sp. gr.
1 ft³	= 62.4 lb × sp. gr.
1 lb	= 0.12 U.S. gal ÷ sp. gr.
	= 0.016 ft³ ÷ sp. gr.
Flow	
1 gal/min (gpm)	= 0.134 ft³/min
	= 500 lb/h × sp. gr.
500 lb/h	= 1 gal/min ÷ sp. gr.
1 ft³/min (cfm)	= 448.8 gal/h (gph)
Work	
1 Btu (mean)	= 778 ft·lb
	= 0.293 Wh
	= 1/180 of heat required to change temp of 1 lb water from 32°F to 212°F
1 hp·h	= 2545 Btu (mean)
	= 0.746 kWh
1 kWh	= 3413 Btu (mean)
	= 1.34 hp·h

TABLE 4.9 Mass Equivalents, USCS and Metric*†

One	Is Equal to	One	Is Equal to
Ounce	16 drams	Ounce	0.02835 kilogram
Pound	16 ounces	Pound	0.4536 kilogram
Pound	7000 grains	Short ton	907.2 kilograms
Short ton	2000 pounds	Long ton	1016 kilograms
Long ton	2240 pounds	Metric ton	1000 kilograms
Metric ton	2205 pounds	Pound per square foot	4.8824 kilograms per square meter

*USCS units are based on avoirdupois weight.

†Density: One pound per cubic foot = 16.02 kilograms per cubic meter.

SOURCE: A. M. Khashab, *Heating, Ventilating, and Air Conditioning Systems Estimating Manual,* 2d ed., McGraw-Hill, New York, © 1984. Used with permission of the publisher.

TABLE 4.10   Velocity Equivalents, USCS and Metric

One	Is Equal to	One	Is Equal to
Foot per second	60 feet per minute	Foot per minute	0.00508 meter per second
Mile per hour	88 feet per minute	Mile per hour	1.609 kilometers per hour

SOURCE: A. M. Khashab, *Heating, Ventilating, and Air Conditioning Systems Estimating Manual*, 2d ed., McGraw-Hill, New York, © 1984. Used with permission of the publisher.

TABLE 4.11   Pressure Equivalents, USCS and Metric

One	Is Equal to	One	Is Equal to
Inch of water (at 60°F)	0.03609 psi*	Inch of water (at 60°F)	248.84 Pa
		Millimeter of water (at 4°C)	9.806 Pa
Foot of water (at 39.2°F)	0.43310 psi*	Foot of water (at 39.2°F)	2988.98 Pa
Inch of mercury (at 32°F)	0.4912 psi*	Inch of mercury (at 32°F)	3386.39 Pa
		Millimeter of mercury (at 0°C)	133.32 Pa
Atmosphere	14.70 psi*	psi*	6894.76 Pa
Atmosphere	1.0333 kgf† per square centimeter	kgf† per square centimeter	98,066.50 Pa

*psi = pound-force per square inch.
†kgf = kilogram-force = 9.81 newtons.

SOURCE: A. M. Khashab, *Heating, Ventilating, and Air Conditioning Systems Estimating Manual*, 2d ed., McGraw-Hill, New York, © 1984. Used with permission of the publisher.

TABLE 4.12 Temperature Equivalents, USCS and Metric

To Convert from	To	Use	To Convert from	To	Use
Degree Fahrenheit	Degree Rankine	°R = °F + 459.67	Degree Celsius	Kelvin	K = °C + 273.15
Degree Fahrenheit	Degree Celsius	°C = $\frac{5}{9}$(°F − 32)	Degree Rankine	Kelvin	K = $\frac{5}{9}$(°R)

SOURCE: A. M. Khashab, *Heating, Ventilating, and Air Conditioning Systems Estimating Manual*, 2d ed., McGraw-Hill, New York, © 1984. Used with permission of the publisher.

TABLE 4.13 Conversion Factors for Length, Torque, Rotation, Moment of Inertia, Power, and Temperature, USCS and Metric

	MULTIPLY	BY	TO OBTAIN
Length	Meters	3.281	Feet
	Meters	39.37	Inches
	Inches	.0254	Meters
	Feet	.3048	Meters
	Millimeters	.0394	Inches
Torque	Newton-Meters	.7376	Lb-Ft
	Lb-Ft	1.3558	Newton-Meter
	Lb-In	.0833	Lb-Ft
	Lb-Ft	12.00	Lb-In
Rotation	RPM	6.00	Degrees/Sec.
	RPM	.1047	Rad./Sec.
	Degrees/Sec.	.1667	RPM
	Rad./Sec.	9.549	RPM
Moment of Inertia	Newton-Meters^2	2.42	Lb-Ft^2
	Oz-In^2	.000434	Lb-Ft
	Lb-In^2	.00694	Lb-Ft^2
	Slug-Ft^2	32.17	Lb-Ft^2
	Oz-In-Sec^2	.1675	Lb-Ft^2
	Lb-In-Sec^2	2.68	Lb-Ft^2
Power	Watts	.00134	HP
	Lb-Ft/Min	.0000303	HP
Temperature	Degree C = (Degree F -32) × 5/9 Degree F = (Degree C × 9/5) + 32		

SOURCE: Boston Gear, Quincy, MA.

TABLE 4.14   DC and AC Power Circuits

POWER IN DC CIRCUITS:

$$\text{Watts} = \text{Volts} \times \text{Amperes} \qquad \text{Horsepower} = \frac{\text{Volts} \times \text{Amperes}}{746}$$

$$\text{Kilowatts} = \frac{\text{Volts} \times \text{Amperes}}{1000}$$

$$\text{Kilowatt-Hours} = \frac{\text{Volts} \times \text{Amperes} \times \text{Hours}}{1000}$$

POWER IN AC CIRCUITS:

Kilovolt - Amperes (KVA)

$$\text{KVA (Single-Phase)} = \frac{\text{Volts} \times \text{Amperes}}{1000}$$

$$\text{KVA (Three-Phase)} = \frac{\text{Volts} \times \text{Amperes} \times 1.73}{1000}$$

Kilowatt (Kw)

$$\text{Kw (Single-Phase)} = \frac{\text{Volts} \times \text{Amperes} \times \text{Power Factor}}{1000}$$

$$\text{Kw (Two-Phase)} = \frac{\text{Volts} \times \text{Amperes} \times \text{Power Factor} \times 1.42}{1000}$$

$$\text{Kw (Three-Phase)} \quad \frac{\text{Volts} \times \text{Amperes} \times \text{Power Factor} \times 1.73}{1000}$$

$$\text{Power Factor} = \frac{\text{Kilowatts}}{\text{Kilovolts} \times \text{Amperes}}$$

SOURCE: Boston Gear, Quincy, MA.

TABLE 4.15   Temperature Conversion Chart, between Celsius (Centigrade) and Fahrenheit

°C		°F		°C		°F
−17.7	0	32		− 3.3	26	78.8
−17.2	1	33.8		− 2.8	27	80.6
−16.6	2	35.6		− 2.2	28	82.4
−16.1	3	37.4		− 1.6	29	84.2
−15.5	4	39.2		− 1.1	30	86.0
−15.0	5	41.0		− 0.6	31	87.8
−14.4	6	42.8		0	32	89.6
−13.9	7	44.6		0.5	33	91.4
−13.3	8	46.4		1.1	34	93.2
−12.7	9	48.2		1.6	35	95.0
−12.2	10	50.0		2.2	36	96.8
−11.6	11	51.8		2.7	37	98.6
−11.1	12	53.6		3.3	38	100.4
−10.5	13	55.4		3.8	39	102.2
−10.0	14	57.2		4.4	40	104.0
− 9.4	15	59.0		4.9	41	105.8
− 8.8	16	61.8		5.5	42	107.6
− 8.3	17	63.6		6.0	43	109.4
− 7.7	18	65.4		6.6	44	111.2
− 7.2	19	67.2		7.1	45	113.0
− 6.6	20	68.0		7.7	46	114.8
− 6.1	21	69.8		8.2	47	116.6
− 5.5	22	71.6		8.8	48	118.4
− 5.0	23	73.4		9.3	49	120.2
− 4.4	24	75.2		9.9	50	122.0
− 3.9	25	77.0		10.4	51	123.8

Note: The center numbers refer to the temperature either in degrees Celsius or degrees Fahrenheit which it is desired to convert into the other scale. If converting from Fahrenheit to Celsius, the equivalent temperature will be found in the left column, while if converting from Celsius to Fahrenheit, the answer will be found in the column on the right.

TABLE 4.15 Temperature Conversion Chart, between Celsius (Centigrade) and Fahrenheit (*continued*)

°C		°F	°C		°F
11.1	52	125.6	25.5	78	172.4
11.5	53	127.4	26.2	79	174.2
12.1	54	129.2	26.8	80	176.0
12.6	55	131.0	27.3	81	177.8
13.2	56	132.8	27.7	82	179.6
13.7	57	134.6	28.2	83	181.4
14.3	58	136.4	28.8	84	183.2
14.8	59	138.2	29.3	85	185.0
15.6	60	140.0	29.9	86	186.8
16.1	61	141.8	30.4	87	188.6
16.6	62	143.6	31.0	88	190.4
17.1	63	145.4	31.5	89	192.2
17.7	64	147.2	32.1	90	194.0
18.2	65	149.0	32.6	91	195.8
18.8	66	150.8	33.3	92	197.6
19.3	67	152.6	33.8	93	199.4
19.9	68	154.4	34.4	94	201.2
20.4	69	156.2	34.9	95	203.0
21.0	70	158.0	35.5	96	204.8
21.5	71	159.8	36.1	97	206.6
22.2	72	161.6	36.6	98	208.4
22.7	73	163.4	37.1	99	210.2
23.3	74	165.2	37.7	100	212.0
23.8	75	167.0	43	110	230
24.4	76	168.8	49	120	248
25.0	77	170.6			

Note: The center numbers refer to the temperature either in degrees Celsius or degrees Fahrenheit which it is desired to convert into the other scale. If converting from Fahrenheit to Celsius, the equivalent temperature will be found in the left column, while if converting from Celsius to Fahrenheit, the answer will be found in the column on the right.

TABLE 4.16   Steam Table

Gauge press., lb	Temp., °F	Gauge press., lb	Temp., °F	Gauge press., lb	Temp., °F	Gauge press., lb	Temp., °F
5	228	55	302	110	344	210	391
6	230	56	303	112	345	212	392
7	233	57	304	114	346	214	393
8	235	58	305	116	348	216	394
9	237	59	306	118	349	218	394
10	240	60	307	120	350	220	395
11	242	61	308	122	351	222	396
12	244	62	309	124	352	224	397
13	246	63	310	126	353	226	397
14	248	64	311	128	354	228	398
15	250	65	312	130	355	230	399
16	252	66	312	132	356	232	400
17	254	67	313	134	357	234	400
18	255	68	314	136	359	235	401
19	257	69	315	138	360	237	401
20	259	70	316	140	361	239	402
21	261	71	317	142	362	241	403
22	262	72	317	144	363	243	403
23	264	73	318	146	364	245	404
24	265	74	319	148	365	247	405
25	267	75	320	150	366	249	405
26	268	76	321	152	367	251	406
27	270	77	321	154	368	253	407
28	271	78	322	156	368	255	407
29	273	79	323	158	369	257	408
30	274	80	324	160	370	259	409
31	275	81	324	162	371	261	409

32	277	82	325	164	372	263	410
33	278	83	326	166	373	265	411
34	279	84	327	168	374	267	411
35	281	85	327	170	375	269	412
36	282	86	328	172	376	271	413
37	283	87	329	174	377	273	414
38	284	88	330	176	378	275	415
39	285	89	330	178	378	277	415
40	287	90	331	180	379	279	416
41	288	91	332	182	380	281	416
42	289	92	332	184	381	283	417
43	290	93	333	186	382	285	420
44	291	94	334	188	383	295	423
45	292	95	334	190	384	305	437
46	293	96	335	192	384	355	442
47	294	97	336	194	385	375	444
48	295	98	337	196	386	385	449
49	297	99	338	198	387	405	460
50	298	100	339	200	388	455	472
51	299	102	340	202	388	510	481
52	300	104	341	204	389	560	560
53	301	106	342	206	390	585	585
54	302	108	343	208	391		

TABLE 4.17   Specific Gravity of Liquids

Liquid	Temp., °F	Specific gravity
Water (1 cu. ft. weighs 62.41 lb)	50	1.00
Brine (sodium chloride 25%)	32	1.20
Pennsylvania crude oil	80	0.85
Fuel oil Nos. 1 and 2	85	0.95
Gasoline	80	0.74
Kerosene	85	0.82
Lubricating oil SAE 10-20-30	115	0.94

TABLE 4.18   Specific Gravity of Gases (at 60°F and 29.9 inHg)

Gases	Specific gravity
Dry air (1 ft³ at 60°F and 29.92 inHg weighs 0.07638 lb)	1.000
Acetylene (C_2H_2)	0.91
Ethane (C_2H_6)	1.05
Methane (CH_4)	0.554
Ammonia (NH_3)	0.596
Carbon dioxide (CO_2)	1.53
Carbon monoxide (CO)	0.967
Butane (C_4H_{10})	2.067
Butene (C_4H_8)	1.93
Chlorine (Cl_2)	2.486
Helium (He)	0.138
Hydrogen (H_2)	0.0696
Nitrogen (N_2)	0.9718
Oxygen (O_2)	1.1053

TABLE 4.19 Coefficients of Expansion of Various Materials for 100 Degrees

Materials	Linear Expansion		Materials	Linear Expansion	
	Centigrade	Fahrenheit		Centigrade	Fahrenheit
METALS AND ALLOYS			STONE AND MASONRY		
Aluminum, wrought	.00231	.00128	Ashlar masonry	.00063	.00035
Brass	.00188	.00104	Brick masonry	.00061	.00034
Bronze	.00181	.00101	Cement, portland	.00126	.00070
Copper	.00168	.00093	Concrete	.00099	.00055
Iron, cast, gray	.00106	.00059	Granite	.00080	.00044
Iron, wrought	.00120	.00067	Limestone	.00076	.00042
Iron, wire	.00124	.00069	Marble	.00081	.00045
Lead	.00286	.00159	Plaster	.00166	.00092
Magnesium, various alloys	.0029	.0016	Rubble masonry	.00063	.00035
Nickel	.00126	.00070	Sandstone	.00097	.00054
Steel, mild	.00117	.00065	Slate	.00080	.00044
Steel, stainless, 18-8	.00178	.00099			
Zinc, rolled	.00311	.00173			
TIMBER			TIMBER		
Fir	.00037	.00021	Fir	.0058	.0032
Maple) parallel to fiber	.00064	.00036	Maple) perpendicular to	.0048	.0027
Oak)	.00049	.00027	Oak) fiber	.0054	.0030
Pine)	.00054	.00030	Pine)	.0034	.0019

EXPANSION OF WATER
Maximum Density = 1

C°	Volume	C°	Volume	C°	Volume	C°	Volume	C°	Volume	C°	Volume
0	1.000126	10	1.000257	30	1.004234	50	1.011877	70	1.022384	90	1.035829
4	1.000000	20	1.001732	40	1.007627	60	1.016954	80	1.029003	100	1.043116

SOURCE: American Institute of Steel Construction.

TABLE 4.20   Hardness Conversion Table

Hardened Steel and Hard Alloys

C	A	15-N	30-N	Brinnell
150	60	15	30	3000
80	92.0	96.5	92.0	—
78	91.0	96.0	91.0	—
76	90.0	95.5	90.0	—
74	89.0	95.0	88.5	—
72	88.0	94.5	87.0	—
70	86.5	94.0	86.0	—
68	85.5	—	84.5	—
66	84.5	92.5	83.0	—
64	83.5	—	81.0	—
62	82.5	91.0	79.0	—
60	81.0	90.0	77.5	614
58	80.0	—	75.5	587
56	79.0	88.5	74.0	560
54	78.0	87.5	72.0	534
52	77.0	86.5	70.5	509
50	76.0	85.5	68.5	484
49	75.5	85.0	67.5	472
48	74.5	84.5	66.5	460
47	74.0	84.0	66.0	448
46	73.5	83.5	65.0	437
45	73.0	83.0	64.0	426
44	72.5	82.5	63.0	415
43	72.0	82.0	62.0	404
42	71.5	81.5	61.5	393
41	71.0	81.0	60.5	382
40	70.5	80.5	59.5	372
39	70.0	80.0	58.5	362
38	69.5	79.5	57.5	352
37	69.0	79.0	56.5	342
36	68.5	78.5	56.0	332
35	68.0	78.0	55.0	322
34	67.5	77.0	54.0	313
33	67.0	76.5	53.0	305
32	66.5	76.0	52.0	297
31	66.0	75.5	51.5	290
30	65.5	75.0	50.5	283
29	65.0	74.5	49.5	276
28	64.5	74.0	48.5	270
27	64.0	73.5	47.5	265
26	63.5	72.5	47.0	260
25	63.0	72.0	46.0	255
24	62.5	71.5	45.0	250
23	62.0	71.0	44.0	245
22	61.5	70.5	43.0	240
21	61.0	70.0	42.5	235
20	60.5	69.5	41.5	230

Unhardened Steel, Soft Tempered Steel Gray and Malleable Cast Iron and most Non-ferrous metals

B	Brinnell	
100	500	3000
100	201	240
98	189	228
96	179	216
94	171	205
92	163	195
90	157	185
88	151	176
86	145	169
84	140	162
82	135	156
80	130	150
78	126	144
76	122	139
74	118	135
72	114	130
70	110	125
68	107	121
66	104	117
64	101	114
62	98	110
60	95	107
58	92	104
56	90	101
54	87	—
52	85	—
50	83	—
48	81	—
44	78	—
42	76	—
38	73	—
34	70	—
30	67	—
28	66	—
26	65	—
25	64	—
23	63	—
21	62	—
19	61	—
17	60	—
15	59	—
13	58	—
10	57	—
7	56	—
5	55	—
2	54	—
0	53	—

SOURCE: Boston Gear, Quincy, MA.

TABLE 4.21   Horsepower Required to Pump 1 Gallon of Water per Minute at 87% Pump Efficiency

Working pressure	HP per gal pumped	Working pressure	HP per gal pumped
100	0.067	1500	1.000
150	0.10	1600	.067
200	0.133	1800	1.20
300	0.20	2000	1.33
400	0.267	2250	1.50
500	0.33	2500	1.67
600	0.40	2750	1.83
700	0.467	3000	2.00
800	0.533	3500	2.33
900	0.60	4000	2.67
1000	0.667	4500	3.00
1200	0.80	5000	3.33
1300	0.867	5500	3.67
1400	0.933	6000	4.00

4.1 HOW TO CALCULATE HORSEPOWER, ACCELERATING FORCE FOR LINEAR MOTION, ACCELERATING TORQUE FOR ROTARY MOTION, AND SHAFT INERTIA[1]

Horsepower

For rotating objects:

$$HP = \frac{TN}{63,000}$$

where T = torque (lb-in)
N = speed (RPM)

or:

$$HP = \frac{TN}{5250}$$

where T = torque (lb-ft)
N = speed (RPM)

For objects in linear motion:

$$HP = \frac{FV}{396,000}$$

where F = force (lb)
V = velocity (IPM)

or:

$$HP = \frac{FV}{33,000}$$

[1]Text, art, and tables in this subsection adapted from material provided by Boston Gear, Quincy, MA. Used with permission.

where F = force (lb)
    V = velocity (FPM)

For pumps:

$$\text{HP} = \frac{(\text{GPM}) \times (\text{head in feet}) \times (\text{specific gravity})}{3950 \times (\text{efficiency of pump})}$$

For fans and blowers:

$$\text{HP} = \frac{\text{CFM} \times (\text{pressure in pounds/sq ft})}{33,000 \times \text{efficiency}}$$

When calculated horsepower falls between standard motor ratings, select the next higher rating.

Accelerating Force for Linear Motion

The following formula can be used to calculate the approximate accelerating force required for linear motion. However, before sizing the drive, add the torque required to accelerate the motor armature, gears, pulleys, etc. to the linear-motion accelerating force converted to torque.

$$\text{Acceleration force } (F) = \frac{W (\Delta V)}{1933t}$$

where W = weight (lb)
    ΔV = change in velocity (FPM)
    t = time (seconds) to accelerate weight

Accelerating Torque for Rotary Motion

When, in addition to the selection of a motor with proper torque capacity to start and maintain machine motion, a desired time for acceleration is involved and the required torque value may be affected, an additional formula must be considered. This formula makes it possible to calculate the average torque required over the complete range of speed change to accelerate a known inertia (WK^2).

On high inertia loads, accelerating torque may be the major factor in the drive selection.

The formula to calculate acceleration torque (torque required above load torque) of a rotating member:

$$T = \frac{(WK^2)(\Delta N)}{308t}$$

where T = acceleration torque (lb-ft)
WK^2 = total system inertia (lb-ft^2)
        that the motor must accelerate.
        This value includes motor armature,
        reducer and load.
ΔN = change in speed required (RPM)
t = time to accelerate total system
        load (seconds)

The same formula can also be used to determine the minimum acceleration time of a given drive, or if it can accomplish the desired change in speed within the required time period.

$$t = \frac{(WK^2)(\Delta N)}{308T}$$

Inertia (WK^2)

The factor WK^2 is the weight (lb) of an object multiplied by the square of the radius of gyration (K). The unit measurement of the radius of gyration is expressed in feet.

For solid or hollow cylinders (Fig. A), inertia may be calculated by the equations shown below.

The inertia of solid steel shafting per inch of shaft length is given in Table A. To calculate for hollow shafts, take the difference between

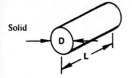

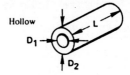

FIG. A

TABLE A Inertia of Steel Shafting (per Inch of Length)

DIAM. (IN.)	WK^2 (=FT.^{2})	DIAM. (IN.)	WK^2 (=FT.^{2})
3/4	.00006	10-1/2	2.35
1	.0002	10-3/4	2.58
1-1/4	.0005	11	2.83
1-1/2	.001	11-1/4	3.09
1-3/4	.002	11-1/2	3.38
2	.003	11-3/4	3.68
2-1/4	.005	12	4.00
2-1/2	.008	12-1/4	4.35
2-3/4	.001	12-1/2	4.72
3	.016	12-3/4	5.11
3-1/2	0.029	13	5.58
3-3/4	0.038	13-1/4	5.96
4	0.049	13-1/2	6.42
4-1/4	0.063	13-3/4	6.91
4-1/2	0.079	14	7.42

(*continued*)

TABLE A   Inertia of Steel Shafting (per Inch of Length) (*continued*)

DIAM. (IN.)	WK² (=FT.²)	DIAM. (IN.)	WK² (=FT.²)
5	0.120	14-1/4	7.97
5-1/2	0.177	14-1/2	8.54
6	0.250	14-3/4	9.15
6-1/4	0.296	15	9.75
6-1/2	0.345	16	12.59
6-3/4	0.402	17	16.04
7	0.464	18	20.16
7-1/4	0.535	19	25.03
7-1/2	0.611	20	30.72
7-3/4	0.699	21	37.35
8	0.791	22	44.99
8-1/4	0.895	23	53.74
8-1/2	1.00	24	63.71
8-3/4	1.13	25	75.02
9	1.27	26	87.76
9-1/4	1.41	27	102.06
9-1/2	1.55	28	118.04
9-3/4	1.75	29	135.83
10	1.93	30	155.55
10-1/4	2.13	—	—

the inertia values for the O.D. and I.D. as the value per inch. For shafts of materials other than steel, multiply the value for steel by the factors in Table B.

Simplified, the WK^2 for a solid cylinder or disc $= W (r_2/2)$, where r = radius in feet and W is weight in pounds.

For a hollow cylinder:

$$WK^2 = W \times \frac{r_1^2 + r_2^2}{2}$$

TABLE B   Shaft Material Factors

SHAFT MATERIAL	FACTOR
Rubber	.121
Nylon	.181
Aluminum	.348
Bronze	1.135
Cast Iron	.922

where r_1 is ID and r_2 is OD.

The inertia of complex concentric rotating parts is calculated by breaking the part up into simple rotating cylinders, calculating their inertia, and summing their values, as shown in Fig. B.

WK^2 of Rotating Elements

In practical mechanical systems, all the rotating parts do not operate at the same speed. The WK^2 of all moving parts operating at each

$$WK^2 = WK_1^2 + WK_2^2 + WK_3^2$$

FIG. B

speed must be reduced to an equivalent WK^2 at the motor shaft, so that they can all be added together and treated as a unit, as follows:

$$\text{Equivalent } WK^2 = WK^2 \left[\frac{N}{N_M} \right]^2$$

where WK^2 = inertia of the moving part
$\quad\quad\quad N$ = speed of the moving part (RPM)
$\quad\quad N_M$ = speed of the driving motor (RPM)

When using speed reducers, and the machine inertia is reflected back to the motor shaft, the equivalent inertia is equal to the machine inertia divided by the square of the drive reduction ratio:

$$\text{Equivalent } WK^2 = \frac{WK^2}{(DR)^2}$$

where DR = drive reduction = $\dfrac{N_M}{N}$

WK^2 of Linear Motion

Not all driven systems involve rotating motion. The equivalent WK^2 of linearly moving parts can also be reduced to the motor shaft speed as follows:

$$\text{Equivalent } WK^2 = \frac{W(V)^2}{39.5 \, (N_M)^2}$$

where W = weight of load (lbs)
$\quad\quad V$ = linear velocity of rack and load or conveyor
$\quad\quad\quad\quad$ and load (FPM)
$\quad\quad N_M$ = speed of the driving motor (RPM)

NOTE: This equation can only be used where the linear speed bears a continuous fixed relationship to the motor speed, such as with a conveyor.

Electrical Formulas

Ohm's law:

$$\text{Amperes} = \frac{\text{volts}}{\text{ohms}} \qquad \text{Ohms} = \frac{\text{volts}}{\text{amperes}}$$

$$\text{Volts} = \text{amperes} \times \text{ohms}$$

TABLE 4.22   Horsepower and Torque Capacity of Shafting

Shaft Dia.	Shaft Horsepower Based on Pure Torsion at 10,000 PSI Maximum Shear Stress							Torque Capacity (Lb. Ins.) Based on 10,000 PSI Shear Stress
	Revolutions Per Minute							
	30	50	100	175	690	1150	1750	
3/8	0.049	0.082	0.164	0.287	1.13	1.88	2.87	103
7/16	0.078	0.130	0.261	0.456	1.79	2.99	4.56	164
1/2	0.117	0.194	0.389	0.681	2.68	4.47	6.80	245
9/16	0.166	0.277	0.554	0.969	3.82	6.36	9.69	349
5/8	0.228	0.380	0.760	1.32	5.24	8.73	13.2	479
11/16	0.303	0.506	1.01	1.76	6.97	11.6	17.6	637
3/4	0.394	0.656	1.31	2.29	9.05	15.0	22.9	827
13/16	0.501	0.834	1.66	2.92	11.5	19.1	29.2	1052
7/8	0.625	1.04	2.08	3.64	14.3	23.9	36.4	1314
15/16	0.769	1.28	2.56	4.48	17.6	29.4	44.3	1616
1	0.933	1.55	3.11	5.44	21.4	35.7	54.4	1961
1-1/16	1.12	1.86	3.73	6.53	25.7	42.9	65.3	2352
1-1/8	1.32	2.21	4.43	7.75	30.5	50.9	77.5	2792
1-3/16	1.56	2.60	5.21	9.11	35.9	59.9	91.1	3283
1-1/4	1.82	3.03	6.07	10.6	41.9	69.8	106.	3830
1-5/16	2.11	3.51	7.03	12.3	48.5	80.	123.	4433
1-3/8	2.42	4.04	8.08	11.1	55.8	93.	141.	5097
1-7/16	2.77	4.62	9.24	16.1	63.7	106.	161.	5824
1-1/2	3.15	5.25	10.5	18.3	72.4	120.	183.	6618
1-9/16	3.56	5.93	11.8	20.7	81.8	136.	207.	7480
1-5/8	4.00	6.67	13.3	23.3	92.1	153.	233.	8414
1-11/16	4.48	7.47	14.9	26.1	103.1	171.	261.	9422
1-3/4	5.00	8.33	16.6	29.1	115.0	191.	291.	10509
1-13/16	5.55	9.26	18.5	32.4	127.8	213.	324.	11675
1-7/8	6.15	10.2	20.5	35.8	141.5	235.	358.	12925
1-15/16	6.78	11.3	22.6	39.6	156.1	260.	396.	14261
2	7.46	12.4	24.8	43.5	171.7	286.	435.	15686
2-1/16	8.18	13.6	27.2	47.7	188.3	313.	477.	17203
2-1/8	8.95	14.9	29.8	52.2	206.0	343.	522.	18815
2-3/16	9.77	16.2	32.5	56.9	224.7	374.	569.	20525
2-1/4	10.6	17.7	35.4	62.0	244.5	407.	620.	22335
2-5/16	11.5	19.2	38.4	67.3	265.4	442.	673.	24248
2-3/8	12.5	20.8	41.6	72.9	287.6	479.	729.	26268
2-7/16	13.5	22.5	45.0	78.8	310.9	518.	788.	29396
2-1/2	14.5	24.3	48.6	85.0	335.1	559.	850.	30637
2-9/16	15.7	26.1	52.3	91.6	361.2	602.	916.	32993
2-5/8	16.8	28.1	56.2	98.4	388.3	647.	984.	35466

The above table is computed based on a torsional stress of 10,000 psi. For applications involving bending moments (gears, sprockets, etc.) the horsepower capacity must be reduced accordingly.

The stress level of 10,000 PSI is representative of medium carbon steel shafting. For other materials, a correction must be made accordingly.

SOURCE: Boston Gear, Quincy, MA.

4.2   HORSEPOWER AND TORQUE[2]

Power is the rate of doing work.

Work is the exerting of a *force* through a *distance*. One foot-pound is a unit of work. It is the work done in exerting a force of one pound through a distance of one foot.

The amount of work done (foot-pounds) is the force (pounds) exerted multiplied by the distance (feet) through which the force acts.

The amount of power used (foot-pounds per minute) is the work (foot-pounds) done divided by the time (minutes) required.

$$\text{Power (foot-pounds per minute)} = \frac{\text{work (ft-lb)}}{\text{time (minutes)}}$$

Power is usually expressed in terms of horsepower.

Horsepower is power (foot-pounds per minute) divided by 33,000.

Horsepower (HP)

$$= \frac{\text{power (ft-lb per minute)}}{33,000}$$

$$= \frac{\text{work (ft-pounds)}}{33,000 \times \text{time (min.)}}$$

$$= \frac{\text{force (lb)} \times \text{distance (feet)}}{33,000 \times \text{time (min.)}}$$

Horsepower (HP)

$$= \frac{\text{force (lb)} \times \text{distance (feet)}}{33,000 \times \text{time (min.)}}$$

[2]Text and unnumbered illustrations in this subsection adapted from material provided by Boston Gear, Quincy, MA. Used with permission.

ILLUSTRATION OF HORSEPOWER

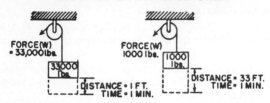

$$HP = \frac{33,000 \times 1}{33,000 \times 1} = 1\ HP \qquad\qquad HP = \frac{1000 \times 33}{33,000 \times 1} = 1\ HP$$

Torque (T) is the product of a force (W) in pounds, times a radius (R) in inches from the center of shaft (lever arm) and is expressed in inch-pounds.

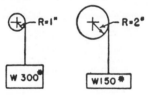

T=WR=300 × 1=300 In. Lbs.   T=WR=150 × 2=300 In. Lbs.

If the shaft is revolved, the force (W) is moved through a distance, and work is done:

$$Work\ (ft\text{-}pounds) = W \times \frac{2\pi R}{12} \times No.\ of\ revolutions\ of\ shaft$$

When this work is done in a specified time, power is used:

$$\text{Power (ft-pounds per min.)} = W \times \frac{2\pi R}{12} \times \text{RPM}$$

Since horsepower = 33,000 foot-pounds per minute

$$\text{Horsepower (HP)} = W \times \frac{2\pi R}{12} \times \frac{\text{RPM}}{33,000} = \frac{W \times R \times \text{RPM}}{63,025}$$

but torque (inch-pounds) = force (W) × radius (R), therefore

$$\text{Horsepower (HP)} = \frac{\text{torque } (T) \times \text{RPM}}{63,025}$$

Where total reductions are small, 50 to 1 or less, HP figures are commonly used. Higher reductions require that torque figures be used to select drive components, because with large reductions, a small motor can produce extremely high torque at the final low speed. For example, $\frac{1}{12}$ HP reduced to 1 RPM using the formula below and neglecting friction:

$$\text{HP} = \frac{\text{torque} \times \text{RPM}}{63,025} \quad \text{or} \quad \text{Torque} = \frac{63,025 \times \text{HP}}{\text{RPM}}$$

$$\text{Torque} = \frac{63,025 \times 1/12}{1} = 5252 \quad \text{in-lb}$$

Therefore, motors for use with large reductions should be carefully selected. Even a small motor, if stalled, can produce enough torque to ruin the drive, unless it is protected by a shear pin or some similar device.

Neglecting frictional losses, the sketch below illustrates the manner in which torque increases as speed decreases.

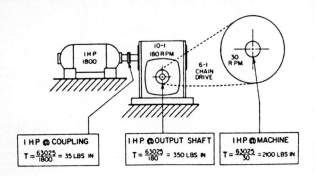

I H P @ COUPLING
$T = \frac{63025}{1800} = 35$ LBS IN

I H P @ OUTPUT SHAFT
$T = \frac{63025}{180} = 350$ LBS IN

I H P @ MACHINE
$T = \frac{63025}{30} = 2100$ LBS IN

TABLE 4.23 Pulley Pitch and Outside Diameters

	.080 PITCH		.0816 PITCH			
	FOR FIBREX BELTS		FOR FIBREX BELTS		FOR POLYURETHANE BELTS	
No. of Grooves	Pitch Diameter	Outside Diameter	Pitch Diameter	Outside Diameter	Pitch Diameter	Outside Diameter
10	.255	.235	.260	.240	.260	.246
11	.280	.260	.286	.266	.286	.272
12	.306	.286	.312	.292	.312	.298
13	.331	.311	.338	.318	.338	.324
14	.357	.337	.364	.344	.364	.350
15	.382	.362	.390	.370	.390	.376
16	.407	.387	.415	.395	.416	.402
17	.433	.413	.442	.422	.442	.428
18	.458	.438	.467	.447	.468	.454
19	.484	.464	.494	.474	.494	.480
20	.509	.489	.519	.499	.520	.506
21	.535	.515	.546	.526	.546	.532
22	.560	.540	.571	.551	.572	.558
23	.580	.566	.598	.578	.598	.584
24	.611	.591	.623	.603	.624	.610
25	.637	.617	.650	.630	.650	.636
26	.662	.642	.675	.655	.676	.662
27	.687	.667	.701	.681	.702	.688
28	.713	.693	.727	.707	.728	.714
29	.738	.718	.753	.733	.754	.740
30	.764	.744	.779	.759	.780	.766
31	.789	.769	.805	.785	.806	.792
32	.815	.795	.831	.811	.832	.818
33	.840	.820	.857	.837	.858	.844
34	.866	.846	.883	.863	.884	.870
35	.891	.871	.909	.889	.910	.896
36	.917	.897	.935	.915	.936	.922
37	.942	.922	.961	.941	.962	.948
38	.968	.948	.987	.967	.988	.974
39	.993	.973	1.013	.993	1.014	1.000
40	1.019	.999	1.039	1.019	1.040	1.026
41	1.044	1.024	1.065	1.045	1.066	1.052
42	1.070	1.050	1.091	1.071	1.092	1.078
43	1.095	1.075	1.117	1.097	1.118	1.104
44	1.120	1.100	1.142	1.122	1.144	1.130
45	1.146	1.126	1.169	1.149	1.170	1.156
46	1.171	1.151	1.194	1.174	1.196	1.182
47	1.197	1.177	1.221	1.201	1.222	1.208
48	1.222	1.202	1.246	1.226	1.248	1.234
49	1.248	1.228	1.273	1.253	1.274	1.260
50	1.273	1.253	1.298	1.278	1.300	1.286
51	1.299	1.279	1.325	1.305	1.326	1.312
52	1.324	1.304	1.350	1.330	1.352	1.338
53	1.350	1.330	1.377	1.357	1.378	1.364
54	1.375	1.355	1.403	1.383	1.404	1.390
55	1.401	1.381	1.429	1.409	1.430	1.416
56	1.426	1.406	1.455	1.435	1.456	1.442

(continued)

SOURCE: Stock Drive Products, New Hyde Park, NY.

TABLE 4.23 Pulley Pitch and Outside Diameters (*continued*)

| | .080 PITCH | | .0816 PITCH | | | |
| | FOR FIBREX BELTS | | FOR FIBREX BELTS | | FOR POLYURETHANE BELTS | |
No. of Grooves	Pitch Diameter	Outside Diameter	Pitch Diameter	Outside Diameter	Pitch Diameter	Outside Diameter
57	1.451	1.431	1.480	1.460	1.482	1.468
58	1.477	1.457	1.506	1.486	1.508	1.494
59	1.502	1.482	1.532	1.512	1.534	1.520
60	1.528	1.508	1.559	1.539	1.560	1.546
61	1.553	1.533	1.584	1.564	1.586	1.572
62	1.579	1.559	1.611	1.591	1.612	1.598
63	1.604	1.584	1.636	1.616	1.638	1.624
64	1.630	1.610	1.663	1.643	1.664	1.650
65	1.655	1.635	1.688	1.668	1.690	1.676
66	1.681	1.661	1.715	1.695	1.716	1.702
67	1.706	1.686	1.740	1.720	1.742	1.728
68	1.731	1.711	1.766	1.746	1.768	1.754
69	1.757	1.737	1.792	1.772	1.794	1.780
70	1.783	1.763	1.819	1.799	1.820	1.806
71	1.808	1.788	1.845	1.825	1.846	1.832
72	1.833	1.813	1.870	1.850	1.872	1.858
73	1.859	1.839	1.896	1.876	1.898	1.884
74	1.884	1.864	1.922	1.902	1.924	1.910
75	1.910	1.890	1.948	1.928	1.950	1.936
76	1.935	1.915	1.974	1.954	1.976	1.962
77	1.961	1.941	2.000	1.980	2.002	1.988
78	1.986	1.966	2.026	2.006	2.028	2.014
79	2.012	1.992	2.052	2.032	2.054	2.040
80	2.037	2.017	2.078	2.058	2.080	2.066
81	2.063	2.043	2.104	2.084	2.106	2.092
82	2.088	2.068	2.130	2.110	2.132	2.118
83	2.114	2.094	2.156	2.136	2.158	2.144
84	2.139	2.119	2.182	2.162	2.184	2.170
85	2.165	2.145	2.208	2.188	2.210	2.196
86	2.190	2.170	2.234	2.214	2.236	2.222
87	2.215	2.195	2.259	2.239	2.262	2.248
88	2.241	2.221	2.286	2.266	2.288	2.274
89	2.266	2.246	2.311	2.291	2.314	2.300
90	2.292	2.272	2.338	2.318	2.340	2.326
91	2.317	2.297	2.363	2.343	2.366	2.352
92	2.343	2.323	2.390	2.370	2.392	2.378
93	2.368	2.348	2.415	2.395	2.418	2.404
94	2.394	2.374	2.442	2.422	2.444	2.430
95	2.419	2.399	2.467	2.447	2.470	2.456
96	2.445	2.425	2.494	2.474	2.496	2.482
97	2.470	2.450	2.519	2.499	2.522	2.508
98	2.496	2.476	2.546	2.526	2.548	2.534
99	2.521	2.501	2.571	2.551	2.574	2.560
100	2.546	2.526	2.597	2.577	2.600	2.586
101	2.572	2.552	2.623	2.603	2.626	2.612
102	2.597	2.577	2.649	2.629	2.652	2.638
103	2.623	2.603	2.675	2.655	2.678	2.664

TABLE 4.23 Pulley Pitch and Outside Diameters (*continued*)

No. of Grooves	.080 PITCH FOR FIBREX BELTS		.0816 PITCH FOR FIBREX BELTS		FOR POLYURETHANE BELTS	
	Pitch Diameter	Outside Diameter	Pitch Diameter	Outside Diameter	Pitch Diameter	Outside Diameter
104	2.648	2.628	2.701	2.681	2.704	2.690
105	2.674	2.654	2.727	2.707	2.730	2.716
106	2.699	2.679	2.753	2.733	2.756	2.742
107	2.725	2.705	2.780	2.760	2.782	2.768
108	2.750	2.730	2.805	2.785	2.808	2.794
109	2.776	2.756	2.832	2.812	2.834	2.820
110	2.801	2.781	2.857	2.837	2.860	2.846
111	2.827	2.807	2.884	2.864	2.886	2.872
112	2.852	2.832	2.909	2.889	2.912	2.898
113	2.878	2.858	2.936	2.916	2.938	2.924
114	2.903	2.887	2.961	2.641	2.964	2.950
115	2.928	2.908	2.987	2.967	2.990	2.976
116	2.954	2.934	3.013	2.993	3.016	3.002
117	2.979	2.959	3.039	3.019	3.042	3.028
118	3.005	2.985	3.065	3.045	3.068	3.054
119	3.030	3.010	3.091	3.071	3.094	3.080
120	3.056	3.036	3.117	3.097	3.120	3.106
121	3.081	3.061	3.143	3.123	3.146	3.132
122	3.107	3.087	3.169	3.149	3.172	3.158
123	3.132	3.112	3.195	3.175	3.198	3.184
124	3.158	3.138	3.221	3.201	3.224	3.210
125	3.183	3.163	3.247	3.227	3.250	3.236
126	3.209	3.189	3.273	3.253	3.276	3.262
127	3.234	3.214	3.299	3.279	3.302	3.288
128	3.259	3.239	3.324	3.304	3.328	3.314
129	3.285	3.265	3.351	3.331	3.354	3.340
130	3.310	3.290	3.376	3.346	3.380	3.366
131	3.336	3.316	3.403	3.383	3.406	3.392
132	3.361	3.341	3.428	3.408	3.432	3.418
133	3.387	3.367	3.455	3.435	3.458	3.444
134	3.412	3.392	3.480	3.460	3.484	3.470
135	3.438	3.418	3.507	3.487	3.510	3.496
136	3.463	3.443	3.532	3.512	3.536	3.522
137	3.489	3.469	3.559	3.539	3.562	3.548
138	3.514	3.494	3.584	3.564	3.588	3.574
139	3.540	3.520	3.611	3.591	3.614	3.600
140	3.565	3.545	3.636	3.616	3.640	3.626
141	3.591	3.571	3.663	3.643	3.666	3.652
142	3.616	3.596	3.688	3.668	3.692	3.678
143	3.641	3.621	3.714	3.694	3.718	3.704
144	3.667	3.647	3.740	3.720	3.744	3.730
145	3.692	3.672	3.766	3.746	3.770	3.756
146	3.718	3.698	3.792	3.772	3.796	3.782
147	3.743	3.723	3.818	3.798	3.822	3.808
148	3.769	3.749	3.844	3.824	3.848	3.834
149	3.794	3.774	3.870	3.850	3.874	3.860
150	3.820	3.800	3.896	3.876	3.900	3.886

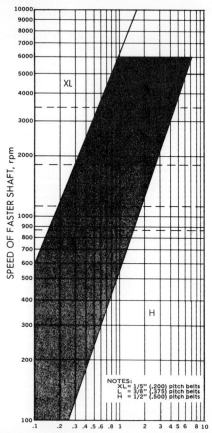

FIG. 4.1 Belt pitch selection. (*Courtesy Stock Drive Products, New Hyde Park, NY.*)

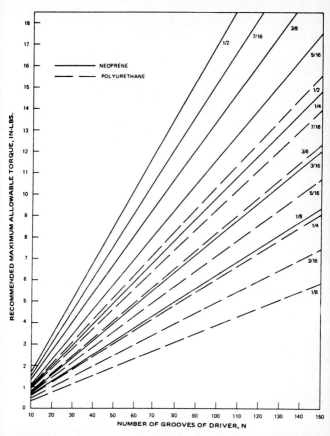

FIG. 4.2 Belt width selection. (*Courtesy Stock Drive Products, New Hyde Park, NY.*)

4.3 CENTER-DISTANCE DETERMINATION IN CUSTOM-DESIGNED BELT DRIVES[3]

The following nomenclature and basic equations are involved (Fig. C illustrates the notation):

C = center distance, inches

L = belt length, inches = PNB

P = circumferential pitch of belt, inches

NB = number of teeth on belt = L/P

$N1$ = number of teeth (grooves) on larger pulley

$N2$ = number of teeth (grooves) on smaller pulley

ϕ = one half angle of wrap on smaller pulley, radians

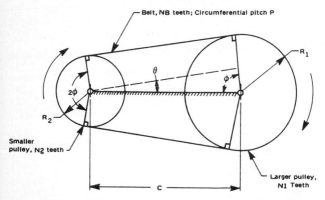

FIG. C Belt geometry.

[3]Text and illustration used in this subsection adapted from material provided by Stock Drive Products, New Hyde Park, NY. Used with permission.

$\theta = \dfrac{\pi}{2} - \phi$ = angle between straight portion of belt and line of centers, radians

R_1 = pitch radius of larger pulley, inches = $N1P/2\pi$

R_2 = pitch radius of smaller pulley, inches = $N2P/2\pi$

π = 3.14159 (ratio of circumference to diameter of circle)

The basic equation for the determination of center distance is:

$$2C \sin \phi = L - \pi(R_1 + R_2) - (\pi - 2\phi)(R_1 - R_2)$$

where $C \cos \phi = R_1 - R_2$

These equations can be combined in different ways to yield various equations for the determination of center distance.

4.4   ANALYSIS OF RADIAL BEARING LOADS FOR UNMOUNTED AND MOUNTED ROLLING ELEMENTS[4]

Radial Load

Radial bearing loads are determined by analysis of all the forces applied to a shaft. In many instances this becomes a complex analysis and should be performed with expertise. However, many applications involve simple loading and may be calculated with basic information.

Many shafts are supported by two bearings, with a load "L" applied either between two bearings, as in Fig. D; or with load overhung, as in Fig. E. In either case, the reaction on the bearing is dependent upon:

a. The point of load application
b. The magnitude of the load
c. The distance between the bearing centers

With the above information known, the reactions, due to the load, on the bearings may be calculated.

The loading of a shaft usually is the result of forces generated by gearing, sprockets or pulleys, the weight of these parts and friction. Normally the weight of the parts and friction are ignored. However, if the weight of these parts is large, they should be considered.

In this text we are mainly considering radial loading of the shaft. Each load should be calculated individually as the sum of these will be used to calculate the load imposed on the bearings.

Load Connection Factor

Loads applied by various types of drives may be calculated with use of the following load connection factors and formula:

---

[4]Text and illustrations in this subsection adapted from material provided by Boston Gear, Quincy, MA. Used with permission.

When the applied load is located between the two bearings, it is commonly referred to as "Straddle" loading.

FIG. D

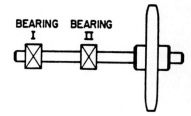

When the applied load is located outside the bearings, it is commonly referred to as "Overhung" loading.

FIG. E

$$L = \frac{2TK}{D}$$

L = load (lb)

T = torque (lb-in) $T = \dfrac{(63025)(HP)}{RPM}$

K = load connection factor

D = P.D. of sprocket, pinion, or pulley (in)

Load connection factors (K) are:

Sprocket or timing belt	1.0
Pinion and gear drive	1.25
Pulley and V-belt drive	1.50
Pulley and flat-belt drive	2.50

EXAMPLE "A": We have the following:

Load smooth and steady 8 hours per day
#40 chain drive
30 tooth sprocket
4.783 sprocket P.D.
2 HP
500 RPM
⅝ shaft diameter

With the above information the load can be calculated as follows:

$$L = \frac{2TK}{D}$$

$$T \doteq \frac{63025 \times 2}{500} = 252 \text{ in-lb}$$

K = 1.0 from load connection factor table

D = 4.783

$$L = \frac{2 \times 252 \times 1.0}{4.783}$$

L = 105 lb radial load

Magnitude of Load Acting on Bearings

Once the applied load or loads that act on the shaft is determined, we may now apply it to the bearings.

There are many types of loadings that can be imposed on a bearing:

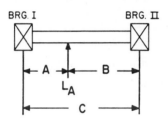

STRADDLE LOADED BEARINGS
Radial Applied Load Acting On Shaft

Load bearing I = $L_I = \dfrac{L_A \times B}{C}$

Load bearing II = $L_{II} = \dfrac{L_A \times A}{C}$

Check $L_I + L_{II} = L_A$

OVERHUNG LOADED BEARINGS
Radial Applied Load Acting On Shaft

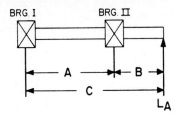

Load bearing I $= L_I = \dfrac{L_A \times B}{A}$

Load bearing II $= L_{II} = \dfrac{L_A \times C}{A}$

Check $L_{II} - L_I = L_A$

EXAMPLE "B": We have the following:

Load given in Example "A" = 105 lb.
Is in overhung condition, as shown.

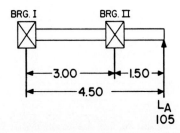

Load bearing I:

$$L_I = \frac{L_A \times B}{A}$$

$$L_I = \frac{105 \times 1.50}{3.00}$$

$$L_I = 52.5 \text{ lb}$$

Load bearing II:

$$L_{II} = \frac{L_A \times C}{A}$$

$$L_{II} = \frac{105 \times 4.5}{3.0}$$

$$L_{II} = 157.5 \text{ lb}$$

Check:

$$L_{II} - L_I = L_A$$
$$157.5 - 52.5 = 105$$
$$105 = 105$$

4.5   AIR VALVE SIZING[5]

Most manufacturers' catalogs give flow rating C_v for the valve, which was established using proposed NFPA standard T3.21.3. The following tables and formulas will enable you to quickly size a valve properly. The traditional, often used, approach of using the valve size equivalent to the port in the cylinder can be very costly. Cylinder speed, not port size, should be the determining factor!

The following C_v calculations are based upon simplified formulas which yield results with acceptable accuracy under the following standard condition:

Air at a temperature of 68°F

Absolute downstream or secondary pressure must be 53% of absolute inlet or primary pressure or greater. Below 53%, the air velocity may become sonic and the C_v formula does not apply. To calculate air flow to atmosphere, enter outlet pressure p_2 as 53% of absolute inlet pressure p_1. Pressure drop Δp would be 47% of absolute inlet pressure. These valves have been calculated for a $C_v = 1$ in Table E.

Nomenclature:

B	Pressure drop factor
C	Compression factor
C_v	Flow factor
D	Cylinder Diameter, in
F	Cylinder Area, sq in
L	Cylinder Stroke, in
p_1	Inlet or primary pressure, psig
p_2	Outlet or secondary pressure, psig
Δp	Pressure differential ($p_1 - p_2$), psid

---

[5]Text and tables in this subsection adapted from material provided by Boston Gear, Quincy, MA. Used with permission.

TABLE C   Cylinder Push Bore Area F for Standard-Size Cylinders

Bore Size D (In.)	Push Bore F (Sq. In.)	Bore Size D (In.)	Push Bore F (Sq. In.)
3/4	.44	4	12.57
1	.79	4-1/2	15.90
1-1/8	.99	5	19.64
1-1/4	1.23	6	28.27
1-1/2	1.77	7	38.48
1-3/4	2.41	8	50.27
2	3.14	10	78.54
2-1/2	4.91	12	113.10
3-1/4	8.30	14	153.94

q Air flow at actual condition, cfm

Q Air flow of free air, scfm

t Time to complete one cylinder stroke, sec

T Absolute temperature at operating pressure, $°R = °F + 460$

Valve Sizing for Cylinder Actuation— Direct Formula

C_v

$$= \frac{\overset{\text{cylinder area}}{\underset{\text{(see Table C)}}{\text{(sq in)}}} F \times \overset{\text{cylinder stroke}}{\underset{\text{(in)}}{}} L \times \overset{\text{compression factor}}{\underset{\text{(see Table D)}}{}} C}{\underset{\substack{\text{factor} \\ \text{(see Table D)}}}{\overset{\text{pressure drop}}{}} B \times \underset{\substack{\text{cylinder stroke}}}{\overset{\text{time to complete}}{}} t \times 29}$$

EXAMPLE: Cylinder size 4″ Dia. × 10″ stroke. Time to extend: 2 seconds. Inlet pressure 90 psig. Allowable pressure drop 5 psid. Determine C_v.

SOLUTION: Table C: $F = 12.57$ sq in

TABLE D   Compression Factor *C* and Pressure Drop Factor *B*

Inlet Pressure (PSIG)	Compression Factor C	Pressure Drop Factor B For Various Pressure Drops Δp				
		2 PSID	5 PSID	10 PSID	15 PSID	20 PSID
10	1.7	6.5				
20	2.4	7.8	11.8			
30	3.0	8.9	13.6	18.0		
40	3.7	9.9	15.3	20.5	23.6	
50	4.4	10.8	16.7	22.6	26.4	29.0
60	5.1	11.7	18.1	24.6	29.0	32.0
70	5.8	12.5	19.3	26.5	31.3	34.8
80	6.4	13.2	20.5	28.2	33.5	37.4
90	7.1	13.9	21.6	29.8	35.5	39.9
100	7.8	14.5	22.7	31.3	37.4	42.1
110	8.5	15.2	23.7	32.8	39.3	44.3
120	9.2	15.8	24.7	34.2	41.0	46.4
130	9.8	16.4	25.6	35.5	42.7	48.4
140	10.5	16.9	26.5	36.8	44.3	50.3
150	11.2	17.5	27.4	38.1	45.9	52.1
160	11.9	18.0	28.2	39.3	47.4	53.9
170	12.6	18.5	29.0	40.5	48.9	55.6
180	13.2	19.0	29.8	41.6	50.3	57.2
190	13.9	19.5	30.6	42.7	51.7	58.9
200	14.6	20.0	31.4	43.8	53.0	60.4
210	15.3	20.4	32.1	44.9	54.3	62.0
220	16.0	20.9	32.8	45.9	55.6	63.5
230	16.7	21.3	33.5	46.9	56.8	64.9
240	17.3	21.8	34.2	47.9	58.1	66.3
250	18.0	22.2	34.9	48.9	59.3	67.7

Table D: $C = 7.1$
$\qquad B = 21.6$

$$C_v = \frac{12.57 \times 10 \times 7.1}{21.6 \times 2 \times 29} = .7$$

Select a valve that has a C_v factor of .7 or higher. In most cases a $\frac{1}{4}''$ valve would be sufficient.

It is considered good engineering practice to limit the pressure drop Δp to approximately 10% of primary pressure p_1. The smaller the allowable pressure drop, the larger the required valve will become.

After the minimum required C_v has been calculated, the proper size valve can be selected from the catalog.

Valve Sizing with C_v = 1 Table

(For nomenclature see above.) This method can be used if the required air flow is known or has been calculated with the formulas as shown below.
Formula 1:

$$Q = .0273 \frac{D^2 L}{t} \times \frac{p_2 + 14.7}{14.7} \quad \text{(scfm)}$$

Conversion of cfm to scfm (formula 2):

$$Q = q \times \frac{p_2 + 14.7}{14.7} \times \frac{528}{T} \quad \text{(scfm)}$$

Flow Factor C_v (standard conditions) (formula 3):

$$C_v = \frac{1.024 \times Q}{\sqrt{\Delta p \times (p_2 + 14.7)}} \qquad \begin{array}{l} \text{Proposed NFPA} \\ \text{Standard T3.21.3} \end{array}$$

Maximum pressure drop Δp across the valve should be less than 10% of inlet pressure p_1.

EXAMPLE 1: Find air flow Q(scfm) if C_v is known. C_v (from valve catalog) = 1.8.

Primary pressure p_1 = 90 psig
Pressure drop across valve Δp = 5 psid

TABLE E   Air Flow Q (SCFM) for $C_v = 1$

Inlet Pressure (PSIG)	Air Flow Q (SCFM) For Various Pressure Drops Δ_p At A $C_v = 1$					Air Flow Q (SCFM) To Atmosphere
	2 PSID	5 PSID	10 PSID	15 PSID	20 PSID	
10	6.7					12.0
20	7.9	11.9				16.9
30	9.0	13.8	18.2			21.8
40	9.9	15.4	20.6	23.8		26.6
50	10.8	16.9	22.8	26.7	29.2	31.5
60	11.6	18.2	24.8	29.2	32.3	36.4
70	12.3	19.5	26.7	31.6	35.1	41.2
80	13.0	20.7	28.4	33.8	37.7	46.1
90	13.7	21.8	30.0	35.8	40.2	51.0
100	14.4	22.9	31.6	37.8	42.5	55.9
110	15.0	23.9	33.1	39.6	44.7	60.7
120	15.6	24.9	34.5	41.4	46.8	65.6
130	16.1	25.8	35.8	43.1	48.8	70.5
140	16.7	26.7	37.1	44.7	50.7	75.3
150	17.2	27.6	38.4	46.3	52.5	80.2
160	17.7	28.4	39.6	47.8	54.3	85.1
170	18.2	29.3	40.8	49.3	56.0	90.0
180	18.7	30.1	42.0	50.7	57.7	94.8
190	19.2	30.9	43.1	52.1	59.4	99.7
200	19.6	31.6	44.2	53.4	60.9	104.6
210	20.1	32.4	45.2	54.8	62.5	109.4
220	20.5	33.1	46.3	56.1	64.0	114.3
230	21.0	33.8	47.3	57.3	65.5	119.2
240	21.4	34.5	48.3	58.6	66.9	124.0
250	21.8	35.2	49.3	59.8	68.3	128.9

Flow through valve from Table E for $C_v = 1$:21.8 scfm

$$Q = C_v \text{ of valve} \times \text{air flow at } C_v = 1 \text{ (scfm)}$$
$$= 1.8 \times 21.8 = 39.2 \text{ scfm}$$

EXAMPLE 2: Find C_v if air flow Q(scfm) is given.

Primary pressure $p_1 = 90$ psig

Pressure drop $\Delta p = 10$ psid

Air flow $Q = 60$ scfm

Flow through valve from Table E for $C_v = 1:30$ scfm

$$C_v = \frac{\text{air flow } Q(\text{scfm})}{\text{air flow at } C_v = 1 \text{ (scfm)}}$$

$$= \frac{60 \text{ scfm}}{30} = 2.0$$

A valve with a C_v of minimum 2 should be selected.

EXAMPLE 3: Find C_v if air flow Q(scfm) to atmosphere is given (from catalog).

Primary pressure $p_1 = 90$ psig

Air flow to atmosphere $Q = 100$ scfm

Flow to atmosphere through valve from Table E for $C_v = 1:51$ scfm

$$C_v = \frac{\text{air flow to atmosphere } Q(\text{scfm})}{\text{air flow to atmosphere at } C_v = 1 \text{ (scfm)}}$$

$$= \frac{100}{51} = 2.0$$

Flow given in catalog is equivalent to a valve with $C_v = 2$. This conversion is often necessary to size a valve properly, since some manufacturers do not show the standard C_v to allow a comparison.

EXAMPLE 4: Find C_v if cylinder size and stroke speed are known using the formulas 1 and 3.

Primary pressure = 90 psig

Pressure drop across valve = 5 psid

Cylinder size is 4" diameter × 10" stroke

Time to complete stroke = 2 sec

$$Q = 0.273 \frac{4^2 \times 10}{2} \times \frac{85 + 14.7}{14.7} = 14.81 \text{ scfm}$$

$$C_v = \frac{1.024 \times 14.81}{\sqrt{5 \times (85 + 14.7)}} = .7$$

NOTES

NOTES

HEATING FUELS
AND SOURCES

TABLE 5.1   Heat Values of Wood Fuel

Type of wood	Average wt/cord,* lb	Btu/cord†	Btu/lb	Comments
Hickory	3595	30,600,000	8510	Highest heat value
Hard maple	3075	29,000,000	9430	High heat value
Beech	3240	27,800,000	8580	
White oak	3750	27,700,000	7380	
Red oak	3240	26,300,000	8110	
Birch	3000	26,200,000	8730	
Elm	2750	24,500,000	8900	Hard to split
Tamarack	2500	24,010,000	9600	
Soft Maple	2500	24,000,000	9600	Acceptable
Cherry	2550	23,500,000	9210	Difficult to find
Ash	2950	22,600,000	7660	Marginal heat value
Spruce	2100	18,100,000	8610	
Hemlock	2100	17,910,000	8520	
White Pine	1800	17,900,000	9940	
Aspen	1900	17,700,000	9310	
Basswood	1900	17,001,000	8940	

*Wood cut for fuel or pulpwood (128 ft³) as arranged in a stack 4 ft × 4 ft × 8 ft. "Fireplace" or "face" cords are only about one-third of a true cord.

†Btu: The British thermal unit (Btu) is the quantity of heat required to raise the temperature of 1 lb of water 1°F.

TABLE 5.2 Typical Btu Values of Fuels

ASTM rank solids	Btu values per lb
Anthracite Class I	11,230
Bituminous Class II Group 1	14,100
Bituminous Class II Group 3	13,080
Subbituminous Class III Group 1	10,810
Subbituminous Class III Group 2	9,670

Liquids	Btu values per gal
Fuel oil No. 1	138,870
Fuel oil No. 2	143,390
Fuel oil No. 4	144,130
Fuel oil No. 5	142,720
Fuel oil No. 6	137,275

Gases	Btu values per ft^3
Natural gas	1030 to 1132
Producers gas	163
Illuminating gas	534
Mixed (coke oven and water gas)	545

Coal, lignite, peat, etc.	Btu values per lb
Eastern coal	13,250
Western coal	9,000
Lignite	7,000
Peat (sods or milled) (@ 30% M.C.)*	6,000
Agri fuel pellets or briquettes (@ 8% M.C.)*	8,000
Wood chips (@ 45% M.C.)*	4,700

*M.C. = Moisture Content

TABLE 5.3   Heat Values of Petrofuels

Type	Btu/lb	Btu/gal or ft³
Fuel oil No. 2	20,571	144,000/gal
Natural gas	18,000	1,030/ft³
Propane	21,564	2,572/ft³
Butane	21,440	3,200/ft³

TABLE 5.4 Physical Properties of Propane

Formula	C_3H_8
Btu/gal	91,500
Btu/ft^3 of gas at 60°F, atmospheric pressure	2,520
Btu/lb of gas	21,560
Range of inflammability: percent of gas in gas-air mixture	2.15 to 9.6%
Vapor pressure, psig at 60°F	92
Vapor pressure, psig at 100°F	172
Pounds per gal of liquid at 60°F, storage tank pressure	4.23
Specific gravity of liquid at 60°F (water = 1)	0.51
Boiling point of liquid at atmospheric pressure	−44°F
Cubic feet of gas per lb of liquid (at 60°F, atmospheric pressure)	8.59
Cubic feet of gas per gal of liquid (at 60°F, atmospheric pressure)	36.5
Specific gravity of gas (air = 1)	1.53

TABLE 5.5   Typical Composition of Commercial Fuel Gases

Constituent	Blast Furnace Gas	Blue Water Gas From Coke	Carbureted Water Gas	Coke Oven Gas	Natural Gas	Oil Gas	Producer Gas
			Percent by Volume				
H_2	2.9	47.3	37.0	52.3	See	48.7	14.9
CO	25.3	37.0	33.5	6.0	page	12.3	27.4
CH_4	0.5	1.3	13.5	30.6		24.8	2.1
CO_2	12.8	5.4	3.4	1.9	24-2	4.7	4.5
N_2	58.5	8.3	4.8	4.9		4.8	50.5
O_2	——	0.7	0.5	0.5		0.5	0.3
Illum.	——	——	7.3	3.8		4.2	0.3
Specific Gravity Air = 1.0	1.02	0.57	0.65	0.41	0.55 to 1.00	0.48	0.85
HHV BTU/cu ft.	94	287	525	578	950 to 1150	541	161

Each column is an average of dry gas analyses from several sources.

SOURCE: Ingersoll-Rand Co. Used with permission.

TABLE 5.6 Heating Efficiency Comparison

Fuel	Heating equipment efficiency percentage*
Electricity	95
No. 2 oil	80
Propane	78
Natural gas	80
Firewood (air-dried)	55
No. 5 & No. 6 oil (low sulfur)	80
Agri fuel pellets or briquettes (@ 8% M.C.)	78
Eastern coal	78
Western coal	75
Wood chips (@ 45% M.C.)	65
Lignite	75
Peat (@ 30% M.C.)	68

*These percentages are general. For specific equipment, see manufacturers' literature and certified test data.

TABLE 5.7   Properties of Saturated Steam—Pressure Table

Abs press in. Hg.	Temp °F	Specific Volume cu ft per lb		Enthalpy (Total Heat) Btu per lb		Entropy Btu/°F.lb	
		Sat Liquid	Sat Vapor	Sat Liquid	Sat Vapor	Sat Liquid	Sat Vapor
0.25	40.23	0.01602	2423.7	8.28	1079.4	0.0166	2.1589
0.50	58.80	0.01604	1256.4	26.86	1087.5	0.0532	2.0985
0.75	70.43	0.01606	856.1	38.47	1092.5	0.0754	2.0635
1.00	79.03	0.01608	652.3	47.05	1096.3	0.0914	2.0387
1.5	91.72	0.01611	444.9	59.71	1101.7	0.1147	2.0041
2.0	101.14	0.01614	339.2	69.10	1105.7	0.1316	1.9797
2.5	108.71	0.01616	274.9	76.65	1108.9	0.1449	1.9609
3.0	115.06	0.01618	231.6	82.99	1111.6	0.1560	1.9456
4.0	125.43	0.01622	176.7	93.34	1116.0	0.1738	1.9214
5	133.76	0.01626	143.25	101.66	1119.4	0.1879	1.9028
6	140.78	0.01630	120.72	108.67	1122.3	0.1996	1.8877
7	146.86	0.01633	104.46	114.75	1124.8	0.2097	1.8750
8	152.24	0.01635	92.16	120.13	1127.0	0.2186	1.8640
9	157.09	0.01638	82.52	124.97	1129.0	0.2264	1.8543
10	161.49	0.01640	74.76	129.38	1130.8	0.2335	1.8456
11	165.54	0.01642	68.38	133.43	1132.4	0.2400	1.8378
12	169.28	0.01644	63.03	137.18	1133.9	0.2460	1.8307
13	172.78	0.01646	58.47	140.68	1135.3	0.2516	1.8241
14	176.05	0.01648	54.55	143.96	1136.6	0.2568	1.8181
15	179.14	0.01650	51.14	147.06	1137.8	0.2616	1.8125
16	182.05	0.01652	48.14	149.98	1138.9	0.2662	1.8072
17	184.82	0.01654	45.48	152.75	1140.0	0.2705	1.8023
18	187.45	0.01655	43.11	155.39	1141.1	0.2746	1.7977
19	189.96	0.01657	40.99	157.91	1142.1	0.2784	1.7933
20	192.37	0.01658	39.07	160.33	1143.0	0.2822	1.7891
21	194.68	0.01660	37.32	162.65	1143.9	0.2857	1.7851
22	196.90	0.01661	35.73	164.87	1144.7	0.2891	1.7814
23	199.03	0.01663	34.28	167.02	1145.5	0.2923	1.7779
24	201.09	0.01664	32.94	169.09	1146.3	0.2955	1.7744
25	203.08	0.01666	31.70	171.09	1147.0	0.2985	1.7711
26	205.00	0.01667	30.56	173.02	1147.8	0.3014	1.7679
27	206.87	0.01668	29.50	174.90	1148.5	0.3042	1.7649
28	208.67	0.01669	28.52	176.72	1149.2	0.3069	1.7619
29	210.43	0.01671	27.60	178.48	1149.9	0.3096	1.7591
30	212.13	0.01671	26.74	180.19	1150.5	0.3122	1.7564
lb/in²							
0.20	53.14	0.01603	1526.0	21.21	1085.0	0.0422	2.1163
0.25	59.30	0.01604	1235.3	27.36	1087.7	0.0542	2.0970
0.30	64.47	0.01605	1039.5	32.52	1090.0	0.0641	2.0812
0.35	68.93	0.01605	898.5	36.97	1091.9	0.0725	2.0678
0.40	72.86	0.01606	791.9	40.89	1093.6	0.0799	2.0563
0.45	76.38	0.01607	708.5	44.41	1095.1	0.0865	2.0462
0.50	79.58	0.01608	641.4	47.60	1096.4	0.0924	2.0372
0.60	85.21	0.01609	540.0	53.21	1098.9	0.1028	2.0216
0.70	90.08	0.01610	466.9	58.07	1101.0	0.1117	2.0085
0.80	94.38	0.01612	411.7	62.36	1102.8	0.1194	1.9971
0.90	98.24	0.01613	368.4	66.21	1104.5	0.1263	1.9871
1.0	101.74	0.01614	333.6	69.70	1106.0	0.1326	1.9782
1.2	107.92	0.01616	280.9	75.87	1108.6	0.1435	1.9628

SOURCE: Ingersoll-Rand Co. Used with permission.

TABLE 5.7   *(continued)*

Abs press lb/in²	Temp °F	Specific Volume cu ft per lb		Enthalpy (Total Heat) Btu per lb		Entropy Btu°F.lb	
		Sat Liquid	Sat Vapor	Sat Liquid	Sat Vapor	Sat Liquid	Sat Vapor
1.4	113.26	0.01618	243.0	81.20	1110.8	0.1528	1.9498
1.6	117.99	0.01620	214.3	85.91	1112.8	0.1610	1.9386
1.8	122.23	0.01621	191.8	90.14	1114.6	0.1683	1.9288
2.0	126.08	0.01623	173.73	93.99	1116.2	0.1749	1.9200
2.2	129.62	0.01624	158.85	97.52	1117.7	0.1809	1.9120
2.4	132.89	0.01626	146.38	100.79	1119.1	0.1864	1.9047
2.6	135.94	0.01627	135.78	103.78	1120.3	0.1916	1.8981
2.8	138.79	0.01629	126.65	106.68	1121.5	0.1963	1.8920
3.0	141.48	0.01630	118.71	109.37	1122.6	0.2008	1.8863
3.5	147.57	0.01633	102.72	115.46	1125.1	0.2109	1.8735
4.0	152.97	0.01636	90.63	120.86	1127.3	0.2198	1.8625
4.5	157.83	0.01638	81.16	125.71	1129.3	0.2276	1.8528
5.0	162.24	0.01640	73.52	130.13	1131.1	0.2347	1.8441
5.5	166.30	0.01643	67.24	134.19	1132.7	0.2411	1.8363
6.0	170.06	0.01645	61.98	137.96	1134.2	0.2472	1.8292
6.5	173.56	0.01647	57.50	141.47	1135.6	0.2528	1.8227
7.0	176.85	0.01649	53.64	144.76	1136.9	0.2581	1.8167
7.5	179.94	0.01651	50.29	147.86	1138.1	0.2629	1.8110
8.0	182.86	0.01653	47.34	150.79	1139.3	0.2674	1.8057
8.5	185.64	0.01654	44.73	153.57	1140.4	0.2718	1.8008
9.0	188.28	0.01656	42.40	156.22	1141.4	0.2759	1.7962
9.5	190.80	0.01658	40.31	158.75	1142.3	0.2798	1.7918
10	193.21	0.01659	38.42	161.17	1143.3	0.2835	1.7876
11	197.75	0.01662	35.14	165.73	1145.0	0.2903	1.7800
12	201.96	0.01665	32.40	169.96	1146.6	0.2967	1.7730
13	205.88	0.01667	30.06	173.91	1148.1	0.3027	1.7665
14	209.56	0.01670	28.04	177.61	1149.5	0.3083	1.7605
14.696	212.00	.01672	26.80	180.07	1150.4	.3120	1.7566
15	213.03	.01672	26.29	181.11	1150.8	.3135	1.7549
20	227.96	.01683	20.089	196.16	1156.3	.3356	1.7319
25	240.07	.01692	16.303	208.42	1160.6	.3533	1.7139
30	250.33	.01701	13.746	218.82	1164.1	.3680	1.6993
35	259.28	.01708	11.898	227.91	1167.1	.3807	1.6870
40	267.25	.01715	10.498	236.03	1169.7	.3919	1.6763
45	274.44	.01721	9.401	243.36	1172.0	.4019	1.6669
50	281.01	.01727	8.515	250.09	1174.1	.4110	1.6585
55	287.07	.01732	7.787	256.30	1175.9	.4193	1.6509
60	292.71	.01738	7.175	262.09	1177.6	4270	1.6438
65	297.97	.01743	6.655	267.50	1179.1	.4342	1.6374
70	302.92	.01748	6.206	272.61	1180.6	.4409	1.6315
75	307.60	.01753	5.816	277.43	1181.9	.4472	1.6259
80	312.03	.01757	5.472	282.02	1183.1	.4531	1.6207
85	316.25	.01761	5.168	286.39	1184.2	.4587	1.6158
90	320.27	.01766	4.896	290.56	1185.3	.4641	1.6112
95	324.12	.01770	4.652	294.56	1186.2	.4692	1.6068
100	327.81	.01774	4.432	298.40	1187.2	.4740	1.6026
105	331.36	.01778	4.232	302.10	1188.1	.4787	1.5986
110	334.77	.01782	4.049	305.66	1188.9	.4832	1.5948

in. Hg x 25.4 = mm Hg; °C = 0.5555 (°F-32); ft³/lb x 0.0624 = m³/kg; Btu/lb x 2326 = J/kg; Btu°F.lb x 4184 = J/kg.K

(continued)

TABLE 5.7 (*continued*)

Abs press lb/in²	Temp °F	Specific Volume cu ft per lb		Enthalpy (Total Heat) Btu per lb		Entropy Btu/°F lb	
		Sat Liquid	Sat Vapor	Sat Liquid	Sat Vapor	Sat Liquid	Sat Vapor
115	338.07	.01785	3.882	309.11	1189.7	.4875	1.5912
120	341.25	.01789	3.728	312.44	1190.4	.4916	1.5878
125	344.33	.01792	3.587	315.68	1191.1	.4956	1.5844
130	347.32	.01796	3.455	318.81	1191.7	.4995	1.5812
135	350.21	.01800	3.333	321.85	1192.4	.5032	1.5781
140	353.02	.01802	3.220	324.82	1193.0	.5069	1.5751
145	355.76	.01806	3.114	327.70	1193.5	.5104	1.5722
150	358.42	.01809	3.015	330.51	1194.1	.5138	1.5694
155	361.01	.01812	2.922	333.24	1194.6	.5172	1.5666
160	363.53	.01815	2.834	335.93	1195.1	.5204	1.5640
165	366.00	.01818	2.752	338.54	1195.6	.5235	1.5615
170	368.41	.01822	2.675	341.09	1196.0	.5266	1.5590
175	370.76	.01825	2.601	343.59	1196.5	.5296	1.5566
180	373.06	.01827	2.532	346.03	1196.9	.5325	1.5542
185	375.31	.01831	2.466	348.94	1197.3	.5354	1.5518
190	377.57	.01833	2.404	350.79	1197.6	.5381	1.5497
195	379.67	.01836	2.344	353.10	1198.0	.5409	1.5474
200	381.79	.01839	2.288	355.36	1198.4	.5435	1.5453
205	383.86	.01842	2.234	357.58	1198.7	.5461	1.5432
210	385.90	.01844	2.183	359.77	1199.0	.5487	1.5412
215	387.89	.01847	2.134	361.91	1199.3	.5512	1.5392
220	389.86	01850	2.087	364.02	1199.6	.5537	1.5372
225	391.79	.01852	2.042	366.09	1199.9	.5561	1.5353
230	393.68	.01854	1.999	368.13	1200.1	.5585	1.5334
235	395.54	.01857	1.9579	370.14	1200.4	.5608	1.5316
240	397.37	.01860	1.9183	372.12	1200.6	.5631	1.5298
245	399.18	.01863	1.8803	374.08	1200.9	.5653	1.5280
250	400.95	.01865	1.8438	376.00	1201.1	.5675	1.5263
260	404.42	.01870	1.7748	379.76	1201.5	.5719	1.5229
270	407.78	.01875	1.7107	383.42	1201.9	.5760	1.5196
280	411.05	.01880	1.6511	386.98	1202.3	.5801	1.5164
290	414.23	.01885	1.5954	390.46	1202.6	.5841	1.5133
300	417.33	.01890	1.5433	393.84	1202.8	.5879	1.5104
320	423.29	.01899	1.4485	400.39	1203.4	.5952	1.5046
340	428.97	.01908	1.3645	406.66	1203.7	.6022	1.4992
360	434.40	.01917	1.2895	412.67	1204.1	.6090	1.4941
380	439.60	.01925	1.2222	418.45	1204.3	.6153	1.4891
400	444.59	.0193	1.1613	424.0	1204.5	.6214	1.4844
420	449.39	.0194	1.1061	429.4	1204.6	.6272	1.4799
440	454.02	.0195	1.0556	434.6	1204.6	.6329	1.4755
460	458.50	.0196	1.0094	439.7	1204.6	.6383	1.4713
480	462.82	.0197	.9670	444.6	1204.5	.6436	1.4673
500	467.01	.0197	.9278	449.4	1204.4	.6487	1.4634
550	476.93	.0199	.8422	460.8	1203.9	.6608	1.4542
600	486.21	.0201	.7698	471.6	1203.2	.6720	1.4454
650	494.89	.0203	.7084	481.8	1202.3	.6826	1.4373
700	503.10	.0205	.6554	491.5	1201.2	.6925	1.4296
750	510.85	.0207	.6093	500.8	1199.1	.7019	1.4223

TABLE 5.8 Tank Sizing (in Gallons) for Vapor Withdrawal

Maximum gas needed to vaporize*	Tank size (in gallons) required if lowest outdoor temperature (average for 24 h) reaches:						
	32°F	20°F	10°F	0°F	−10°F	−20°F	−30°F
125,000 Btu/h (50 ft³/h)	115	115	115	250	250	400	600
250,000 Btu/h (100 ft³/h)	250	250	250	400	500	1000	1500
375,000 Btu/h (150 ft³/h)	300	400	500	500	1000	1500	2500
500,000 Btu/h (200 ft³/h)	400	500	750	1000	1200	2000	3500
750,000 Btu/h (300 ft³/h)	750	1000	1500	2000	2500	4000	5000

* Average rate of withdrawal during an 8-h phase.

TABLE 5.9   Heating Values of Substances Occurring in Common Fuels*

Substance	Molecular Symbol	Higher Heating Values, †		Lower Heating Values, †	
		kJ/kg	Btu/lb	kJ/kg	Btu/lb
Carbon (to CO)	C	9 188	3 950	9 188	3 950
Carbon (to CO$_2$)	C	32 780	14 093	32 780	14 093
Carbon Monoxide	CO	10 111	4 347	10 111	4 347
Hydrogen	H$_2$	142 107	61 095	118 680	51 023
Methane	CH$_4$	55 533	23 875	49 997	21 495
Ethane	C$_2$H$_6$	51 923	22 323	47 492	20 418
Propane	C$_3$H$_8$	50 402	21 669	46 373	19 937
Butane	C$_4$H$_{10}$	49 593	21 321	45 771	19 678
Ethylene	C$_2$H$_4$	50 325	21 636	47 160	20 275
Propylene	C$_3$H$_6$	48 958	21 048	45 792	19 687
Acetylene	C$_2$H$_2$	50 028	21 508	48 309	20 769
Sulfur (to SO$_2$)	S	2 957	3 980	9 257	3 980
Sulfur (to SO$_3$)	S	13 816	5 940	13 816	5 940
Hydrogen Sulfide	H$_2$S	16 508	7 097	15 205	6 537

*Fuel factors include available (including dependability of) supply, convenience of use and storage, economy, and cleanliness.

†All values corrected to 15.6°C, 101.325 kPa (60°F, 30 inHg), dry. For gases saturated with water vapor at 15.6°C (60°F), deduct 1.74% of the value.

SOURCE: Reprinted by permission from *ASHRAE Handbook—1981 Fundamentals*.

TABLE 5.10 Typical Gravity and Heating Value of Standard Grades of Fuel Oil

Grade No.	Gravity API	Mass (Weight) kg/L	Mass (Weight) lb/gal	Heating Value kJ/L	Heating Value Btu/gal
1	38 to 45	0.833 to 0.800	6.950 to 6.675	38 180 to 37 040	137 000 to 132 900
2	30 to 38	0.874 to 0.834	7.296 to 6.960	39 520 to 38 180	141 800 to 137 000
4	20 to 28	0.933 to 0.886	7.787 to 7.396	41 280 to 39 880	148 100 to 143 100
5L	17 to 22	0.951 to 0.921	7.940 to 7.686	41 810 to 40 920	150 000 to 146 800
5H	14 to 18	0.968 to 0.945	8.080 to 7.890	42 360 to 41 640	152 000 to 149 400
6	8 to 15	1.012 to 0.965	8.448 to 8.053	43 450 to 42 170	155 900 to 151 300

SOURCE: Reprinted by permission from *ASHRAE Handbook—1981 Fundamentals.*

TABLE 5.11   Sulfur Content of Marketed Fuel Oils

Grade of Oil	No. 1	No. 2	No. 4	No. 5 (Light)	No. 5 (Heavy)	No. 6
Total Fuel Samples	123	158	13	15	16	96
Sulfur Content % Wt Min	0.002	0.03	0.46	0.90	0.57	0.32
Max	0.380	0.64	1.44	3.50	2.92	4.00
No. Samples with S over 0.3%	1	32	13	15	16	96
over 0.5%	0	1	11	15	16	93
over 1.0%	0	0	3	9	11	60
over 3.0%	0	0	0	2	0	8

SOURCE: Reprinted by permission from *ASHRAE Handbook—1981 Fundamentals*.

TABLE 5.12 Combustion Reactions of Common Fuel Constituents

Constituent	Molecular Symbol	Combustion Reactions	Stoichiometric Oxygen and Air Requirements			
			kg/kg (lb/lb) Fuel*		m³/ft³ Fuel	
			O_2	Air	O_2	Air
Carbon (to CO)	C	$C + 0.5o_2 \rightarrow CO$	1.33	5.75	—	—
Carbon (to CO_2)	C	$C + O_2 \rightarrow CO_2$	2.66	11.51	—	—
Carbon Monoxide	CO	$CO + 0.5o_2 \rightarrow CO_2$	0.57	2.47	0.50	2.39
Hydrogen	H_2	$H_2 + 0.5o_2 \rightarrow H_2O$	7.94	34.28	0.50	2.39
Methane	CH_4	$CH_4 + 2O_2 \rightarrow CO_2 + 2H_2O$	3.99	17.24	2.00	9.57
Ethane	C_2H_6	$C_2H_6 + 3.5O_2 \rightarrow 2CO_2 + 3H_2O$	3.72	16.09	3.50	16.75
Propane	C_3H_8	$C_3H_8 + 5O_2 \rightarrow 3CO_2 + 4H_2O$	3.63	15.68	5.00	23.95
Butane	C_4H_{10}	$C_4H_{10} + 6.5O_2 \rightarrow 4CO_2 + 5H_2O$	3.58	15.47	6.50	31.14
	C_nH_{2n+2}	$C_nH_{2n+2} + (1.5n + 0.5)O_2 \rightarrow nCO_2 + (n+1)H_2O$	—	—	$1.5n + 0.5$	$7.18n + 2.39$
Ethylene	C_2H_4	$C_2H_4 + 3O_2 \rightarrow 2CO_2 + 2H_2O$	3.42	14.78	3.00	14.38
Propylene	C_3H_6	$C_3H_6 + 4.5O_2 \rightarrow 3CO_2 + 3H_2O$	3.42	14.78	4.50	21.53
	C_nH_{2n}	$C_nH_{2n} + 1.5nO_2 \rightarrow nCO_2 + nH_2O$	3.42	14.78	$1.50n$	$7.18n$
Acetylene	C_2H_2	$C_2H_2 + 2.5O_2 \rightarrow 2CO_2 + H_2O$	3.07	13.27	2.50	11.96
	C_nH_{2m}	$C_nH_{2m} + (n + 0.5m)O_2 \rightarrow nCO_2 + mH_2O$	—	—	$n + 0.5m$	$4.78n + 2.39m$
Sulfur (to SO_2)	S	$S + O_2 \rightarrow SO_2$	1.00	4.31	—	—
Sulfur (to SO_3)	S	$S + 1.5O_2 \rightarrow SO_3$	1.50	6.47	—	—
Hydrogen Sulfide	H_2S	$H_2S + 1.5O_2 \rightarrow SO_2 + H_2O$	1.41	6.08	1.50	7.18

*Atomic mass (weights): H = 1.008; C = 12.01; O = 16.00; S = 32.06.

SOURCE: Reprinted by permission from *ASHRAE Handbook—1981 Fundamentals.*

5.1 SOLAR ANGLES AND SUN PATH DIAGRAMS[1]

Solar Angles

The position of the sun in relation to specific geographic locations, seasons, and times of day can be determined by several different methods. Model measurements, by means of sun machines or shade dials, have the advantage of direct visual observations. Tabulative and calculative methods have the advantage of exactness. However, graphic projection methods are usually preferred by architects as they are easily understood and can be correlated to both radiant energy and shading calculations.

Sun-Path Diagrams

The most practical graphic projection is the Sun-Path Diagram method. Such diagrams depict the path of the sun within the sky-vault as projected onto a horizontal plane. The horizon is represented as a circle with the observation point in the center. The sun's position at any date and hour can be determined from the diagram in terms of its altitude (α) and bearing angle (β). (See figure below.) The graphs are constructed in equidistant projection. The altitude angles are represented at 10° intervals by equally spaced concentric circles; they range from 0° at the outer circle (horizon) to 90° at the center point. These intervals are graduated along the south meridian. Bearing angles are represented at 10° intervals by equally spaced radii; they range from 0° at the south meridian to 180° at the north meridian. These intervals are graduated along the periphery. The sun's bearing

[1]Text, unnumbered table, and unnumbered illustrations in this subsection adapted from Joseph N. Boaz (ed.), Charles G. Ramsey's and Harold R. Sleeper's *Architectural Graphic Standards,* 6th ed., John Wiley & Sons, New York, © 1970. Reprinted by permission of the publisher.

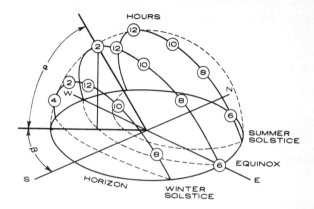

will be to the east during morning hours, and to the west during afternoon hours.

The earth's axis is inclined 23°27′ to its orbit around the sun and rotates 15° hourly. Thus, from all points on the earth, the sun appears to move across the skyvault on various parallel circular paths with maximum declinations of ±23°27′. The declination of the sun's path changes in a cycle between the extremes of the summer solstice and winter solstice. Thus, the sun follows the same path on two corresponding dates each year. Due to irregularities between the calendar year and the astronomical data, here a unified calibration is adapted. The differences, as they do not exceed 41′, are negligible for architectural purposes.

The elliptical curves in the diagrams represent the horizontal projections of the sun's path. They are given on the 21st day of each month. Roman numerals designate the months. A cross grid of curves graduates the hours indicated in arabic numerals. Eight sun-path diagrams are shown at 4° intervals from 24° N to 52° N latitude.

DECLINATION OF THE SUN

Date	Declination	Corresp. Date	Declination	Unified Calibr.
June 21	+23°27′			+23°27′
May 21	+20°09′	July 21	+20°31′	+20°20′
Apr. 21	+11°48′	Aug. 21	+12°12′	+12°00′
Mar. 21	+ 0°10′	Sep. 21	+ 0°47′	+ 0°28′
Feb. 21	−10°37′	Oct. 21	−10°38′	−10°38′
Jan. 21	−19°57′	Nov. 21	−19°53′	−19°55′
Dec. 21	−23°27′			−23°27′

EXAMPLE: Find the sun's position in Columbus, Ohio, on February 21st at 2 P.M.:

STEP 1. Locate Columbus on the map. The latitude is 40° N.

STEP 2. In the 40° sun-path diagram select the February path (marked with II), and locate the 2-hour line. Where the two lines cross is the position of the sun.

STEP 3. Read the altitude on the concentric circles (32°) and the bearing angle along the outer circle (35°30′W).

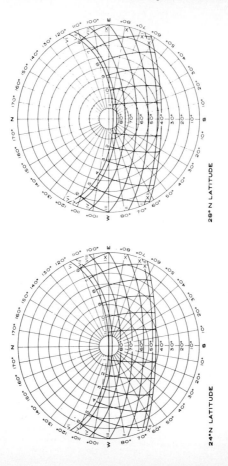

28° N LATITUDE

24° N LATITUDE

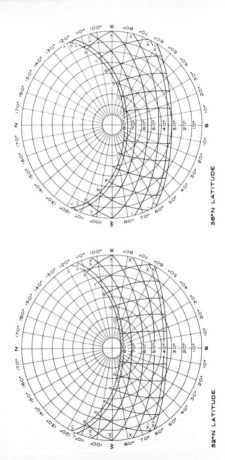

36°N LATITUDE

32°N LATITUDE

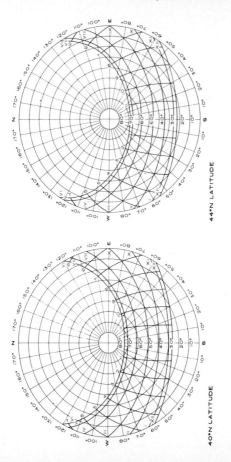

44°N LATITUDE

40°N LATITUDE

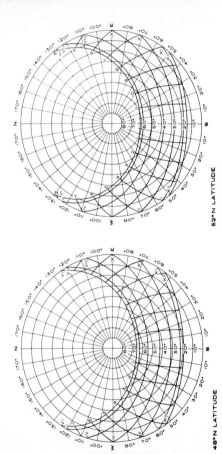

5.2 CALCULATION OF SOLAR POSITION[2]

One can calculate accurately the solar position to any locale and time by relating the spherical triangle formed by the observer's celestial meridian, the meridian of the sun, and the great circle passing through zenith and the sun. The following formulas can be used to find the altitude and bearing angles:

$$\sin A = \sin d \sin l + \cos d \cos l \cos t \qquad (1)$$

$$\cos B = \frac{\sin d \cos l - \cos d \sin l \cos t}{\cos A} \qquad (2)$$

where: A = altitude of the sun in degrees, measured from the horizontal.

d = declination of the sun (see page on solar angles) at the desired date. (North declinations are conventionally positive; south declinations negative.)

l = latitude of the locale; conventionally negative in the southern hemisphere.

t = hour angle of the sun in degrees, measured counterclockwise from north towards east. At solar noon it is zero and changes 15° per hour.

B = bearing angle of the sun in degrees; here measured clockwise from north towards east.

[2]Text in this subsection adapted from Joseph N. Boaz (ed.), Charles G. Ramsey's and Harold R. Sleeper's *Architectural Graphic Standards,* 6th ed., John Wiley & Sons, New York, © 1970. Reprinted by permission of the publisher.

5.3   CALCULATION OF SOLAR RADIATION[3]

To evaluate the importance of solar shading one has to know the amount of solar energy falling on the exposed surface. As the primary protection of the shading devices is from the direct solar radiation, only these energy calculations are described here.

The magnitude of solar radiation depends, first of all, on the sun's altitude. The tabulated values indicate direct radiation energies received under clear atmospheric conditions at normal incidence (ID): Solar altitude in degrees:

5	10	15	20	25	30	35	40	45	50	60	70	80	90
67	123	166	197	218	235	248	258	266	273	283	289	292	294

Btu/sq ft/hour

The energy received on a surface depends also on the cosine of the angle of incidence. As this is a spacial angle, it is conventional on vertical surfaces to substitute it with the functions of the altitude ($\angle A$) and the azimuth ($\angle a$) angles related to the normal of the surface in question. Thus, the direct radiation on vertical surfaces (R) can be defined as:

$$R = I_D \times \cos A \cos a$$

For horizontal surfaces the received direct radiation energy will be:

$$R = I_D \times \sin A$$

In the following tables calculated values of solar position in degrees, and direct radiation energies in Btu/sq ft/hour values, are shown at different orientations. The following tables indicate from 24°N to 46°N latitude at 4° intervals.

[3]Text and unnumbered tables in this subsection adapted from Joseph N. Boaz (ed.), Charles G. Ramsey's and Harold R. Sleeper's *Architectural Graphic Standards*, 6th ed., John Wiley & Sons, New York, © 1970. Reprinted by permission of the publisher.

26° N. LATITUDE

JUNE 22

AM	ALT	BEAR	S	SE	E	NE	N	SW	HOR	PM
6 a.m.	10.05	111.30		49	113	111	44		22	6 p.m.
7	22.82	105.97		93	185	168	53		81	5
8	35.93	100.15		113	199	168	39		147	4
9	49.24	94.45		111	176	176	20		206	3
10	62.69	88.83	9	94	131	90			253	2
11	76.15	82.61	19	55	69	43			283	1
12	87.45	0.00	13	9				9	293	12
	α	β	S	SE	E	NE	N	SW	HOR	PM

MARCH 21, SEPT 24

AM	ALT	BEAR	S	SE	E	NE	N	SW	HOR	PM
6 a.m.	0.00	90.00								6 p.m.
7				17	147				36	5
8				49	172	194			101	4
9				80	185	182			163	3
10			104	171	137				213	2
11			120	137	73				246	1
12			125	88					257	12
	α	β	S	SW	E	NE	N	SE	HOR	PM

DECEMBER 22

AM	ALT	BEAR	S	SE	E	N	NW	SW	HOR	PM
7 a.m.	2.23	62.48	14	87	149	9			1	5 p.m.
8	13.76	54.88	138	196	123	26			37	4
9	24.12	45.30	171	199	111	42			88	3
10	32.66	33.01	190	190	61	92			131	2
11	38.45	17.05	197	139					159	1
12	40.55	0.00	197	139			139		168	12
	α	β	S	SW	W	N	NE	SE	HOR	PM

30° N. LATITUDE

JUNE 22

AM	ALT	BEAR	S	SE	E	NE	N	SW	HOR	PM
6 a.m.	11.48	110.59		55	124	121	47		27	6 p.m.
7	23.87	104.30		100	168	162	48		86	5
8	36.60	98.24		121	200	162	29		150	4
9	49.53	91.79	15	121	177	129	6		207	3
10	62.50	83.46	15	103	131			24	252	2
11	75.11	67.48	29	69	69			44	281	1
12	83.45	0.00	29	24				100	291	12
	α	β	S	SE	E	NE	N	SW	HOR	PM

MARCH 21, SEPT 24

AM	ALT	BEAR	S	SE	E	NE	N	SW	HOR	PM
6 a.m.	0.00	90.00								6 p.m.
7	12.95	82.37		19	115	143	88		33	5
8	25.66	73.90		55	174	191	96		95	4
9	37.76	63.44		90	190	179	63		155	3
10	48.59	49.11	117	179	136	13			203	2
11	56.77	28.19	135	147	72				234	1
12	60.00	0.00	142	100					245	12
	α	β	S	SW	E	NE	N	SE	HOR	PM

DECEMBER 22

AM	ALT	BEAR	S	SE	E	N	NW	SW	HOR	PM
7 a.m.	0.38	62.40	2	5	5					5 p.m.
8	11.44	54.15	78	131	108	21			27	4
9	21.77	44.12	135	189	131		3		73	3
10	29.28	31.73	173	197	107		47		114	2
11	34.64	16.77	195	179	59		96		140	1
12	36.55	0.00	202	143			143		150	12
	α	β	S	SW	W	N	NE	SE	HOR	PM

34° N. LATITUDE

JUNE 22

AM		ALT.	BEAR.	S	SE	E	NE	N	SW	HOR.
5 a.m.	7 p.m.	1.47	117.57		6	17	19	9	9	1
6	6	12.86	109.78		61	135	130	49		33
7	5	24.80	102.54		106	192	166	43		91
8	4	37.07	95.28		129	200	155	19		152
9	3	49.49	87.10	9	131	177	119			207
10	2	61.79	76.00	32	115	130	69			250
11	1	73.17	55.11	48	83	69	15			278
12		79.45	0.00	53	38				38	287
PM		α	β	S	SW	W	NW	N	SE	HOR.

MARCH 21, SEPT. 24

AM		ALT.	BEAR.	S	SE	E	NE	N	SW	HOR.
6 a.m.	6 p.m.	0.00	90.00							
7	5	12.39	81.48	21	113	139	83			31
8	4	24.49	72.11	60	175	187	90			89
9	3	35.89	60.79	99	195	177	55			146
10	2	45.89	45.92	129	186	134	3			192
11	1	53.21	25.60	149	156	71				221
12		56.00	0.00	156	110				55	231
PM		α	β	S	SW	W	NW	N	SE	HOR.

DECEMBER 22

AM		ALT.	BEAR.	S	SE	E	NE	N	SW	HOR.
8 a.m.	4 p.m.	9.08	53.57	66	110	90	17			18
9	3	18.38	43.12	129	177	121			6	59
10	2	25.86	30.65	171	193	101			49	96
11	1	30.81	16.05	196	178	56			99	121
12		32.55	0.00	204	144				144	130
PM		α	β	S	SW	W	NW	N	SE	HOR.

38° N. LATITUDE

JUNE 22

AM		ALT.	BEAR.	S	SE	E	NE	N	SW	HOR.
5 a.m.	7 p.m.	3.32	118.42		13	39	42	20		3
6	6	14.18	108.87		68	146	138	50		39
7	5	25.60	100.70		112	195	164	37		83
8	4	37.33	93.25		136	201	148	8		153
9	3	49.13	82.47	23	141	176	108			206
10	2	60.58	69.06	50	127	130	57			246
11	1	70.61	45.67	67	96	69	1			273
12		75.45	0.00	73	52				52	281
PM		α	β	S	SW	W	NW	N	SE	HOR.

MARCH 21, SEPT. 24

AM		ALT.	BEAR.	S	SE	E	NE	N	SW	HOR.
6 a.m.	6 p.m.	0.00	90.00							
7	5	11.77	80.63	22	110	134	79			28
8	4	23.20	70.43	65	175	182	83			83
9	3	33.86	58.38	107	198	173	47			137
10	2	43.03	42.16	140	192	131			6	179
11	1	49.57	23.52	162	164	71			65	207
12		52.00	0.00	169	120				120	217
PM		α	β	S	SW	W	NW	N	SE	HOR.

DECEMBER 22

AM		ALT.	BEAR.	S	SE	E	NE	N	SW	HOR.
8 a.m.	4 p.m.	6.69	53.12	51	84	68	12			10
9	3	15.44	42.30	120	162	109			8	45
10	2	22.40	29.75	166	185	95			50	79
11	1	26.96	15.45	193	174	53			99	102
12		28.55	0.00	202	143				143	110
PM		α	β	S	SW	W	NW	N	SE	HOR.

42° N. LATITUDE

JUNE 22

AM	ALT.	BEAR.	S	SE	E	NE	N	SW	HOR.
5 a.m. / 7 p.m.	5.15	117.16							6
6 / 6	15.44	107.87		21	61	65	31		45
7 / 5	26.28	98.78		74	155	145	50		98
8 / 4	37.38	89.19	3	118	197	161	30		153
9 / 3	48.45	77.96	37	144	201	140			203
10 / 2	58.95	62.79	67	151	176	98			242
11 / 1	67.64	38.62	85	138	129	44		12	266
12	71.45	0.00	92	109	68			65	274
PM	α	β	S	SW	W	NW	N	SE	HOR.

MARCH 21, SEPT. 24

AM	ALT.	BEAR.	S	SE	E	NE	N	SW	HOR.
6 a.m. / 6 p.m.	0.00	90.00		23	107	128	74		25
7 / 5	11.09	79.84		68	174	177	77		76
8 / 4	21.81	68.88	23	113	200	169	40		126
9 / 3	31.70	57.81	68	150	197	129		15	166
10 / 2	40.06	40.79	113	173	171	69		73	192
11 / 1	46.88	21.82	150	181	128			128	201
12	48.00	0.00	173						201
PM	α	β	S	SW	W	NW	N	SE	HOR.

DECEMBER 22

AM	ALT.	BEAR.	S	SE	E	NE	N	SW	HOR.
8 a.m. / 4 p.m.	4.28	52.82	35	57	46	8			4
9 / 3	12.46	41.63	105	141	94				31
10 / 2	18.91	29.01	157	173	87			50	62
11 / 1	23.09	14.96	187	167	50			97	82
12	24.55	0.00	197	139				139	90
PM	α	β	S	SW	W	NW	N	SE	HOR.

46° N. LATITUDE

JUNE 22

AM	ALT.	BEAR.	S	SE	E	NE	N	SW	HOR.
5 a.m. / 7 p.m.	6.97	116.78							11
6 / 6	16.63	106.77		28	79	84	40		50
7 / 5	26.82	96.80		80	162	149	49		101
8 / 4	37.22	86.15	14	124	199	157	24		153
9 / 3	47.47	73.66	51	151	201	132			199
10 / 2	56.95	57.25	83	160	175	87			235
11 / 1	64.40	33.33	103	149	128	32		25	258
12	67.45	0.00	110	121	78			78	265
PM	α	β	S	SW	W	NW	N	SE	HOR.

MARCH 21, SEPT. 24

AM	ALT.	BEAR.	S	SE	E	NE	N	SW	HOR.
6 a.m. / 6 p.m.	0.00	90.00		23	103	122	70		23
7 / 5	10.36	79.09		71	172	172	71		69
8 / 4	20.32	67.45		119	200	165	33		114
9 / 3	29.42	54.27	23	157	200	126		22	152
10 / 2	36.98	38.75	103	182	176	68		81	175
11 / 1	42.14	20.43	122	190	126			134	184
12	44.00	0.00	165						184
PM	α	β	S	SW	W	NW	N	SE	HOR.

DECEMBER 22

AM	ALT.	BEAR.	S	SE	E	NE	N	SW	HOR.
8 a.m. / 4 p.m.	1.86	52.65	15	25	20	3			1
9 / 3	9.46	41.12	87	115	76				19
10 / 2	15.41	28.41	143	156	77			8	45
11 / 1	19.23	14.56	176	156	46			46	63
12	20.55	0.00	187	132				132	70
PM	α	β	S	SW	W	NW	N	SE	HOR.

Note: direction columns are in BTU/Sq. ft./Hour.

5.4 RADIATION CALCULATOR[4]

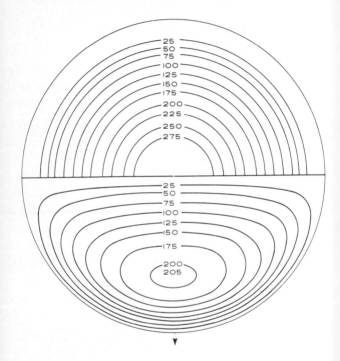

[4]Text and unnumbered illustration in this subsection adapted from Joseph N. Boaz (ed.), Charles G. Ramsey's and Harold R. Sleeper's *Architectural Graphic Standards,* 6th ed., John Wiley & Sons, New York, © 1970. Reprinted by permission of the publisher.

Radiation calculations can be performed by graphical means. The upper half of the above shown direct radiation calculator charts the energies falling on a horizontal plain under clear sky conditions. The equi-intensity radiation lines are indicated at 25-Btu/sq ft/hour intervals. The lower half circle shows the amount of direct radiation falling on a vertical surface. The calculator can be used at any latitude and at any orientation. The calculator is in the same scale and projection as the sun-path diagrams shown in the material on solar angles. Transfer the calculator diagram to a transparent overlay, and superimpose it on a sun-path diagram in the desired orientation: the radiation values can be read directly.

5.5 SHADING: DEVICES, MASKS, PROTRACTOR[5]

Shading Devices

The effect of shading devices can be plotted in the same manner as the sun-path was projected. The diagrams show which part of the sky-vault will be obstructed by the devices and are projections of the surface covered on the sky-vault as seen from an observation point at the center of the diagram. These projections also represent those parts of the sky-vault from which no sunlight will reach the observation point; if the sun passes through such an area, the observation point will be shaded.

Shading Masks

Any building element will define a characteristic form in these projection diagrams, known as "shading masks." Masks of horizontal devices (overhangs) will create a segmental pattern; vertical intercepting elements (fins) produce a radial pattern; shading devices with horizontal and vertical members (eggcrate type) will make a combinative pattern. A shading mask can be drawn for any shading device, even for very complex ones, by geometric plotting. As the shading masks are geometric projections, they are independent of latitude and exposed directions; therefore they can be used in any location and at any orientation. By overlaying a shading mask in the proper orientation on the sun-path diagram, one can read off the times when the sun rays will be intercepted. Masks can be drawn for full shade (100% mask) when the observation point is at the lowest point of the surface needing shading; or for 50% shading when the observation point is placed at the halfway mark on the surface. It is

[5]Text and illustration in this subsection adapted from Joseph N. Boaz (ed.), Charles G. Ramsey's and Harold R. Sleeper's *Architectural Graphic Standards*, 6th ed., John Wiley & Sons, New York, © 1970. Reprinted by permission of the publisher.

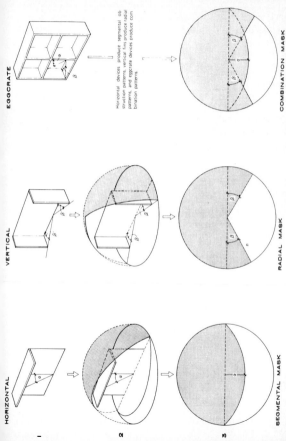

FIG. 5.1 Shading geometry. [*From Joseph N. Boaz (ed.), Charles G. Ramsey's and Harold R. Sleeper's Architectural Graphic Standards, 6th ed., John Wiley & Sons, New York, © 1970. Reproduced with permission of the publisher.*]

customary to design a shading device in such a way that as soon as shading is needed on a surface the masking angle should exceed 50%. Solar calculations should be used to check the specific loads. Basic shading devices are shown in Fig. 5.1, with their obstruction effect on the sky-vault and with their projected shading masks.

Shading Mask Protractor

The half of the protractor showing segmental lines is used to plot lines parallel and normal to the observed vertical surface. The half showing bearing and altitude lines is used to plot shading masks of vertical fins or any other obstruction objects. The protractor is in the same projection and scale as the sun-path diagrams (see material on solar angles); therefore it is useful to transfer the protractor to a transparent overlay to read the obstruction effect.

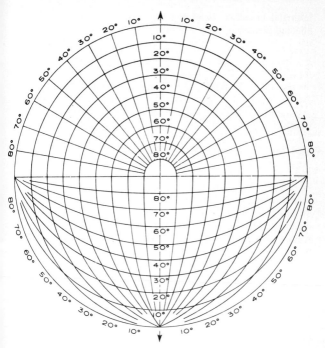

SHADING MASK PROTRACTOR

Examples of Various Types of Shading Devices

The illustrations in Fig. 5.2 show a number of basic types of devices, classified as horizontal, vertical, and eggcrate types. The dash lines shown in the section diagram in each case indicate the sun angle at the time of 100% shading. The shading mask for each device is also shown, the extent of 100% shading being indicated by the gray area.

General rules can be deduced for the types of shading devices to be used for different orientations. Southerly orientations call for shading devices with segmental mask characteristics, and horizontal devices work in these directions efficiently. For easterly and westerly orientations vertical devices serve well, having radial shading masks. If slanted, they should incline toward the north, to give more protection from the southern positions of the sun. The eggcrate type of shading device works well on walls facing southeast, and is particularly effective for southwest orientations. Because of this type's high shading ratio and low winter heat admission, its best use is in hot climate regions. For north walls, fixed vertical devices are recommended; however, their use is needed only for large glass surfaces, or in hot regions. At low latitudes on both south and north exposures eggcrate devices work efficiently.

Whether the shading devices be fixed or movable, the same recommendations apply in respect to the different orientations. The movable types can be most efficiently utilized where the sun's altitude and bearing angles change rapidly: on the east, southeast, and especially, because of the afternoon heat, on the southwest and west.

Horizontal Types: (1) Horizontal overhangs are most efficient toward south, or around southern orientations. Their mask characteristics are segmental. (2) Louvers parallel to wall have the advantage of permitting air circulation near the elevation. Slanted louvers will have the same characteristics as solid overhangs, and can be made retractable. (4) When protection is needed for low sun angles, louvers hung from solid horizontal overhangs are efficient. (5) A solid, or perforated screen strip parallel to wall

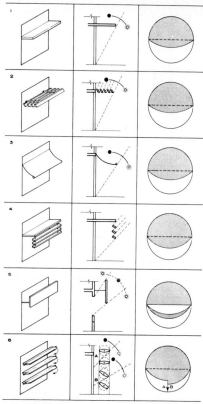

(continued)

FIG. 5.2 Types of shading devices. [*From Joseph N. Boaz (ed.), Charles G. Ramsey's and Harold R. Sleeper's Architectural Graphic Standards, 6th ed., John Wiley & Sons, New York, © 1970. Reproduced with permission of the publisher.*]

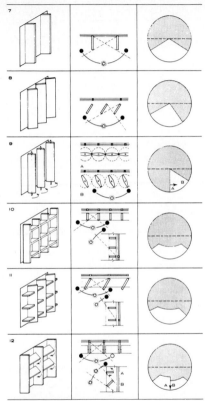

FIG. 5.2—(continued)

cuts out the lower rays of the sun. (6) Movable horizontal louvers change their segmental mask characteristics according to their positioning.

Vertical Types: (7) Vertical fins serve well toward the near east and near west orientations. Their mask characteristics are radial. (8) Vertical fins oblique to wall will result in asymmetrical mask. Separation from wall will prevent heat transmission. (9) Movable fins can shade the whole wall, or open up in different directions according to the sun's position.

Eggcrate Types: (10) Eggcrate types are combinations of horizontal and vertical types, and their masks are superimposed diagrams of the two masks. (11) Solid eggcrate with slanting vertical fins results in asymmetrical mask. (12) Eggcrate device with movable horizontal elements shows flexible mask characteristics. Because of their high shading ratio, eggcrates are efficient in hot climates.

5.6   EXAMPLE OF SHADING DEVICE CALCULATION[6]

The structure in this example is located in New York, N.Y. Two sides of the building (here called north and south) are fully glazed. The two other sides are closed.

STEP 1. Position structure to true orientation (see material on orientation). The long axis of the building lies 15° north of east (see Fig. A below).

STEP 2. To evaluate the need for solar control, the amount of solar energy falling on the exposed glass surfaces should be calculated. New York lies nearest to the 40°N latitude. The most penetrating sun angles occur at June 21st. Superimpose over the 40° sun-path diagram (see material on solar angles) the radiation calculator (see material on solar energy calculation), and turning the calculator 15° east of south, read along the June 21st sun-path the hourly radiation impacts. Figure B below shows the Btu/sq ft/hour sun energy values falling on the south side. One can see from it that this surface receives an eight-hour insolation, with energies over 90 Btu/sq ft/hour around 11 A.M. Figure C below shows the sun energies impinging on the north side. One can see from it that the early morning impact is negligible, but around 6 P.M. a considerable amount of energy falls on the surface. Conclusion: both exposed sides should be protected by shading devices.

STEP 3. To determine the times when shading is needed: during cool times of the year (called "underheated period") the warming

---

[6]Text and unnumbered illustrations in this subsection adapted from Joseph N. Boaz (ed.), Charles G. Ramsey's and Harold R. Sleeper's *Architectural Graphic Standards*, 6th ed., John Wiley & Sons, New York, © 1970. Reprinted by permission of the publisher.

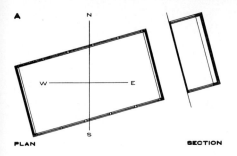

A

PLAN   SECTION

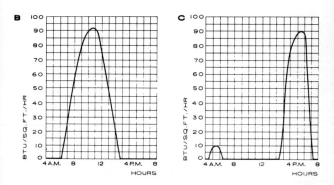

B

BTU/SQ.FT./HR

4 A.M.   8   12   4 P.M.   8

HOURS

C

BTU/SQ.FT./HR

4 A.M.   8   12   4 P.M.   8

HOURS

effect of the sun is desirable. During the warm times (called "overheated period") shading is needed to approach comfort conditions. For practical use the 70°F temperature can be accepted as a dividing line between these two conditions.

Figure D below illustrates the New York (40°N latitude) sun-path diagram on which are charted all conditions throughout the

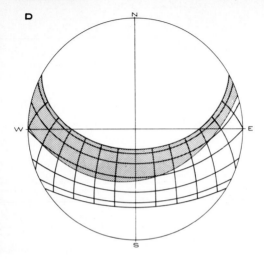

year when the temperatures equal or exceed 70°F. In these overheated times, illustrated with the shaded area on the graph, shading will be needed.

STEP 4. Construction of shading mask: lay the "shading mask protractor" (see material on shading devices) over the overheated period diagram in the proper orientation, as Fig. E below illustrates. Here the contours of the overheated period are shown by the dotted line. From the shading mask lines one can see that towards the south, devices with segmental character (overhang types) will cover conveniently the overheated period. At the north side the application of devices having radial mask patterns (fin types) will be effective.

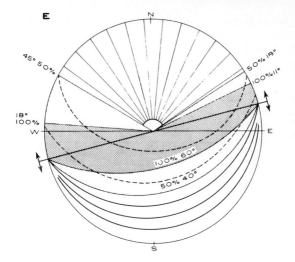

In Fig. D in darker tone is shown the 100% shading effect when the total wall surface is in shadow. In lighter tone is shown the 50% shading effect when only half of the surface will be in shade.

STEP 5. Design of shading devices from shading masks; the mask defines the type and the angles of the devices only, and possibilities remain for various design arrangements. In the example at the north side vertical fins will serve effectively. As the required angles in the shading mask are different towards the west (18° for 100% and 45° for 50% shading effect) than for the easterly direction (11° for 100% and 18° for 50% shading effect), the device shall be oblique to the wall surface. Figure F below illustrates an application for the north side. The necessary shading angle is measured in the plan from the middle of the shading fins; the full shade giving angles are measured from the inside corners of the shading elements.

On the south side one could apply a 60° solid overhang. However, this might be too long to cantilever. Instead, the solution here is a combination of horizontal and vertical elements, which corresponds to the same shading mask (see material on Shading Devices). In the section the 50% shading effect is measured from the middle of the glass pane, the 100% shading effect from the bottom. The horizontal part of the shading device could be solid; however, it is constructed here with louver elements to secure ventilation. The critical angle for the louvers is 73½ °, to correspond to the sun's highest altitude angle at this latitude. (See section in Fig. F below.)

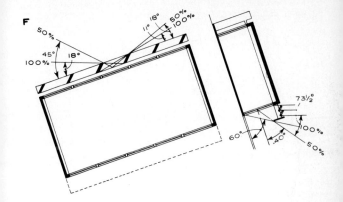

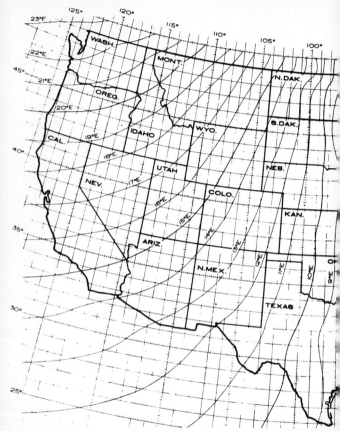

FIG. 5.3 Isogonic chart of the United States, from U.S. Department of Commerce, Coast and Geodetic Survey, 1965. [*From Joseph N. Boaz (ed.), Charles G. Ramsey's and Harold R. Sleeper's Architectural Graphic Standards, 6th ed.,*

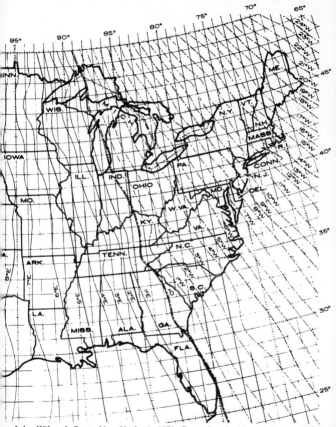

NOTES

Section 6

HEAT LOSS

TABLE 6.1   Recommended Design for Indoor Winter Temperatures, by Type of Building

Type of building	Temp., °F
Schools:	
Classrooms	72
Assembly rooms, dining rooms	72
Playrooms, gymnasiums	65
Swimming pool	75
Locker rooms	70
Hospitals:	
Private rooms	72
Operating rooms	75
Wards	70
Toilets	70
Bathrooms	75
Kitchens and laundries	66
Theaters	72
Hotels:	
Bedrooms	70
Ballrooms	68
Residences	72
Stores	68
Offices	72
Factories	65

TABLE 6.2 Recommended Design for Outdoor Winter Temperatures, for Various Locations

State	City	Temp., °F
Ala.	Birmingham	10
Ariz.	Flagstaff	− 10
Ariz.	Phoenix	25
Ark.	Little Rock	5
Calif.	Los Angeles	35
Calif.	San Francisco	35
Colo.	Denver	− 10
Conn.	Hartford	0
D.C.	Washington	0
Fla.	Jacksonville	25
Fla.	Miami	35
Ga.	Atlanta	10
Idaho	Boise	− 10
Ill.	Chicago	− 10
Ind.	Indianapolis	− 10
Iowa	Des Moines	− 15
Kans.	Topeka	− 10
Ky.	Louisville	0
La.	New Orleans	20
Maine	Portland	− 5
Md.	Baltimore	0
Mass.	Boston	0
Mich.	Detroit	− 10
Minn.	Minneapolis	− 20
Miss.	Vicksburg	10
Mo.	St. Louis	0
Mont.	Helena	− 20
Nebr.	Lincoln	− 10
Nev.	Reno	− 5
N.H.	Concord	− 15
N.J.	Trenton	0
N.Mex.	Albuquerque	0
N.Y.	New York	0
N.C.	Greensboro	10
N.Dak.	Bismarck	− 30
Ohio	Cincinnati	0
Okla.	Tulsa	0
Ore.	Portland	10
Pa.	Philadelphia	0
R.I.	Providence	0
S.C.	Charleston	15
S.Dak.	Rapid City	− 20
Tenn.	Nashville	0
Tex.	Dallas	0
Tex.	Houston	20
Utah	Salt Lake City	− 10
Vt.	Burlington	− 10
Va.	Richmond	15

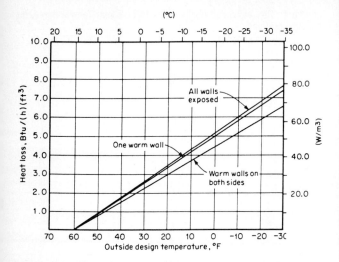

FIG. 6.1 Estimating heat loss at 60°F (15.6°C) for a one-story building with a flat roof (for skylight in roof add 7½%). (*From A. M. Khashab, Heating, Ventilating, and Air Conditioning Systems Estimating Manual, 2d ed., McGraw-Hill, New York, © 1984. Reproduced with permission of the publisher.*)

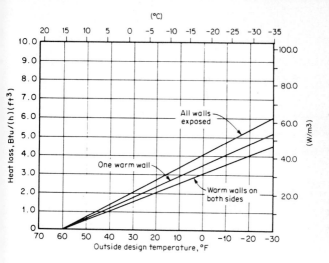

FIG. 6.2 Estimating heat loss at 60°F (15.6°C) for a one-story building with heated space above. (*From A. M. Khashab, Heating, Ventilating, and Air Conditioning Systems Estimating Manual, 2d ed., McGraw-Hill, New York, © 1984. Reproduced with permission of the publisher.*)

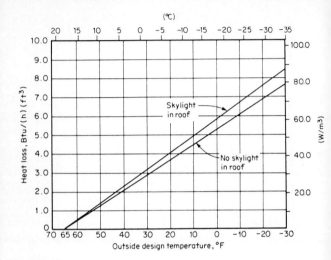

FIG. 6.3 Estimating heat loss at 65°F (18.3°C) for a one-story building. (*From A. M. Khashab, Heating, Ventilating, and Air Conditioning Systems Estimating Manual, 2d ed., McGraw-Hill, New York, © 1984. Reproduced with permission of the publisher.*)

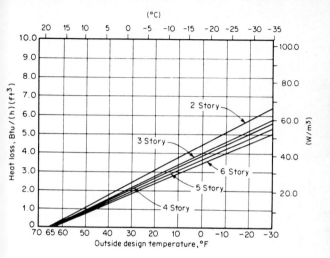

FIG. 6.4 Estimating heat loss at 65°F (18.3°C) for a multistory building. (*From A. M. Khashab, Heating, Ventilating, and Air Conditioning Systems Estimating Manual, 2d ed., McGraw-Hill, New York, © 1984. Reproduced with permission of the publisher.*)

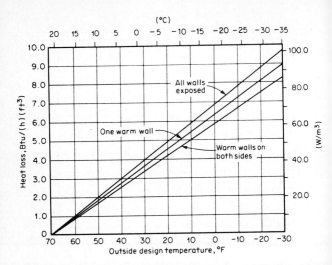

FIG. 6.5 Estimating heat loss at 70°F (21.1°C) for a one-story building with a flat roof. (*From A. M. Khashab, Heating, Ventilating, and Air Conditioning Systems Estimating Manual, 2d ed., McGraw-Hill, New York, © 1984. Reproduced with permission of the publisher.*)

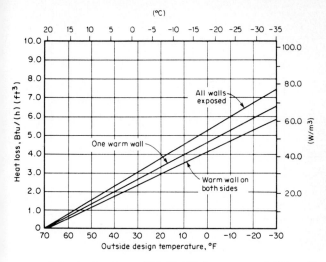

FIG. 6.6 Estimating heat loss at 70°F (21.1°C) for a one-story building with heated space above. (*From A. M. Khashab, Heating, Ventilating, and Air Conditioning Systems Estimating Manual, 2d ed., McGraw-Hill, New York, © 1984. Reproduced with permission of the publisher.*)

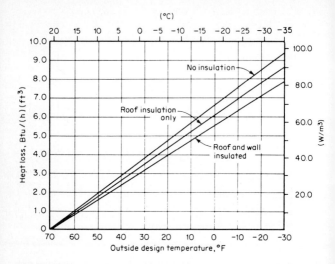

FIG. 6.7 Estimating heat loss at 70°F (21.1°C) for a multistory building (add 10% for bad north and west exposures). (*From A. M. Khashab, Heating, Ventilating, and Air Conditioning Systems Estimating Manual, 2d ed., McGraw-Hill, New York, © 1984. Reproduced with permission of the publisher.*)

TABLE 6.3   Estimated Heat Lost from Building by Infiltration

Room or Building Type	No. of Walls with Windows	Temp. Difference, deg F			
		25	50	75	100
	None	0.23	0.45	0.68	0.90
	1	0.34	0.68	1.02	1.36
A	2	0.68	1.35	2.02	2.70
	3 or 4	0.90	1.80	2.70	3.60
B	Any	1.35	2.70	4.05	5.40
C	Any	0.90-1.35	1.80-270	2.70-4.05	3.60-5.40
D	Any	0.45-0.68	0.90-1.35	1.35-2.02	1.80-2.70
E	Any	0.68-1.35	1.35-2.70	2.03-4.05	2.70-5.40

A = Offices, apartments, hotels, multistory buildings in general.

B = Entrance halls or vestibules.

C = Industrial buildings.

D = Houses, all types, all rooms except vestibules.

E = Public or institutional buildings.

SOURCE: Reprinted by permission from *ASHRAE Handbook—1981 Fundamentals.*

TABLE 6.4   Below-Grade Heat Losses

Ground water temp., °F	Basement floor loss,* Btu/h/ft^2	Below-grade wall loss, Btu/h/ft^2
40	3.0	6.0
50	2.0	4.0
60	1.0	2.0

*Based on basement temperature of 70°F.

DESIGN DATA: -25° to 67° F.	HEAT LOSS CALCULATIONS													UNHEATED	
ITEM	"U"	1 AREA	1 BTU	2 AREA	2 BTU	3 AREA	3 BTU	4 AREA	4 BTU	5 AREA	5 BTU	6 AREA	6 BTU	7 AREA	7 BTU
A GROSS WALL	---	1131		170		270		90		233		231			
B WINDOWS	0.61	96	59	16	10	8	5	0		0		0			
C DOORS	0.07	40	3	0		0		0		0		60	4		
D WALLS	0.026	995	26	154	4	262	7	90	2	233	6	171	5		
E COLD CEILING	0.025	1600	40	320	8	200	5	139	4	139	4	482	12		
F COLD FLOOR	0.04	1600	64	320	13	200	8	139	6	139	6	482	20		
G PERIMETER INFIL PER LIN. FT.	0.45	116	52	20	9	27	12	11	5	24	11	22	10		
H DRS & WINDOWS INFIL/ L.F.	0.45	169	76	23	10	11	5	0		0		42	19		
BTU LOSS PER 1° TEMP. (F)		320		54		42		17		27		70			
TOTAL ROOM LOSS 92°F TD		29,440		4,968		3,864		1,564		2,484		6,440			
CFM TOTAL EXHAUST PER ROOM				250				120		120					
BTU/H/ROOM= 60TC÷55				N/A SUMMER				12,044		12,044					
NOTES															
TOTALS		29,440		4,968		3,864		13,608		14,528		6,440			

THERMAL PERFORMANCE:

MAX.- 12 BTU/H/FT^2 OF ABOVE GRADE EXTERIOR BLDG. ENVELOPE

$$T P = \frac{92(\text{BTU ROOMS})}{A + E \ (\text{AREA})}$$

$$= \frac{92 \times 413}{2125 + 2880}$$

$$= 7.59 \ \text{BTU/H/FT}^2$$

FIG. 6.8 Typical heat loss calculation. (*Courtesy Morgan & Parmley, Ltd.*)

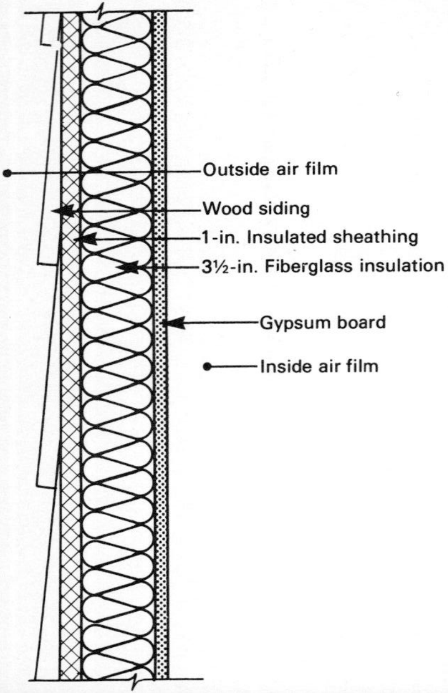

Example

Frame wall with 2x6 studs @ 16″ o/c

	At Framing	Between Framing
outside air	0.17	0.17
wood siding	0.81	0.81
sheathing	1.32	1.32
insulation	0.00	17.20
wood studs	4.38	0.00
gypsum bd.	0.45	0.45
inside air	0.68	0.68
Total Thermal Resist.	7.81	20.63

In a frame wall, studs occupy approximately 20% of the framing with insulation occupying the remaining 80%.

Therefore: $(20\% \times 7.81) + (80\% \times 20.63)$

$= $ composite R

$= 1.562 + 16.50 = 18.06$

$U = 1/R = 0.055 \ Btu/(h)(ft^2)(°F)$

To calculate total heat loss through a given area:

Heat Loss (H) $= U \times A \times \Delta T$

where: U = coefficient of heat transmission
   A = area of section in square feet
   ΔT = difference in temperature on the two sides of assembly in °F

FIG. 6.9   Basic heat loss calculation for a frame wall.

Outside air film
Wood siding
1-in. Insulated sheathing
3½-in. Fiberglass insulation
Gypsum board
Inside air film

NOTES

NOTES

HEATING SYSTEMS

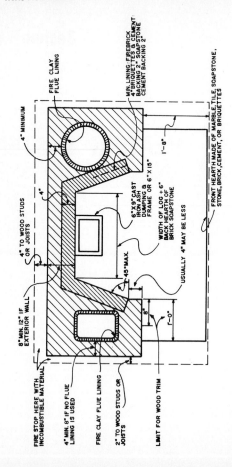

FIG. 7.1 Plan view of a typical fireplace.

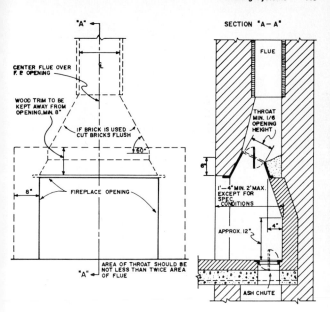

FIG. 7.2   Front elevation of a typical fireplace.

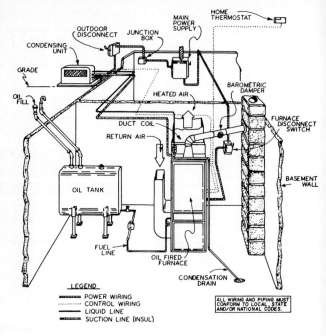

FIG. 7.3 Typical oil-fired forced-air heating system.

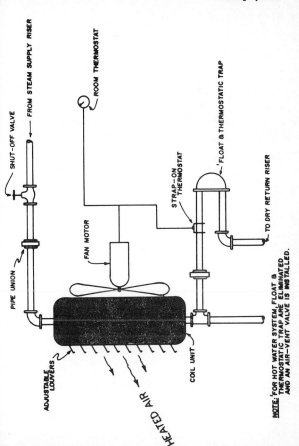

NOTE: FOR HOT WATER SYSTEM, FLOAT &
THERMOSTATIC TRAP ARE ELIMINATED
AND AN AIR-VENT VALVE IS INSTALLED.

FIG. 7.4 Typical steam unit heater installation diagram.

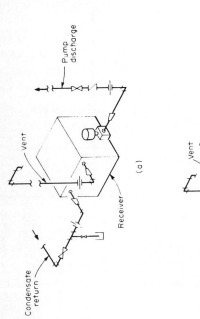

(a)

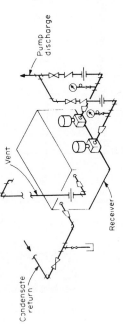

(b)

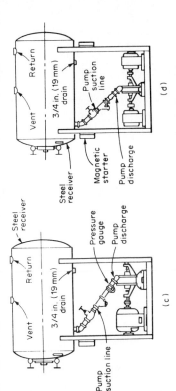

FIG. 7.5 Condensate pump and feedwater packaged unit diagrams: (a) single condensate pump with cast-iron receiver; (b) duplex condensate pump with cast-iron receiver; (c) feedwater packaged unit (single pump); (d) feedwater packaged unit (duplex pump). (From A. M. Khushab, Heating, Ventilating, and Air Conditioning Systems Estimating Manual, 2d ed., McGraw-Hill, New York, © 1984. Reproduced with permission of the publisher.)

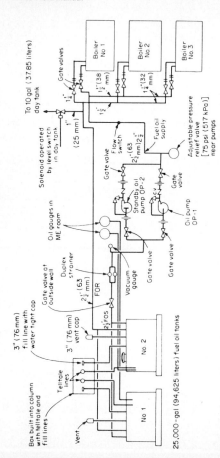

FIG. 7.6 Boiler fuel-oil diagram. (*From A. M. Khashab, Heating, Ventilating, and Air Conditioning Systems Estimating Manual, 2d ed., McGraw-Hill, New York, © 1984. Reproduced with permission of the publisher.*)

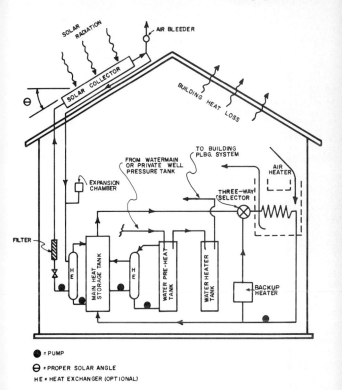

FIG. 7.7 Schematic diagram of a typical residential solar heat system (water-cooled collector).

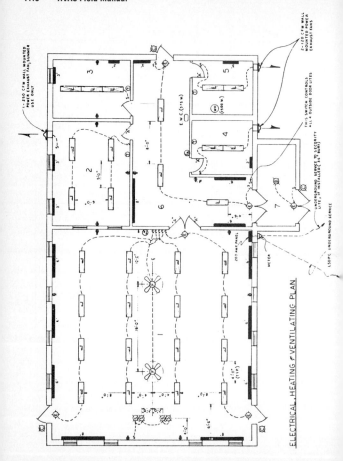

ELECTRICAL HEATING & VENTILATING PLAN

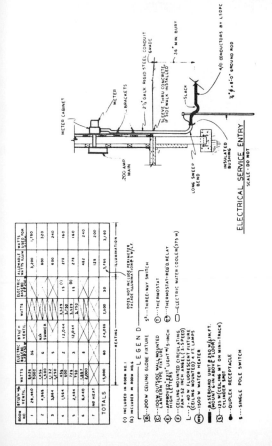

FIG. 7.8 Typical layout of an electrically heated building. (*Courtesy Morgan & Parmley, Ltd.*)

NOTES

PIPING DATA

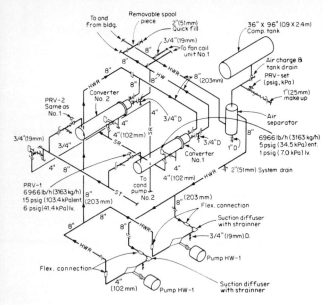

FIG. 8.1 Hot-water piping diagram. (*From A. M. Khashab, Heating, Ventilating, and Air Conditioning Systems Estimating Manual, 2d ed., McGraw-Hill, New York, © 1984. Reproduced with permission of the publisher.*)

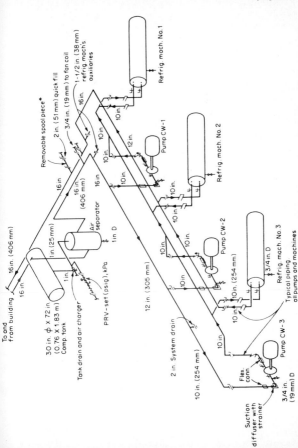

FIG. 8.2 Chilled-water piping diagram (* spool is a piece of pipe flanged at both ends). (*From A. M. Khashab, Heating, Ventilating, and Air Conditioning Systems Estimating Manual, 2d ed., McGraw-Hill, New York, © 1984. Reproduced with permission of the publisher.*)

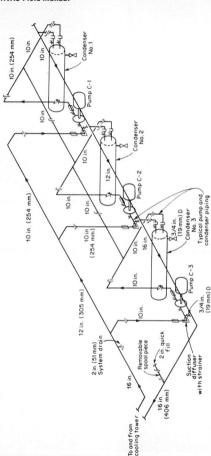

FIG. 8.3 Condenser-water piping diagram. (*From A. M. Khushab, Heating, Ventilating, and Air Conditioning Systems Estimating Manual, 2d ed., McGraw-Hill, New York, © 1984. Reproduced with permission of the publisher.*)

TABLE 8.1 How to Cut Odd-Angle Elbows

	ODD DEGREE LONG RADIUS ELBOWS						
NOM SIZE	OUTSIDE ARC						
	A	B	C	D	E	F	G
2	5/64	3/8	23/32	1 3/32	1 21/32	2 3/4	3 9/32
2 1/2	3/32	7/16	29/32	1 11/32	2 1/32	3 3/8	4 1/16
3	7/64	9/16	1 1/8	1 5/8	2 15/32	4 3/32	4 29/32
3 1/2	1/8	5/8	1 9/32	1 29/32	2 27/32	4 3/4	5 11/16
4	9/64	23/32	1 7/16	2 5/32	3 1/4	5 13/32	6 15/32
5	3/16	29/32	1 25/32	2 11/16	4 1/32	6 23/32	8 1/16
6	7/32	1 1/16	2 5/32	3 7/32	4 27/32	8 1/16	9 21/32
8	9/32	1 7/16	2 27/32	4 9/32	6 13/32	10 11/16	12 13/16
10	11/32	1 25/32	3 9/16	5 11/32	8	13 11/32	16
12	7/16	2 1/8	4 1/4	6 3/8	9 9/16	15 31/32	19 5/32
14	1/2	2 7/16	4 7/8	7 5/16	11	18 5/16	22
16	9/16	2 13/16	5 19/32	8 3/8	12 9/16	20 15/16	25 1/8
18	5/8	3 1/8	6 9/32	9 7/16	14 1/8	23 9/16	28 9/32
20	11/16	3 1/2	7	10 15/32	15 23/32	26 3/16	31 13/32
22	3/4	3 27/32	7 11/16	11 17/32	17 9/32	28 13/16	34 9/16
24	27/32	4 3/16	8 3/8	12 9/16	18 27/32	31 13/32	37 11/16
26	29/32	4 17/32	9 3/32	13 5/8	20 13/32	34 1/32	40 27/32
30	1 1/32	5 1/4	10 15/32	15 3/4	23 9/16	39 1/4	47 1/8
34	1 5/32	5 29/32	11 27/32	17 13/16	26 23/32	44 17/32	53 3/8
36	1 7/32	6 1/4	12 17/32	18 7/8	28 7/32	47	56 17/32
42	1 7/16	7 5/16	14 5/8	22	32 31/32	54 31/32	65 15/16

(continued)

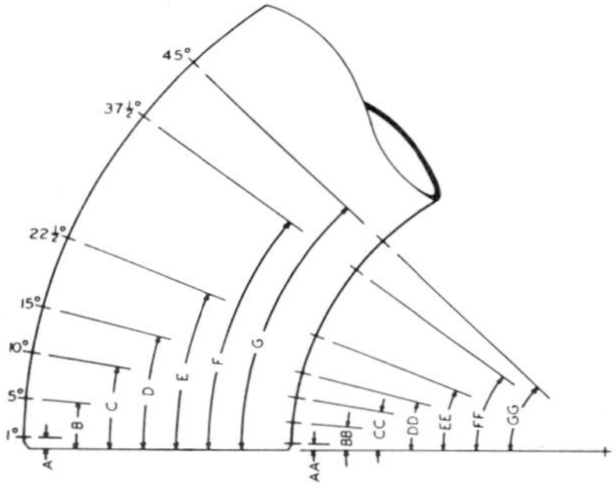

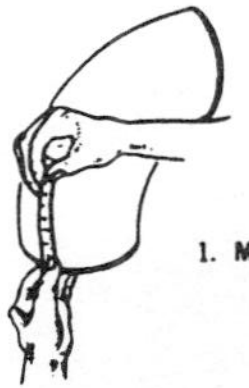

1. MEASURE DISTANCE ON OUTSIDE ARC

TABLE 8.1 How to Cut Odd-Angle Elbows (*continued*)

ODD DEGREE LONG RADIUS ELBOWS

NOM SIZE	INSIDE ARC						
	AA	BB	CC	DD	EE	FF	GG
2	1/32	5/32	5/16	15/32	23/32	1 3/16	1 7/16
2½	3/64	3/16	13/32	19/32	29/32	1½	1 13/16
3	3/64	¼	½	23/32	1 3/32	1 13/16	2 5/32
3½	1/16	9/32	9/16	27/32	1 9/32	2⅛	2 9/16
4	1/16	5/16	21/32	31/32	1 15/32	2 7/16	2 15/16
5	5/64	13/32	13/16	1¼	1 27/32	3 3/32	3 23/32
6	3/32	½	1	1½	2 7/32	3 23/32	4 15/32
8	⅛	11/16	1 11/32	2	3 1/32	5 1/32	6 1/32
10	5/32	27/32	1 11/16	2 17/32	3 25/32	6 5/16	7 9/16
12	7/32	1	2 1/32	3 1/16	4 9/16	7 19/32	9⅛
14	¼	1 7/32	2 7/16	3 21/32	5½	9 5/32	11
16	9/32	1 13/32	2 13/16	4 3/16	6 9/32	10 15/32	12⅝
18	5/16	1 9/16	3⅛	4 23/32	7 1/16	11 25/32	14⅛
20	11/32	1¾	3½	5¼	7 27/32	13 3/32	15 11/16
22	⅜	1 29/32	3 27/32	5¾	8⅝	14⅜	17 9/32
24	13/32	2 3/32	4 3/16	6 9/32	9 7/16	15 11/16	18 27/32
26	15/32	2 9/32	4 17/32	6 13/16	10 7/32	17 1/32	20 13/32
30	17/32	2⅝	5¼	7⅞	11 25/32	19⅝	23 9/16
34	19/32	2 31/32	5 29/32	8 29/32	13⅜	22 9/32	26 11/16
36	⅝	2 13/16	6¼	9 7/16	14⅛	23⅝	28¼
42	23/32	3 21/32	7 5/16	10 19/32	16½	26⅜	32 31/32

2. MEASURE DISTANCE ON INSIDE ARC

3. WRAP TAPE AROUND ELBOW AND MARK CUTTING LINE

SOURCE: Tube Turns, Inc., Louisville, KY. Used with permission.

8.1 PIPE ALIGNMENT[1]

Proper alignment is one of the most important tasks performed by the pipe fitter. If done correctly, welding will be much easier and the piping system will be properly fabricated. If alignment is poor, however, welding will be difficult and the piping system may not function properly.

Many devices are available to aid alignment. Tube Turns manufactures three types of welding rings which not only make alignment easier but also provide the correct gap for welding.

Methods of alignment vary widely throughout the trade. There is no *best* system . . . any number of methods have proven successful. The procedures suggested by this manual are popular with many craft workers and will enable you to quickly obtain good alignment.

PIPE-TO-PIPE

Move pipe lengths together until bevels are nearly abutted, allowing space for welding gap. Center squares on top of both pipes and move pipe up and down until squares are aligned. Tack weld top and bot-

---

[1]Text and unnumbered illustrations in this subsection from material provided by Tube Turns, Inc., Louisville, KY. Used with permission.

tom. Repeat procedure by placing squares on side of pipe. Correct alignment by moving pipe left or right. Tack weld each side.

90° ELBOW-TO-PIPE

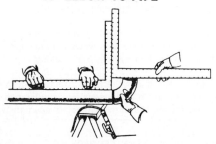

Place fitting bevel in line with bevel of pipe, allowing for welding gap. Tack weld on top. Center square on top of pipe. Center second square on elbow's alternate face. Move elbow until squares are aligned.

45° ELBOW-TO-PIPE

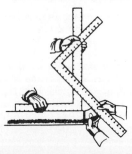

Follow procedure described above except squares will cross. To obtain correct 45° angle, align the same numbers on the inside scale of the tilted square. (*Note:* The numbers 4 and 7 are used in the illustration.)

Alternate Method

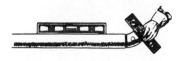

Use same procedure to abut pipe and fitting. Center spirit level on pipe. Next, center 45° spirit level on face of elbow and move elbow until 45° bubble is centered.

TEE-TO-PIPE

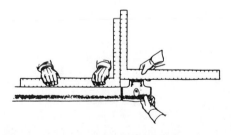

Abut bevels, allowing for welding gap. Tack weld on top. Center square on top of pipe. Place second square on center of branch outlet. Move tee until squares are aligned.

Alternate Method

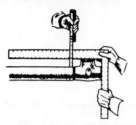

Follow same procedure to abut pipe and fitting. Place square on tee as illustrated. Center rule on top of pipe. Blade of square should be parallel with pipe. Check by measuring with rule at several points along the pipe.

FLANGE-TO-PIPE

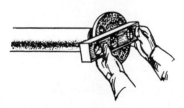

Step 1. Abut flange to pipe. Align top two holes of flange with spirit level. Move flange until bubble is centered. Make one tack weld on top.

Step 2. Center square on face of flange. Center rule on top of pipe. Move flange until square and pipe are parallel. Tack weld bottom.

Step 3. Center square on face of flange. Center rule on side of pipe and align as in Step 2. Tack both sides.

8.2 JIG FOR SMALL-DIAMETER PIPING[2]

Many pipe fitters have found this simple jig to be helpful in aligning small diameter pipe and elbows.

It is made from channel iron approximately 3 ft 9 in long. Use ⅛ in × 1½ in for pipe sizes 1¼ in thru 3 in; ⅛ in × ¾ in for sizes 1 in or smaller.

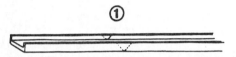

①

Mark 90° notch on side of channel iron about 9 in from end. Make equal notch on other side.

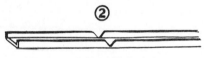

②

Cut out notches with hack saw.

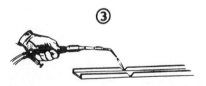

③

Heat bottom of channel iron between the notches.

[2]Text and unnumbered illustrations in this subsection from material provided by Tube Turns, Inc., Louisville, KY. Used with permission.

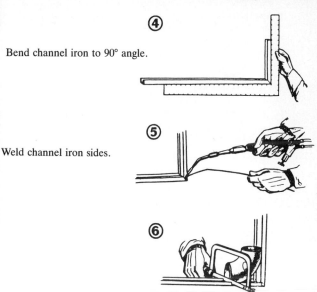

④

Bend channel iron to 90° angle.

⑤

Weld channel iron sides.

⑥

Place Tube Turns 90° Long Radius Elbow in jig. Saw half through both sides of channel iron as shown in illustration. Repeat with other size elbows so jig may be used for several sizes.

⑦

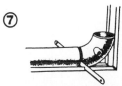

Place used hack saw blade in slot to obtain proper alignment and correct welding gap.

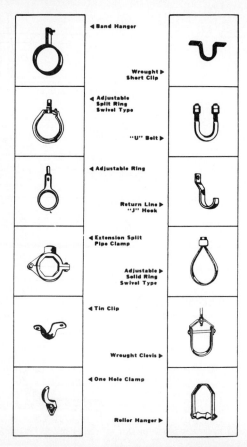

FIG. 8.4 Typical pipe hangers. (*Courtesy Tube Turns, Inc., Louisville, KY.*)

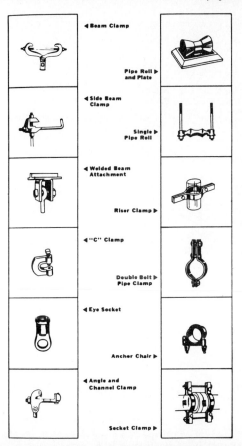

Beam Clamp

Pipe Roll and Plate ▶

◀ Side Beam Clamp

Single ▶ Pipe Roll

◀ Welded Beam Attachment

Riser Clamp ▶

◀ "C" Clamp

Double Bolt ▶ Pipe Clamp

◀ Eye Socket

Anchor Chair ▶

◀ Angle and Channel Clamp

Socket Clamp ▶

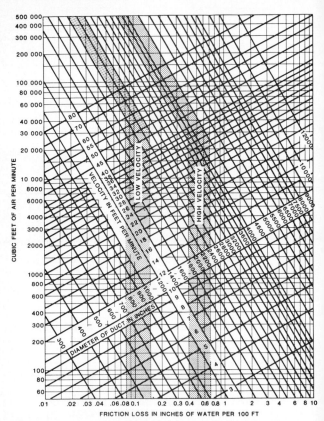

FIG. 8.5 Suggested velocity and friction rate design limits. (*Reprinted by permission from ASHRAE Handbook—1981 Fundamentals.*)

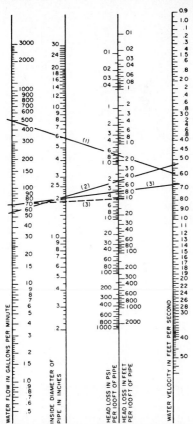

FIG. 8.6 Flow loss characteristics of water flow through rigid plastic pipe. (*Reprinted by permission from ASHRAE Handbook—1981 Fundamentals.*)

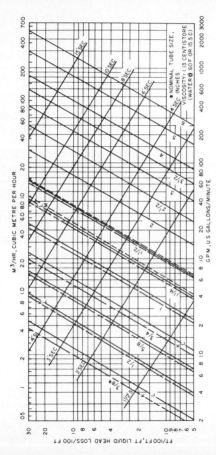

FIG. 8.7 Friction loss due to flow of water in Type L copper tube. (*Reprinted by permission from ASHRAE Handbook—1981 Fundamentals.*)

TABLE 8.2 Fitting Losses in Equivalent Feet of Pipe

Nominal Pipe or Tube Size (in.)	Smooth Bend Elbows						Smooth Bend Tees			
	90° Std*	90° Long Rad.†	90° Street*	45° Std*	45° Street*	180° Std*	Flow Through Branch	Straight-Through Flow		
								No Reduction	Reduced 1/4	Reduced 1/2
3/8	1.4	0.9	2.3	0.7	1.1	2.3	2.7	0.9	1.2	1.4
1/2	1.6	1.0	2.5	0.8	1.3	2.5	3.0	1.0	1.4	1.6
3/4	2.0	1.4	3.2	0.9	1.6	3.2	4.0	1.4	1.9	2.0
1	2.6	1.7	4.1	1.3	2.1	4.1	5.0	1.7	2.2	2.6
1 1/4	3.3	2.3	5.6	1.7	3.0	5.6	7.0	2.3	3.1	3.3
1 1/2	4.0	2.6	6.3	2.1	3.4	6.3	8.0	2.6	3.7	4.0
2	5.0	3.3	8.2	2.6	4.5	8.2	10	3.3	4.7	5.0
2 1/2	6.0	4.1	10	3.2	5.2	10	12	4.1	5.6	6.0
3	7.5	5.0	12	4.0	6.4	12	15	5.0	7.0	7.5
3 1/2	9.0	5.9	15	4.7	7.3	15	18	5.9	8.0	9.0
4	10	6.7	17	5.2	8.5	17	21	6.7	9.0	10
5	13	8.2	21	6.5	11	21	25	8.2	12	13
6	16	10	25	7.9	13	25	30	10	14	16
8	20	13	—	10	—	33	40	13	18	20
10	25	16	—	13	—	42	50	16	23	25
12	30	19	—	16	—	50	60	19	26	30
14	34	23	—	18	—	55	68	23	30	34
16	38	26	—	20	—	62	78	26	35	38
18	42	29	—	23	—	70	85	29	40	42
20	50	33	—	26	—	81	100	33	44	50
24	60	40	—	30	—	94	115	40	50	60

*R/D approximately equal to 1.
†R/D approximately equal to 1.5.

SOURCE: Reprinted by permission from *ASHRAE Handbook—1981 Fundamentals*.

TABLE 8.3 **Special Fitting Losses in Equivalent Feet of Pipe**

Nom. Pipe or Tube Size (in.)	Sudden Enlargement d/D			Sudden Contraction d/D			Sharp Edge		Pipe Projection	
	1/4	1/2	3/4	1/4	1/2	3/4	Entrance	Exit	Entrance	Exit
3/8	1.4	0.8	0.3	0.7	0.5	0.3	1.5	0.8	1.5	1.1
1/2	1.8	1.1	0.4	0.9	0.7	0.4	1.8	1.0	1.8	1.5
3/4	2.5	1.5	0.5	1.2	1.0	0.5	2.8	1.4	2.8	2.2
1	3.2	2.0	0.7	1.6	1.2	0.7	3.7	1.8	3.7	2.7
1 1/4	4.7	3.0	1.0	2.3	1.8	1.0	5.3	2.6	5.3	4.2
1 1/2	5.8	3.6	1.2	2.9	2.2	1.2	6.6	3.3	6.6	5.0
2	8.0	4.8	1.6	4.0	3.0	1.6	9.0	4.4	9.0	6.8
2 1/2	10	6.1	2.0	5.0	3.8	2.0	12	5.6	12	8.7
3	13	8.0	2.6	6.5	4.9	2.6	14	7.2	14	11
3 1/2	15	9.2	3.0	7.7	6.0	3.0	17	8.5	17	13
4	17	11	3.8	9.0	6.8	3.8	20	10	20	16
5	24	15	5.0	12	9.0	5.0	27	14	27	20
6	29	22	6.0	15	11	6.0	33	19	33	25
8	—	25	8.5	—	15	8.5	47	24	47	35
10	—	32	11	—	20	11	60	29	60	46
12	—	41	13	25	—	13	73	37	73	57
14	—	—	16	—	—	16	86	45	86	66
16	—	—	18	—	—	18	96	50	96	77
18	—	—	20	—	—	—	115	58	115	90
20	—	—	—	—	—	20	142	70	142	108
24	—	—	—	—	—	—	163	83	163	130

SOURCE: Reprinted by permission from *ASHRAE Handbook—1981 Fundamentals.*

TABLE 8.4 Equivalent Length for Fittings

SIZE OF PIPE IN INCHES	GATE VALVE	ST'D ELBOW	RED. COUPLING	SIDE OUTLET "T"	ANGLE VALVE	GLOBE VALVE
			EQUIVALENT LENGTH OF PIPE IN FEET			
1/2	0.3	1.3	1.5	3.0	7.0	14.0
3/4	0.4	1.8	2.0	4.0	10.0	18.0
1	0.5	2.2	2.5	5.0	12.0	23.0
1 1/4	0.6	3.0	3.0	6.0	15.0	29.0
1 1/2	0.8	3.5	3.5	7.0	18.0	34.0
2	1.0	4.3	5.0	8.0	22.0	46.0
2 1/2	1.1	5.0	6.0	11.0	27.0	54.0
3	1.4	6.5	7.0	13.0	34.0	66.0
3 1/2	1.6	8.0	9.0	15.0	40.0	80.0
4	1.9	9.0	10.0	18.0	45.0	92.0
5	2.2	11.0	13.0	22.0	56.0	112.0
6	2.8	13.0	15.0	27.0	67.0	136.0
8	3.7	17.0	20.0	35.0	92.0	180.0
10	4.6	21.0	25.0	45.0	112.0	230.0
12	5.5	27.0	30.0	53.0	132.0	270.0
14	6.4	30.0	35.0	63.0	152.0	310.0

TYPE OF FITTING

SOURCE: The Trane Company, LaCross, WI. Reproduced with permission.

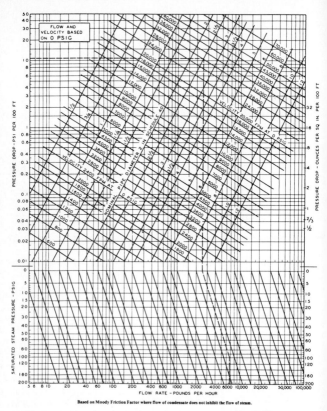

FIG. 8.8 Basic chart for flow rate and velocity of steam in Schedule 40 pipe based on a saturation pressure of 0 psig. (*Reprinted by permission from ASHRAE Handbook—1981 Fundamentals.*)

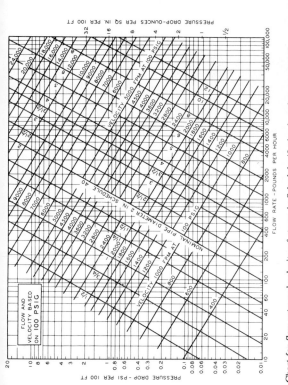

FIG. 8.9 Chart for flow rate and velocity of steam in Schedule 40 pipe based on a saturation pressure of 100 psig (may be used for steam pressures from 85 to 120 psig with an error not exceeding 8%). (*Reprinted by permission from ASHRAE Handbook—1981 Fundamentals.*)

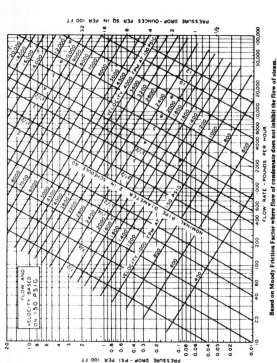

Based on Moody Friction Factor where flow of condensate does not inhibit the flow of steam.

FIG. 8.10 Chart for flow rate and velocity of steam in Schedule 40 pipe based on a saturation pressure of 150 psig (may be used for steam pressures from 127 to 180 psig with an error not exceeding 8%). *(Reprinted by permission from ASHRAE Handbook—1981 Fundamentals.)*

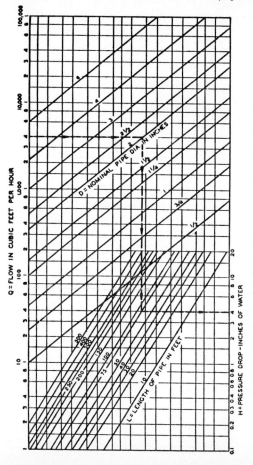

Based on Spitzglass Friction Factor and 0.60 Specific Gravity. To convert for other specific gravity see Table 32.

FIG. 8.11 Flow vs. pressure drop for low-pressure natural gas. (*Reprinted by permission from ASHRAE Handbook—1981 Fundamentals.*)

TABLE 8.5   Wire and Sheet Metal Gages (Diameters and Thicknesses in Decimal Parts of an Inch)

Gage No.	American wire gage, or Brown & Sharpe (for non-ferrous sheet and wire)	Steel wire gage or Washburn & Moen or Roebling (for steel wire)	Birmingham wire gage (B.W.G.) or Stubs iron wire (for steel iron rods or sheets)	Stubs steel wire gage	British Imperial standard wire gage (S.W.G.)	U.S. standard gage for wrought iron sheet (480 lb per cu ft)	U.S. standard gage for steel and open-hearth iron sheet (489.6 lb per cu ft)	British standard for iron and steel sheets and hoops, 1914 (B.G.)
0000000		0.4900			0.500	0.500	0.4902	0.6666
000000		0.4615			0.464	0.469	0.4596	0.6250
00000		0.4305			0.432	0.438	0.4289	0.5883
0000	0.460	0.3938	0.454		0.400	0.406	0.3983	0.5416
000	0.410	0.3625	0.425		0.372	0.375	0.3676	0.5000
00	0.365	0.3310	0.380		0.348	0.344	0.3370	0.4452
0	0.325	0.3065	0.340		0.324	0.312	0.3064	0.3964
1	0.289	0.2830	0.300		0.300	0.281	0.2757	0.3532
2	0.258	0.2625	0.284	0.227	0.276	0.266	0.2604	0.3147
3	0.229	0.2437	0.259	0.219	0.252	0.250	0.2451	0.2804
4	0.204	0.2253	0.238	0.212	0.232	0.234	0.2298	0.2500
5	0.182	0.2070	0.220	0.207	0.212	0.219	0.2145	0.2225
6	0.162	0.1920	0.203	0.204	0.192	0.203	0.1991	0.1981
7	0.144	0.1770	0.180	0.201	0.176	0.188	0.1838	0.1764
8	0.128	0.1620	0.165	0.199	0.160	0.172	0.1685	0.1570
9	0.114	0.1483	0.148	0.197	0.144	0.156	0.1532	0.1398
10	0.102	0.1350	0.134	0.194	0.128	0.141	0.1379	0.1250
11	0.091	0.1205	0.120	0.191	0.116	0.125	0.1225	0.1113
12	0.081	0.1055	0.109	0.188	0.104	0.109	0.1072	0.0991
13	0.072	0.0915	0.095	0.185	0.092	0.094	0.0919	0.0882
14	0.064	0.0800	0.083	0.182	0.080	0.078	0.0766	0.0785
15	0.057	0.0720	0.072	0.180	0.072	0.070	0.0689	0.0699
16	0.051	0.0625	0.065	0.178	0.064	0.062	0.0613	0.0625
17	0.045			0.175				

19	0.036	0.0410	0.042	0.164	0.040	0.0438	0.0429	0.0440
20	0.032	0.0348	0.035	0.161	0.036	0.0375	0.0368	0.0392
21	0.0285	0.0317	0.032	0.157	0.032	0.0344	0.0337	0.0349
22	0.0253	0.0286	0.028	0.155	0.028	0.0312	0.0306	0.0313
23	0.0226	0.0258	0.025	0.153	0.024	0.0281	0.0276	0.0278
24	0.0201	0.0230	0.022	0.151	0.022	0.0250	0.0245	0.0248
25	0.0179	0.0204	0.020	0.148	0.020	0.0219	0.0214	0.0220
26	0.0159	0.0181	0.018	0.146	0.018	0.0188	0.0184	0.0196
27	0.0142	0.0173	0.016	0.143	0.0164	0.0172	0.0169	0.0175
28	0.0126	0.0162	0.014	0.139	0.0148	0.0156	0.0153	0.0156
29	0.0113	0.0150	0.013	0.134	0.0136	0.0141	0.0138	0.0139
30	0.0100	0.0140	0.012	0.127	0.0124	0.0125	0.0123	0.0123
31	0.0089	0.0132	0.010	0.120	0.0116	0.0109	0.0107	0.0110
32	0.0080	0.0128	0.009	0.115	0.0108	0.0102	0.0100	0.0098
33	0.0071	0.0118	0.008	0.112	0.0100	0.0094	0.0092	0.0087
34	0.0063	0.0104	0.007	0.110	0.0092	0.0086	0.0084	0.0077
35	0.0056	0.0095	0.005	0.108	0.0084	0.0078	0.0077	0.0069
36	0.0050	0.0090	0.004	0.106	0.0076	0.0070	0.0069	0.0061
37	0.0045	0.0085		0.103	0.0068	0.0066	0.0065	0.0054
38	0.0040	0.0080		0.101	0.0060	0.0062	0.0061	0.0048
39	0.0035	0.0075		0.099	0.0052	0.0059	0.0057	0.0043
40	0.0031	0.0070		0.097	0.0048	0.0055	0.0054	0.0039
41		0.0066		0.095	0.0044	0.0053	0.0052	0.0034
42		0.0062		0.092	0.0040	0.0051	0.0050	0.0031
43		0.0060		0.088	0.0036	0.0049	0.0048	0.0027
44		0.0058		0.085	0.0032	0.0047	0.0046	0.0024
45		0.0055		0.081	0.0028			0.0022
46		0.0052		0.079	0.0024			0.0019
47		0.0050		0.077	0.0020			0.0017
48		0.0048		0.075	0.0016			0.0015
49		0.0046		0.072	0.0012			0.0014
50		0.0044		0.069	0.0010			0.0012

SOURCE: Ingersoll-Rand Co. Used with permission.

TABLE 8.6   Pipe Sizes and Pump Capacities

Pipe Size, in*	Pump Capacity, gpm	Pump Capacity, lps	Pipe Size, mm*
1	Up to 15	Up to 1.0	25
1¼	5–20	0.5–1.5	32
1½	10–25	1.0–2.0	38
2	20–50	1.5–3.0	51
2½	30–75	2.0–5.0	64
3	55–135	3.5–8.5	76
4	115–275	7.0–17.0	102
5	200–500	12.5–31.5	127
6	330–800	21.0–50.0	152
8	700–1300	44.0–82.0	203
10	1250–1750	79.0–110.0	254
12	1600–2500	100.0–160.0	305

*Pipe sizes in inches represent the "nominal" diameters. Depending on the material, the nominal diameter may refer to inside or outside pipe diameter. The nominal diameter thus becomes a reference size that does not have a true equivalent SI unit conversion. The pipe sizes indicated above, in millimeters, are not the true pipe diameters. They are only given as approximations.

SOURCE: A. M. Khashab, *Heating, Ventilating, and Air Conditioning Systems Estimating Manual*, 2d ed., McGraw-Hill, New York, © 1984. Used with permission of the publisher.

NOTES

NOTES

ROOF SNOW AND ICE-DAM PROBLEMS[1]

[1] The text material and illustrations in this section were originally prepared by Howard L. Grange and Lewis T. Hendricks and published under the title *Roof-Snow Behavior and Ice-Dam Prevention in Residential Housing,* by the Agriculture Extension Service, University of Minnesota (Extension Bulletin 399-1976). Permission has been granted by Minnesota Extension Services–University of Minnesota for its use in this manual.

9.1 INTRODUCTION

The intent of this publication is to awaken public interest in the behavior of roof-snow and the part it plays in ice-dam formation. Each year, ice-dams cause millions of dollars in damage to homes located in the "snow-belt." Yet there is little understanding among homeowners, builders, and suppliers as to the cause of these ice-dams and the remedial measures that can be undertaken to prevent further damages to residential units.

Many areas and facets of roof-snow could not be included in this publication. Further in-depth investigation by better equipped technicians is recommended. The interpretations presented are generally based upon first-hand observations; perhaps higher levels of technical sophistication will be needed to convince reluctant building interests that they are clearly "involved" in the problems of roof-snow behavior. Satisfactory servicing of ice-dammed buildings requires the enthusiastic support of responsible and knowledgeable suppliers of both materials and services.

In general, ice-dams are formed when attic heat moves upward to warm the roof and melt roof-snow at or near ridge areas. Melting of snow occurs at the snow-shingle interface and runs downward (under the snow) as snow-water. At or near the edge of the roof, colder conditions exist that usually result in the freezing of the snow-water, thus forming the ice-dam. Subsequent melting of roof-snow usually accumulates as a pocket of snow-water that eventually backs up under the shingles to cause major damage in the plateline area. This damage can appear in the form of soaked (inefficient) insulation; stained, cracked, and spalled plaster or sheetrock; damp, odorous, and rotting wall cavities; and stained, blistered, and peeling wall paint, both inside and outside the house. Insulation alone will not solve the ice-dam problem nor will contemporary ventilation techniques. Other commonly used techniques such as heating cables and the removal of snow at roof edges are of little value in combating ice-dam problems. Rather, it is the proper use of insulation and ventilation in conjunction with correct house design that offers the best solutions to the problem.

As a final note, the authors do not intend this report to be used as a "book of instructions" or as a maintenance guide. Improvisations by do-it-yourselfers or untrained installers will most likely result in blunders, disappointments, and disillusionment in the bright promises of the observations, comments, and products mentioned in this report.

The explanations, interpretations, and suggestions of this report are founded on a 10-year study of actual roofs and visible roof-snow behavior in several thousand observations as well as color slide documentations in 36 states and 5 Canadian provinces.

Visible evidence everywhere has indicated the frustrations of building owners and the inability, or unwillingness, of building professionals to adequately cope with extensive damages caused by eave ice-dams and blocked snow-water penetrations.

These observations can be verified by streetside observation during almost every winter in the Twin Cities in late February, three winters out of four in Chicago, and at least occasional winters throughout latitudes as far south as the Ohio River.

Although snowloading of structures has been well recognized and documented in architectural texts and structural design data, few references (13, 3, 6, 7) to roof-snow melting and ice-dam behavior have been found in technical literature. Hopefully, this report will bridge a sadly neglected informational gap and also stimulate homeowner response and action to eliminate this needless, but widespread and devastating, fault of building design and performance.

9.2   RECOGNITION OF ICE-DAMS

Our investigations have revealed that many extensively involved and troubled homeowners are reluctant to admit (and discuss) *their in-*

FIG. 9.1 Typical snow-country eave in late winter.

dividual problems with ice-dams. The most common response to our question has been, *"What are ice-dams?"*

Figure 9.1 shows late winter snow on a roof, a bulge of ice attached to the eave, with numerous icicles that may drip from it on warmer days, and stained siding that may be optimistically written off as *"a bit of rust stain from the overloaded rain gutters."* Similarly, a wetted interior ceiling and wall are hopefully nothing more than a *"bad place in the shingles that we will fix next spring."* Perhaps this is true, but the attacks are more likely to be as indicated in Fig. 9.2 revealing the "actions" inside the cornices and attic spaces.

The ice-loaded eave, the puffed snow-blanket, and the dangling icicles of Fig. 9.1 may be all that is apparent; but inside the cornices, within the attic spaces, and within the wall cavity, the destructive actions—seldom detected by direct observation—are illustrated for easy recognition in Fig. 9.2. The existence of these hidden actions is

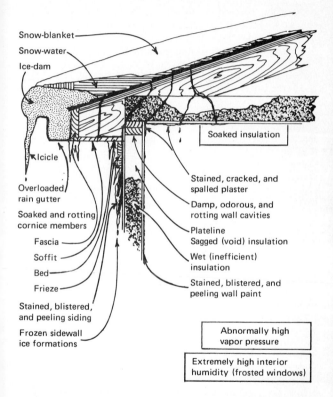

Snow-blanket

Snow-water

Ice-dam

Soaked insulation

Icicle

Overloaded rain gutter

Soaked and rotting cornice members

Fascia

Soffit

Bed

Frieze

Stained, blistered, and peeling siding

Frozen sidewall ice formations

Stained, cracked, and spalled plaster

Damp, odorous, and rotting wall cavities

Plateline

Sagged (void) insulation

Wet (inefficient) insulation

Stained, blistered, and peeling wall paint

Abnormally high vapor pressure

Extremely high interior humidity (frosted windows)

FIG. 9.2 This sketch of the ice-dam problem identifies both the ice-dam and its damages. Of course, all the damages illustrated may (or may not) occur at any one instance. The damages illustrated here are far more common and costly than is generally acknowledged.

occasionally betrayed on the exterior by wetted walls, by massive wall ice formations, by wall stains, and subsequently—usually later in the spring—by blistered and peeling paint.

Inside the house, the distraught housewife may not identify ice-blocked snow-water as the source of wetted ceiling and walls, of stained and peeling paint or blistered, cracked, and spalling plaster, and perhaps the offensive, yet difficult to locate, odor of mildewed and rotting wood.

Unseen, *and detectable only indirectly,* are the damages in the wall, ceiling, and cornice cavities, such as soaked, sagging inefficient insulation; mildewed and rotting wood structural members; rust-marked, stain-oozing degraded sidewalls; and corrosion of nails, flashings, and fasteners. Hidden cavities are also slow to dry out after the roof-snow season. Such severely soaked insulation and wall cavities contribute substantially to house moisture and to damages associated with excessive vapor pressures in residential interiors.

9.3   OCCURRENCE OF ROOF-SNOW AND ICE-DAM PROBLEMS

Ice-dams develop on buildings whenever and wherever roof-snow accumulates on roofs to depths of an inch or two and when the weather turns or remains generally below freezing for 3 or 4 days.

No precise limits of snow-depth and temperature conditions have been determined, nor would such specific limits be significant; wide variations in the range of roof-snow depth and temperature depend on the whims of nature, the direction and velocity of winds, and the shape of the roof.

In general, the deeper the snow, either by single incident snowfall or by accumulations, and the lower (and more persistent) the subfreezing temperatures, the more inevitable and sizeable are the ice formations for typical houses.

A 5-year study (1) of winter conditions in 25 locations sought to develop an index of snow-cold severity for comparisons of relative ice-dam susceptibility in such diverse locations as *Sault Ste. Marie,* Michigan (Ontario); *Minneapolis,* Minn.; *Madison,* Wis.; *Chicago,* Ill.; *Pittsburgh,* Penn.; and *Lexington,* Ky. Snowfalls and roof-damages, as anticipated, proved generally more severe, frequent, and enduring with increasing north latitude locations. Other climatic factors may supersede latitude.

Wind direction and velocity may free some roofs of snow or selectively clear certain planes and drift deeply on another plane or in a susceptible valley.

Heavy snowstorms characteristically move as low pressure frontal systems across the midwest in approximately southwesterly to northeasterly directions. Such blizzard-type winds blow moisture in from the southeast, shifting to the northeast and to heavy snow as the storm front moves in. Snowfalls diminish as winds shift to the north and northwest as the front passes and cold-temperature air follows. Thus, deepest snow and drifts occur normally on northerly and westerly roof planes and valleys. Such selective orientation of deepest snow helps explain why snow depths and snow-water damages are generally geater on northerly and westerly roof planes and valleys. Wind and complex roof-structures may confuse observers of roof-snow actions unless wind factors are carefully recognized and considered.

Farm homes and buildings with wind-exposed roofs are less prone to roof-snow problems. Clearly, roof planes freed of snow will avoid ice-dams and snow-water difficulties.

Affluent suburban homes where trees have been carefully preserved or where trees have been densely re-established have a much higher incidence and severity of eave-ice damming than do rural or windswept (treeless) new developments. High winds, or the lack of winds, that may or may not blow snow from roofs are key factors in the relative depths and damages of roof-snow.

Within metropolitan areas, there may be startling variations in ice-dam susceptibility from dense residential areas to the windswept, thinly built-up, outlying areas.

There are striking local variations, such as Chicago's lakefront in relation to its western suburbs. The moderating effect of large bodies of water distorts the norms of latitude, and the greater incidence of snow in South Bend's Chicago-polluted air indicates need for further study of the snow/cold variables that lead to roof-snow and snow-water damages. Snow density, as it affects both roof loads and snow-water quantity, has received little attention and consideration.

The snow/cold conditions for severe roof-snow difficulties vary little from year to year for the Soo and the Twin Cities, where coping with snow and cold is accepted as part of life. In Madison and Chicago latitudes, severe snow/cold may occur in only 2 years out of 5, and, in more southerly latitudes, damaging roof-snow may occur only once in several years. Curiously, homeowner attitudes appear to be optimistically myopic in locations where chances of ice-dammed snow-water attacks are less than 50 percent; this is sadly shortsighted, because one ice-dammed snow-water attack in a 10-year period can be devastating.

9.4   IDENTIFICATION OF SNOW-WATER DAMAGES

Icicles

Icicles, sparkling in the sun of midwinter days, are only indicators of the interactions of roof-snow, the efficiency of ceiling insulation, and the effects of warmed attic spaces. Decorative icicles hanging beyond the eaves are solidified snow-water that is out and over the dams and are hazardous only to those who carelessly knock them down.

Roof-Ice Glacier Actions

The visible troublemaker on snow-decked buildings is the massive ice formed from successive flows of roof snow-water freezing as it emerges from the protective insulating blanket of roof-snow. The disruptive forces of expanding ice-actions may damage roof shingles and the roof-deck. The glacierlike forces on roofs and in valleys deserve better consideration by building designers.

Rain Gutters

Rain gutters (eaves troughs) are often condemned as responsible for ice-dammed eaves. Ice-filled rain gutters may contribute to concentrated overloads at the eaves. Massive ice in (and around) eaves troughs inevitably slows melting and interferes with desirable roof drainage and runoffs; however, rain gutters, like icicles, are simply symptomatic of snow-water actions, rather than causative.

Snow-Water Penetrations

The deceptive offender on winter roofs is snow-water. Its many actions are disguised in successive transformations. It develops from snow-crystals of varying densities into snow-water, and it may transform on cooling into massive ice-dams that block further runoff at the eave. Roof-snow can sporadically alter its water-ice character with thermal variations in atmosphere and roofdeck thermal relationships. Snow-water deceptions have been inadequately recognized and confronted.

Snow-water blocked by ice-dams may develop to depths sufficient to penetrate cracks and flaws in a deck that was not built to resist water attack from below and beneath it. Alternate freezing and thawing increases both the incidence and the magnitude of roof-deck flaws. Snow-water penetration into the structure may follow paths and channels difficult to analyze as the source of sidewall soaking. Penetrating snow-water often seeps down rafters, along plates, into wall cavities, and saturates sidewall insulation to reduce its efficiency, passes into and through sheathing-siding as either water or vapor, and may stain or blister the thin skins of paint to effect a thoroughly undesirable appearance and a lowered protective performance. The ultimate damage through rotting of structure by successive waves of snow-water attacks may destroy the building, even when cosmetically covered with vapor and water-impervious materials.

The amount of sidewall moisture originating from snow-water is impossible to determine because most of the evidence is hidden initially by the snow-cover and, after penetration, by the roof and sidewall coverings. Actions within the sidewall cavities can be deduced from the sometimes visible wetting and from the occasional frozen sidewall-icicle, the developing stains, and the subsequent paint blisters. Winter snow-water seepages may be difficult to identify, but the musty odor of wet and rotting wall and cornice members may permeate the house during summer. Massive sidewall soaking may saturate cavity insulations and retain moist and rotting conditions for months.

Frozen Sidewall Leaks

Visible evidence of the penetration of ice-dammed snow-water into attic-warmed cornices is not rare in the upper Midwest in December through March. Intensely cold weather may freeze the cornice leaks in a stop-action sidewall ice-formation, as displayed in Fig. 9.3. The visible ice may be apparent for only a few days. With even slight warming, the frozen action may melt to an almost invisible flow. Similarly, ice-blocked snow-water undoubtedly also flows on the interior of the wall cavity where its detection is difficult; evidence of its presence may subsequently develop as stain, blistering, and peeling of the painted siding.

Sidewall Degradation

Snow-water penetrations through roof-decks, into cornices, and seeping generally within and without the side wall cause not only paint failures, but general degradation of the structure. Insulation wetted by water seepages loses most of its efficiency with consequent increases in snow-melting thermals in the attics and cornices.

FIG. 9.3 Frozen sidewall leaks emanating from ice-blocked eave.

Plateline Thermals

Perhaps the most neglected heat losses of well-insulated modern houses occur at the exterior platelines. Sagging sidewall insulation—due to settling, wetting, or careless installation—adds to the overlooked heat losses in the critical plateline region. Such heat losses further aggravate snow-water penetrations by supporting the liquid ice-dammed pool in a most vulnerable position for penetration damage on the roof immediately above the eave platelines. The conditions promote snow-water seepage during extended periods of cold/snowy winters to soak insulation and wall cavities. Complete drying is extremely sluggish, and the rot-promoting conditions may persist for several months.

Other Damages

A re-examination of the damaging effects shown in Fig. 9.2 will identify and explain the damages of ice-dammed snow-water far better than words. All damages illustrated have been observed and recorded in a series of private audio-visual reports (2). These audio-visual reports form much of the data substantiating these comments. The dank and rotting odor of some of the soaked structures is the one feature that cannot be described or sensed with the audio-visual data.

9.5 HOMEOWNER CONFRONTATIONS WITH ROOF-SNOW PROBLEMS

Expert Advice and the State of the Art

With periodic regularity, matched only by falling leaves and snow, newspaper and magazine articles offer warnings, tips, and counsel for homeowners faced with the annual siege of ice-dammed snow-water on roofs. With good intentions, but with a deplorable revelation of the inadequacies of the art, "expert advisors" are prone to contribute to the problem rather than to its solution.

Roof Snow Removals

The most common, and apparently reasonable, advice is to shovel or rake the snow from the roof, since with no snow, there will be no snow-water, or ice-dam, and, ostensibly, no problems. However, homeowner reactions to such news article suggestions often develop even more frustrating side effects than the predictable hazards of falling bodies and broken bones following each snowfall.

Partial Roof-Snow Removals

Most homeowners are seldom equipped with sufficient incentive and enthusiasm to achieve more than partial snow removal. Equipment,

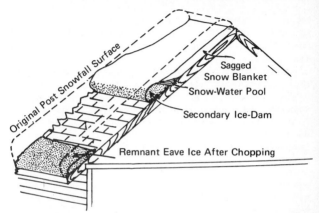

FIG. 9.4 Partial roof-snow removals develop secondary ice-dams.

FIG. 9.5 Secondary dam developed at snow-blanket terminus at the limit of partial snow-removal reach.

energy, dedication, and the many required roof-snow removals appear to limit the effort to a short reach uproof. Such partial snow removals leave substantial snow-blankets on the upper part of the roof slope.

With only partial removals, uproof snow will melt and run downslope to emerge from the protective snow-blanket, freeze, and build into an ice-dam that blocks snow-water uproof in what may be a more vulnerable leak position than at the eaves (Figs. 9.4 and 9.5). Such secondary ice-dams—oftentimes multiple uproof ice-dams—are visible to observers who take the time and winter-cold discomfort to examine the reaction of homeowners to the "shovel-it-off" advice of newspaper authorities.

Complete removal of roof-snow, of course, prevents ice formations and snow-water damages. In many cities that are annually plagued with roof-snow the yellow pages of the phone directory list firms pleased to be paid to remove each successive snowfall. The recurrent expense discourages many homeowners. Many budget-strapped and ill-advised homeowners attack the roof-snow and eave-ice in a desperate do-it-yourself assault of shoveling, chipping, hosing, picking, chiseling, and (believe it or not) blow-torching to rid their eaves of ice-dammed snow-water.

Neither professionals, who are seldom reported falling from the treacherous roof planes, nor the amateurs, who frequently drop into the news as killed or injured, should risk cavorting around on slippery roofs. Nor should prudent caretakers permit such stomping around on winter-embrittled roofs that are not designed to resist such riproaring gyrations. Roofing companies are pleased to shovel off roofsnow as off-season employment; they may score a "double-shot" in returning later to repair or replace the shattered shingles and leaking roof.

Eave Heat Tapes (Electric Cables)

Building owners, under the duress of ice-dam inundations, often install electric cables along the eaves and in the valleys. Unfortunately, the melting effectiveness of such heat tapes is limited to only a few inches from the cable; the typical zigzag pattern of installation betrays this localization of melting effectiveness.

Heat Tape Secondary Ice-Dams

The characteristic sawtooth melting of snow within a few inches of the heat tape develops a limited and selective removal of snow and ice; partial removals, whether by raking, shoveling, or localized heat cables, often develop uproof secondary ice-dams. Heat-tape secondary dams are observable along the terminal lines of the protective snow-blanket in a zigzag pattern prescribed by the limited melting zone of the heat tape.

Figure 9.6 shows the characteristic pattern of limited zone melting of an electric heat-taped eave. Figure 9.7 verifies the formation of the heat-taped secondary dam immediately above the highly localized melt zone of the heat tape.

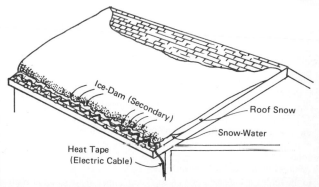

FIG. 9.6 Characteristic zigzag pattern of limited melting zone of eave heat tape.

FIG. 9.7   Secondary ice-dam developing uproof above heat zone of electric cable.

Valley Heat Tapes

Electric cables have been frequently observed in their zigzag pattern in valleys. Limited melt zones near the cable are often flanked with secondary ice-dams that obstruct normal drainage into the valleys. It is difficult to justify the expenses of electric cables, their installation, their recurrent power demands, and their adverse side effects in light of their limited effectiveness. Employing heat to melt ice into water on a troubled roof where heat has already produced the potentially penetrating snow-water suggests such approaches are not only wrong solutions, but are clearly complications of the eave-ice problem. Surprisingly extensive use of eave and valley heat tapes may be attributed to the futility and desperation of frustrated homeowners seeking "push-button gadgets" to alleviate their problems. Heat tapes are more numerous on roofs in latitudes where winters are not severe than where eave-ice is an annual assault and the futility and wasted expenses of localized melting are better recognized.

Counteracting Roof Thermals

Confronting roof-snow with controlled thermal gradients to melt the snow (as with electric heat cables) has been studied and developed for structures in the specific climate of the California High Sierras at Lake Tahoe (13). The energy costs and the adaptation to a specific climate of freeze-at-night and thaw-by-day limit the usefulness of houses designed to establish thermal gradient patterns from eave to ridge to melt troublesome roof-snow. Such counter-gradient installations in the upper Midwest would effect a return to the old *heat waster* (Fig. 9.12) performances of roof-snow melting.

Hosing with Tap Water

Another method of thermal attack on roof-snow has been employed in emergency situations to remove roof-snow and ice. Simply hosing down the roof (on mild days) from ridge to eave with tap water will remove roof-snow with reasonable efficiency; the method might be acceptable *if* the owner has no concern for the ice-formations that coat and overload the shrubbery around the house. Melting snow with flows of water is a method that has most of the hazards to life, limbs, and shingles of the ill-conceived shoveling techniques.

Eave Flashings

In the snow-country near the Canadian border from Maine to British Columbia, many houses can be observed with eave flashings of metal (Fig. 9.8). The 2- to 3-foot widths of eave flashing seldom improve the appearance of the house, but distressed people living in deep snow-country seem willing to forgo the esthetics of appearance for

FIG. 9.8 Eave flashings of aluminum.

some measure of protection from snow-water penetrations.

Flashings of smooth metal (or roll roofing) serve as leak protection devices along the leak-vulnerable eaves. The smooth (usually aluminum) edge-band eases the slough-off of eave snow and ice along the eave projections where attic heat is insufficient to sustain further down-roof flows of snow-water. With the lowest edge of the snow-blanket a few feet above the eave, *secondary ice-dams will form and build above the metallic flashed eave.*

Heat tapes fastened to eave flashings have been observed in several states and provinces. Such improvisations suggest a high degree of owner frustration and compounded desperation in such "systems" of ice-dam prevention. The practical usefulness of such metallic eave flashings appears to be in the leak protection of a continuous (monolithic) sheet of material.

Metal Roofs

Sealed-seam metallic roof coverings are common on homes of deep-snow ski areas from Stowe, Vt., to Aspen, Colo. These interlocked (monolithic) coverings indicate an awareness of the damage threats of roof-snow and ice. Such "total roof flashings" are directed toward leak protection and appear to defy or ignore ice-dams. Metal roofs apparently serve well (especially on steeply pitched roofs) by simply sloughing off the troublesome roof-snow.

Double Roofs

Double-surface roofs constructed with cool-air venting spaces between them are reportedly used in European snow-countries to prevent ice formations (13). The report also suggests this device has found little acceptance in this country because of the excessive costs of an insulated inner roof with a second weather-roof spaced atop by the furring supports. In several thousand observations, only one building (at Lake Louise, Alberta) was detected as newly constructed with an inner roof to retain the attic heat and an offset top surface (of corrugated steel) to protect against snow and rain. Furring strips running from eave to ridge supported the "weather-roof" and provided space between the two roofs for a gravitational flow of air from eave to vented ridge. Presumably, the inner roof was well insulated to retain and conserve attic and room heat. What little heat does escape the inner roof should air-wash upward and out the ridge *without melting the weather-roof snow-blanket from its underside.*

When such vent-spaced double roofs are functioning properly, roof-snow is not melted by room and attic heat, and snow-water does not flow down the roof-slope beneath the snow-cover to freeze into ice formations at the eaves. The double roof is, in effect, a "cold roof" and is a natural method of avoiding destructive ice-dams.

Explanations of roof-snow melting behavior considerations should precede further considerations of cold-roof performances. An understanding of roof-snow behavior (and misbehavior) on familiar roof designs and shapes leads to logical explanations of the errors and follies of our present approaches to ice-dam problems and results in an appreciation of the natural efficiency of a simple adaptation of the "cold-roof concept" of eave-ice prevention.

A following section of this publication shows how the concept of a "cold-roof" can be achieved by practical methods without the excessive expense of the furred-out double roof.

9.6   WHY THE NEGLECT OF ICE-DAMS?

Anyone recognizing the extent, frequency, and severity of roof-snow damages may be puzzled by homeowners' strangely apathetic and passive tolerance toward houses plagued with such a visible deficiency.

Respected technical professionals have generally neglected, ignored, or denied the impacts and importance of roof-snow behavior. It may appear incredible—as it was for us early in this research—that technically competent and responsible professionals in building design and supply have so strangely failed to attend and resolve the destructive impacts of eave ice-dams and roof-snow.

Some possible explanations of this neglect are offered here, not to fuel an argument, but to encourage evaluation of these observations that may initially appear to affront established custom and conflict with the (limited) writings of respected technical professionals.

With ice-dammed snow-water visibly soaking the ceilings and sidewalls of their homes, distraught owners have had no obvious target for their wrath. Homeowners cannot identify any specific material or product manufacturer to blame. Materials suppliers and installers neither recognize nor acknowledge any involvement with (or responsibility for) the ice-dammed destruction. Technically responsible authorities have generally ignored the extent, severity, and increasing intensity of roof-snow damages; such experts must recognize the inadequate state of the art and endorse research and development programs if this correctable fault of building performance in the snow-country is to be eliminated.

The lack (or avoidance) of involvement in snow-water actions by building industry leaders is vividly indicated in an upper Midwest newspaper quote (7) of a building association chief executive in recommending heat tapes and shoveling the snow from the roof: "It's the homeowner's responsibility, just as much as shutting the windows when it rains or removing the snow from the sidewalk." The director of a highly respected government research laboratory confirmed those recommendations as the state of the art in coping with ice-dams.

Such building industry attitudes are in puzzling conflict with private communications with federal housing officials and another metropolitan builders' association executive who said such ice-dammed snow-water "is the greatest single complaint we receive from homeowners."

Homeowners, with little competent technical support to guide them, turn to various makeshift methods, gadgets, attitudes, and escapes from the attacks of snow-water and eave-ice. A few methods are partially helpful, some are monstrously hazardous, some are illusionary, and some are little more than deceptions. Distraught homeowners are reluctant to discuss degradations of their property. They have developed, perhaps with the support of compensatory insurance settlements, an amazingly apathetic, stoical, or calloused acceptance of this "unmentionable" of north country homes.

A fundamental reason for the apparent indifference of technical professionals to the damages of ice-dammed snow-water may be found in the assignment of vapor condensate as the almost sole source of sidewall moisture problems. Sidewall inundations from eave-blocked snow-water penetrations have been ignored or assigned an insignificant role by most technologists in residential housing.

Our observations and recorded color slide data indicate a substantial, if not a major, role of snow-water in sidewall moisture accumulations and damages.

It will be productive to assign competent technical attention and consideration to long-neglected ice-dammed snow-water as a significant source of sidewall moisture and damage. As promised in this report, *eave-ice can be prevented, and with the prevention of eave-ice, homeowners can be assured there will be no sidewall moisture originating from blocked eaves.*

It is not productive to argue the relative importance (or destructions) of vapor condensate versus ice-dammed snow-water as sources of unwanted sidewall moisture. It is important to note that, even with the best recommendations for blocking vapor movement into sidewalls, moisture damages continue to occur with embarrassing frequency in spite of scrupulously specified (and installed) vapor barrier installations.

Vapor Condensate Theory

A classic study and report (5) of the causes of sidewall moisture in the 30s—when houses were initially tightened with insulation and asphalt shingles—identified and emphasized sidewall moisture sources as vapor movement from the tightened interior. Unfortunately, eave-blocked snow-water was apparently neither considered nor researched.

Acceptance of vapor movements as the major (if not the only) troublesome source of sidewall soaking has permeated all segments of the responsible building industry. This is evident in pamphlets from paint, vapor barrier, insulation, design, louvering, and associated manufacturers and associations (see References 4, 5, 6, 8, 10, 11, 12, 14, 16, 17, 22, 23, 24, 25). Four decades of neglect of ice-dammed snow-water as a second (and possibly major) cause of sidewall soaking may be explained, but to continue to ignore snow-water attacks is unconscionable.

Definition of Sidewall Damage Region

The classic definition of geographic limits of sidewall moisture troubles as the "January Mean Temperatures of 35°F," is illustrated in

Fig. 9.9. The origin of this geographic definition of sidewall troubles can best be understood from the following quote from page 15 of FPL Report #1710(11), originally issued in September 1947:

> The Forest Products Laboratory has been receiving reports of condensation in houses from various parts of the country for many years. In normal or mild winters most of these are from areas north of the Ohio River, but after a severe winter, such as occurs every 4 or 5 years, the reports are more numerous and include many from areas farther south. On a basis of these reports, it has been established that condensation problems may be expected in houses in those parts of the country where the average January temperature, according to Weather Bureau reports, is 35°F. or lower.

It is interesting, if not amazingly revealing, to find that the January 35°F mean temperature geographical limit of sidewall moisture prob-

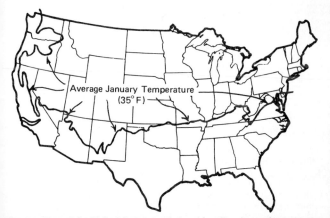

FIG. 9.9 Map of the United States associating sidewall moisture damage susceptibility with temperature (Ref. 6).

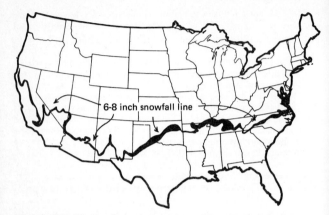

FIG. 9.10 U.S. Weather Bureau map and data (Ref. 18) of mean annual total snowfall with 6- to 8-in line delineated.

lems coincides almost precisely with the U.S. Dept. of Commerce Weather Bureau's map of the 6- to 8-inch Mean Annual Total Snowfall. [See Fig. 9.10 reproduced from U.S. Department of Commerce Weather Bureau Map and Data (18).] Had those geographical limits of trouble been identified by *snowfall criteria* instead of *temperature criteria,* ice-dams (as well as vapor) might have received some, if not a major, consideration in sidewall moisture problems.

Ice-Dams Overlooked

Ice-dams and damages have been almost ignored in reports of sidewall moisture problems throughout several revisions and building industry derivations of the classic vapor study (5); a recent (1972) revision (6) of the original research has added sketches and brief explanations of ice-dams caused by insufficient insulation and ventilation.

In discussing what appears to have been an honest, but possibly embarrassing, oversight of the important factor of roof-snow effects on sidewall degradation, we do not want to be argumentative. We believe it is in the public interest to re-examine all sources of sidewall moisture.

9.7 TYPICAL ROOF-SNOW MELTING ACTIONS

Contrary to popular belief, the radiant energy of the sun does not provide the major source of heat that melts roof-snow on snow-country buildings. During cold winter days, attic air warms the roof-deck, which then melts roof-snow from its underside.

Light and fluffy snow crystals serve as unusually efficient insulation that slows atmospheric melting, supports roof-contact melting, and maintains fluid flows of roof-deck snow-water during subfreezing weather.

In Fig. 9.11, "heat" arrows are used to indicate thermal movement, direction, and concentration. Room heat is shown passing sluggishly (slowed by insulation) into the attic where it warms the attic air. As attic air rises in temperature, it expands, becomes lighter, and rises to accumulate and increase air temperatures in the higher attic spaces at or near the ridge. Attic air warms the roof-deck, making it warmest near the ridge and less warm toward and along the eaves. These thermal gradients of the roof-deck cause the roof-snow cover to melt initially and more intensely at the warmer ridge zone.

Roof-snow, melted from its underside by the heat of the attic-warmed roof-deck, flows downroof as snow-water under the insulative snow-blanket. When typical snowstorms are followed by subfreezing cold, melted snow under the snow-blanket flows slowly downroof to emerge from the snow-cover and freeze into accumulating ice at the edge of the snow-blanket. Normally, ice-dams form and develop along the eavelines where the snow-blanket ceases to protect the fluid

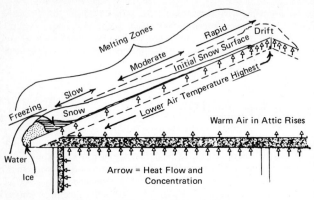

FIG. 9.11 Schematic representation of roof-snow melting.

flow and where the roof-extension beyond the plates has less (or no) attic heat to keep it liquid.

As the snow-water emerges to freeze into ice masses along the edges of the snow-blanket, subsequent water beneath the protective blanket may be blocked from runoff by the "ice-dams"; it may freeze or be held as pools under the insulating blanket of snow. Blocked by massive ice-formations, snow-water pools develop and may rise on roofs to heights sufficient to penetrate between and under the shingles and roof-deck. Snow-water penetrations through spaces, cracks, and flaws of the roof-deck lead to soaking and extensive damages to the roof, cornices, and sidewalls. Such typical roof-snow behavior is illustrated in Fig. 9.11, and damages are shown in Fig. 9.2.

The major destructive agent on snow-covered roofs is the ice-blocked snow-water, which is kept liquid by attic heat and protected from atmospheric freezing cold by the insulative snow-blanket. The roof-snow covers and hides the deceptive liquid–ice–liquid (and vapor) transformations energized by attic-heated roof-decks.

Roof-melted snow-water accumulates and develops ice-formations at any roof location where there ceases to be sufficient attic heat to counteract freezing or where the insulating snow-blanket is terminated. If the snow-blanket is undisturbed, the ice-dam will form at the eave; if snow is removed part way uproof from the eave, the ice-dam will form at that newly established snow/freezing air interface.

Intermittent and sporadic melting and freezing of roof-snow and snow-water is difficult to follow because most of the actions are invisible beneath the covering of roof-snow.

Simple corroboration of these actions of roof-snow can be observed from streetside. Thin coatings of frost or light snow showers reveal many highly localized sources of attic heat, such as: chimney; bathroom; dryer; vent-stacks; voids of carelessly applied ceiling insulation; platelines, because of their higher thermal contribution; and the pattern or ridge area heat concentrations.

Such visible indicators of roof-snow behavior develop more slowly with heavy snow-blankets and during severely cold weather. Attic-heat actions are visibly evident in terms of the diminishing depths of roof-snow near the ridge where the warm air of the attic rises to produce thermal concentration. Roof-snow patterns provide proof that roof-snow melting proceeds primarily from roof-decks warmed by the internally developed attic or building heat.

With very short (or no) cornice projections, massive ice and blocked snow-water pools develop at the platelines where attic heat and sidewall heat leakage are greatest; observations suggest that such short cornice houses have the greater incidence of *visible* sidewall soaking because pools are more likely to develop directly over the wall lines. Cornice projections, increased insulation, steeply pitched roofs, "cathedral" ceiling structures, knee-walled half stories, split-levels, lean-tos, and improved louvering practices have varying influences on the formation of snow-water and eave-ice.

9.8   ROOF-SNOW PERFORMANCES ON PRE-30'S HOUSES

A review of the progress in house construction since the 1930's reveals some of the reasons for increasingly troublesome roof-snow problems. These considerations also disclose inadequacies of present construction methods, materials, codes, and standards in coping with roof-snow.

Typical American houses built before 1930 were constructed with relatively steeply pitched roofs, with "open" (spaced) sheathing, and mostly with wood shingles; these (cedar) shingles had spaces between them that closed when wetted. Such gaps between shingles served to ventilate attics and to cool the roof-decks.

Most "pre-30's" houses had no insulation in either ceilings or walls; indeed, there was little demand for heat conservation when rooms were selectively *warmed* by stoves rather than heated by a central furnace. Some homes had central heating, of course, but only a few had insulation in ceilings and sidewalls. Wood and coal were plentiful; there were no outraged public demands for energy conservation. Nor were paint staining, blistering, and peeling of the traditional wood siding embarrassing and expensive problems. *The houses built in those days were not plagued with paint problems until they were modernized;* "updating" with asphalt shingles, insulation, and other installations designed to "tighten the old heat waster" and conserve heat led to sidewall moisture and paint failures of alarming severity.

Numerous pre-1930 houses in our country have been "modernized" with tight roofing, variable (but almost always inadequate) insulation, ventilation, and vapor barriers. Because of the highly heated air of the attic, the warmed roof melts roof-snow rapidly; in only a few days, roof-snow is transformed to snow-water and/or ice.

Figure 9.12 is a schematic representation of the roof-snow melting of such houses. Heat concentrations and heat escape movements are indicated in the quantity and directions of the arrows.

Notice the hot air concentration (as heat arrows) in the high spaces

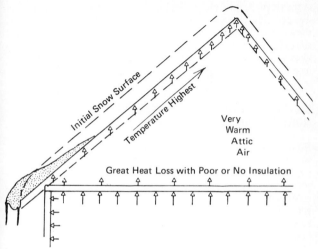

FIG. 9.12 Schematic representation of thermal influences on roof-snow melting of pre-1930's houses ("the old heat wasters").

of the attic where the warmest air rises and is trapped. Such ridge buildup of attic heat initiates and supports the most intense melt actions on the warmer roof-deck nearest the ridge. Roof-snow converted to snow-water flows downroof under the protection of the snow cover. On the old heat wasters, the excessive attic heat supports rapid flows of snow-water downroof to the cooler eave projection and beyond. When air temperatures are less than extremely cold and attic heat is in plentiful supply, rapid melting of roof-snow quickly clears the roof of snow—with only a gutterful or eave-line of ice and icicles.

Such roofs flow fast and furiously; roof-snow disappears initially at the ridge, next from the midrun, and last from the eave. "Old heat wasters" flow comparatively rapidly and copiously with snow-water; their roofs clear of snow more rapidly than do neighboring moderns; snow-water penetrations are of comparatively short duration; and massive eave-ice may be all that remains of the roof-snow until the next snowfall.

9.9 ROOF-SNOW PERFORMANCES ON MODERN HOUSES

Roof-snow blanketing on modern houses, in contrast with "old heat wasters," melts slowly; beneath the cover, water pools may threaten penetration into the house for periods of weeks or months in the snow/cold of the upper Midwest.

Modern houses, as a type apart from the old heat wasters, are generally considered adequately insulated and ventilated since they meet contemporary standards, codes, and recommendations.

With codes usually stated as minimums in thickness (or R-factor), builders tend to insulate (and ventilate) close to the minimums to keep building costs competitive. Few houses, other than those specified for electric heat, have more than the minimums. Generally, there is an unquestioning public acceptance of the minimum standards as "adequate."

Residential areas are dominated by "old heat wasters" in older communities and by many "adequately" insulated and ventilated moderns in the newer developments. Roof-snow melting may be observed as fast and furious for "old heat wasters"; in contrast, modern houses melt roof-snow with comparative slowness, greater deception, and often more insidious damages. Simply increasing insulation will not stop the flow of waste heat into attics; insulation will only slow the movement of thermal energy to economic and ecologic levels of acceptance. Warm, if not hot, attics are an undesirable characteristic of millions of "modern" American homes.

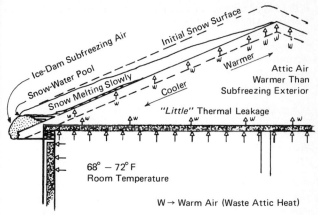

FIG. 9.13 Schematic representation of thermal influences on roof-snow melting of modern (contemporary) houses.

Thermal movements and concentrations for moderns are represented schematically by the arrows in Fig. 9.13. Compared with the "old heat wasters," there are fewer arrows of heat passing through the ceiling into the attic. Because insulation only slows heat escape, some waste heat passes into the attic air in contact with the roof-deck. With less heat entering the attic, there will be less intense heating of the roof; indeed, if heat escape is very slight because of generously insulated ceilings, only slight warming of the roof-deck may be expected. However, with present attic venting standards, the roof is often warmer than the freezing atmospheric air; imperceptibly slow and intermittent melting of the underside of the snow-blanket will (and does) develop roof snow-water that persists and threatens roof penetrations for periods of several weeks or months.

Roof-snow melting of modern houses will usually initiate near the ridge where the thermal concentrations of the attic create the warmest gradients. Streetside observations reveal these thermal gradients in the sagging patterns of roof snow-blankets. Ridge-snow, melted by the warm roof-deck, will flow downroof under the insulative snow-blanket to freeze into ice—either at the edge of the roof snow-blanket with the subfreezing air or where the roof thermals will not maintain the liquid flow. Roof snow-water may sporadically freeze and thaw as it progresses downroof to eventually emerge from the snow-cover where it meets the subfreezing cold air, solidifies, and grows into massive ice-dams with blocked snow-water pools immediately uproof. With long periods of formation, snow-water pools, and potential penetrations, modern houses are more prone to severe damages from snow-water than are the icicle-trimmed old-timers.

The sagging roof-snow pattern of Fig. 9.14 demonstrates the effects and locations of thermal gradients on this 1962 house. Differences in melting actions as delicate as the shingle pattern are evident. The deeper snow along the rake and eave cornices is a result of cool roof projections beyond the attic heat supply. The eave ice-dam, with its probable snow-water pool immediately uproof, is topped and protected with an insulative snow-cover along the 24-inch eave projection.

Thermal patterns on this roof are somewhat more pronounced than on many contemporaries, but the roof-heat pattern is characteristic for houses built under minimum standards.

Observations attest to this slow, deceptive, and insidiously penetrating attack by roof-snow on "adequately" insulated, and even on electrically heated houses. The problems of roof-snow melting and its damages can be seen on almost all buildings: the older poorly or noninsulated houses and on moderns complying with the latest insulation and ventilation building codes and standards. Contemporary specifications may provide only illusions of adequate protection.

Ice-dams and snow-water damages occur on almost all homes, ranging from "old heat wasters" through modernized old timers and contemporary moderns to the thermally efficient electric heated homes:

FIG. 9.14 Telescopic lens photo of shrinking and sagging roof snow on a modern (1962) roof.

all develop ice-dams, either fast and furiously flowing with snow-water or slow insidious seepage for as much as 3 months' duration. The roof performances can be visibly verified almost every year during February and March in the Twin Cities and in those recurring "hard" winters in Chicago's western suburbs.

The present emphasis (and sales promotions) of "thicker" insulation to offset the inflating costs of residential heating should be considered in relation to ice-dam formation. If energy conservation is limited to only greater insulation depths (and R-factor improvements) without recognizing and considering roof-snow behavior, *sidewall soaking and related damages can be expected to intensify* to epidemic proportions, possibly more confusing and devastating than the scourge of baffling sidewall degradations of the 1930's and 1940's.

These explanations of roof-snow behavior ranging from old heat wasters to the "adequate" moderns are intended to stimulate thought and consideration of roof-snow actions; they are also essential for a fair appreciation of the extent of this wintertime scourge of the snow-country where *almost no contemporary dwellings are immune to roof–snow–water damages.*

Observations and recognition of roof-snow actions on contemporary housing point to an appalling inadequacy in present standards and lack of building industry response to this neglected misperformance of snow-country buildings.

9.10   ROOF-SNOW BEHAVIOR ON A COLD-ROOF

Elimination (or control) of undesirable attic heat suggests a fundamental approach to eave-ice prevention; in essence, the object is to develop inexpensive cold-roof houses.

The effectiveness of cold-surface snow behavior to avoid massive ice formations is exhibited everywhere in winterized snow country. There are no massive ice formations on felled timber, rock outcrops, woodpiles, picnic tables, sheds, patios, marquees, and *unheated buildings*!

Cold-surface reaction to covers of snow is the way nature confronts and directs orderly melting and dissipation of snow. Ice may accumulate as glaciers from snow-packs or snow-water accumulations; however, generally, *nature is not troubled with ice-dams.* The human race, with its heated constructions, has created the monster, and as we errant humans are so often reminded, "It is not nice to fool Nature!"

Examination and interpretation of an idealized cold-roof performance should prove helpful, encouraging thought and action for practical application and adaptation of the cold-roof principle of eave ice-

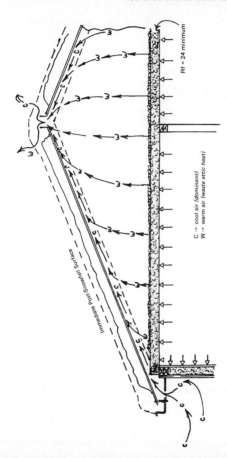

FIG. 9.15 Schematic representation of roof-snow melting behavior on a cold-roof house.

prevention. Such an ideally functioning house is illustrated in Fig. 9.15.

Continuous removal of attic air to maintain the roof-deck at atmospheric air temperatures requires minimizing the amount of heat entering the attic, together with a scavenging free movement of air from eave inlets to uproof (ridge) outlets. The small thin arrows in Fig. 9.15 illustrate escaping room heat passing sluggishly through the insulated ceiling. Arrows are few because heat loss of this very efficiently insulated house is minimal compared to the conventional modern house of Fig. 9.13.

In this idealized sketch, none of the heat arrows contact and warm the roof. When comparatively insignificant amounts of heat are lost through the well-insulated ceiling, the expansion of the warmed attic air causes it to rise slowly and be swept out the ridge vents by the natural flow of air from eave to ridge. Incoming atmospheric air flowing under the roof-deck keeps the roof cool.

For an ideally functioning cold-roof house, the snow-blanketed roof-deck will be maintained near the atmospheric temperature: the roofside of the snow-blanket will remain frozen during cold weather; on warm (or sunny) days, roof-snow melting will progress from the exposed atmospheric side of the roof-snow blanket.

9.11   COLD-ROOF PERFORMANCE IN WARM CLIMATES

The performance and desirability of cold-roof designs for warmer climates and for *better summer characteristics* should be recognized; cold-roof technology should be made available for all homeowners.

Because of this publication's concern for and emphasis on roof-snow, the full performances and benefits of cold-roof house construction may appear beneficial and limited to only deep snow-country buildings.

Attic heat buildups are possibly more undesirable and uneconomical in summer than they are in winter. The cold-roof design, with its attic air-sweep of air, merits its adaptation to all climates.

Attic air temperatures build to the 140–150°F range under the hot summer sun. With a room cooled to 75°F and the attic at 150°F, the heat flow impact is equivalent to the typical winter conditions of the room at 72°F and the attic at −3°F. The need for generous insulation to slow the heat flow is easily recognized. For both summer comfort and air-conditioning (cooling) energy conservation, sales promoters urge homeowners to purchase attic exhaust fans.

Electrically operated exhaust fans consume energy in their operation to conserve energy. Designed cold (atmospheric) roofs flush out the rising (expanded and lightened) air in the free air-flow movements from the lower eave inlets to higher ridge outlets.

No energy-consuming exhaust fans are needed in the cold-roof attics in the summer, either in the north country or in the hot deep South.

9.12 UNIQUE CHARACTERISTICS OF COLD-ROOF SNOW PATTERNS

Roof-snow disappearance patterns display sharply contrasting differences between cold-roofs and others; these contrasts provide visible evidence of a cold-roof performance; on cold roofs, the edges are the first to melt, and the center of the blanket is the last to disappear. Roof-snow disappearance patterns provide a convenient means to evaluate thermal gradients on subject buildings and for monitoring the effectiveness of cold-roof conversions.

The snow on the roof of the cold-roof house of Fig. 9.16 is in a midroof pattern that is in sharp contrast with most (occupied) residential buildings. Notice the rake areas have cleared, and the eave edge is also free of both snow and ice; snow-water drainage has been as free-flowing and orderly as the run-off rains. This performance can be better appreciated in its contrast with the roof-snow pattern

FIG. 9.16 Residual snow pattern on a cold-roof house specifically designed for ice-dam immunity.

of Fig. 9.14, where attic-warmed ridge and midslope areas melt first, the cooler rake edges follow, and the eave-ice persists in its drainage-blocking action until it, at last, melts.

9.13 DETECTION OF COLD-ROOF EFFICIENCIES IN ROOF-SNOW PATTERNS

The distinctive patterns that develop as roof-snow dissipates provide a convenient way to recognize and evaluate a subject roof's characteristics in eave-ice resistance.

The sketches illustrating three types of melt patterns developed from the snowfall of Fig. 9.17 are shown in Figs. 9.18, 9.19, and 9.20.

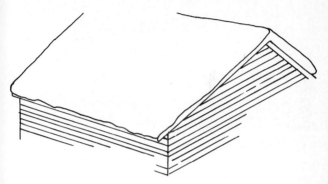

FIG. 9.17 Typical post-snowfall pattern.

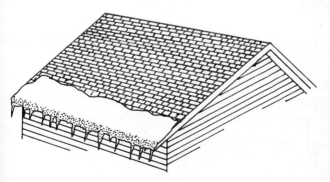

FIG. 9.18 "Old heat waster" after a few days.

FIG. 9.19 Modern house roof after several days or weeks.

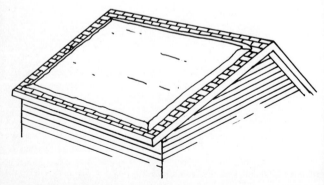

FIG. 9.20 Cold-roof house.

These three house types relate to an "old heat waster," a modern, and a cold-roof. These roof-snow patterns may be correlated to the melting performances earlier described for these three characteristic types of residential housing.

A properly functioning cold-roof can often be visibly identified in the unique ruffling and eroding of the roof-snow edges into the peculiarly concave shapes indicated in Fig. 9.20. Cold-roof snow edges at ridge, rakes, and eaves contrast revealingly with the convex eave and rake edges of roof-snow on almost all contemporary buildings.

When there is no attic (or underroof) heat to melt the snow-cover from its underside, the snow-blanket on the cold-roof will dissipate initially and progressively from the exposed edges at ridge, rakes, and eaves.

On a cold-roof, snow melting and compacting are accompanied by wind actions that activate blow-offs, sublimation, evaporation, melting, and a generally orderly free-flowing dissipation and drainage at the greater surface exposure of the edges.

Deep roof-snow has long been recognized (and proclaimed in insulation promotions) as an indicator of well-insulated houses. Unfortunately, the depth of snow accumulations does not indicate the combined, or balanced, efficiencies of insulation and ventilation.

How roof-snow "shapes up" after a period of exposure to winds, sun, and interior heating energy will provide a simple means to evaluate the insulation-ventilation characteristics of any subject house or building.

As you look for the characteristic edge dissipation patterns of cold roofs, you will find almost none on contemporary housing, *unless* you chance upon a residence where the occupant has vacated and the house-heat is cut back or off.

Simple visible verification of most of the observations and comments of this publication may be monitored in roof-snow patterns, especially in *late* winter.

With reasonable recognition of the erratics and complexities of wind-drifted snows, interpretations of roof-snow behavior should prove stimulating and productive.

9.14 EAVE-ICE PREVENTION TECHNOLOGY

Basic Guidelines

The fundamental concept of a cold-roof is quite simple: adaptation of the basic guidelines to new construction, and to existing buildings, may be accomplished with *apparently* little more than these two modifications of conventional residential construction:

1. *Insulate* room ceilings far more than customary to minimize heat losses and attic temperatures.
2. *Ventilate* profusely at all eaves and ridge for a natural flow of air to sweep out the warmed attic air.

Limitations, Exceptions, Difficulties, and Solutions

Earlier in this publication, better eave and ridge venting was suggested, as well as improved insulation for homes. Such broad suggestions, like our basic guidelines, may be technically profound, but of dubious practical value without further explanations of the details and problems of their applications.

The *apparently* simple application of the basic guidelines for eave-ice prevention may often be difficult, sometimes less than fully satisfactory, and downright impractical for a few building types.

The cold-roof concept is simple; its applications are complex.

The following discussions on the complexities of eave-ice prevention technology are not intended to discourage corrections. The difficulties are described and the limitations are interpreted *to prevent misapplications* and possible disillusionment with the effectiveness and widespread adaptability of *competently installed eave-ice constructions and corrections.* Cautious homeowners should be aware of these difficulties and limitations so they may judge applicator capability as well as the quality of the completed application.

The Trouble with Insulation

"If a little is good, more is better." Certainly this phrase has its limits of judicious application, but in present standards (and existing installations) of ceiling insulation, much, much, *much more* will be better for 99+ percent of residential housing; this is especially important if attics are to be properly vented and eave-ice is to be prevented.

Apologetic marketing of ceiling insulation still persists since the early introductory sales of the 1930's when the additional expense of an unfamiliar item met "dubious-buyer" resistance. Justification and redemption of insulation costs through amortization of "additional outlays" has, until the recent energy conservational promotions, limited installations of ceiling insulation to little, or no more than, the Federal Housing Authority (FHA) minimum standard specifications.

No stipulations of the amount of insulation (or louvering) can be recommended *as sufficient,* partly because of the innumerable variables in residential designs and requirements, but mostly because there should be no interpretation of such a stipulated "sufficiency" that might inhibit an owner's urge to "really insulate (and ventilate) the house!"

A homeowner might be satisfied that he or she has accomplished all that is feasible in ceiling insulation when room to attic heat losses are reduced to the level required for efficient electric heating. Such quality of room heat retention is in harmony with the needs and promises of contemporary promotions for energy conservation.

Budget-strapped homeowners may be somewhat reluctant to purchase these "extra" insulation improvements, but the savings in energy outlays alone should not be difficult to appreciate in this era of runaway fuel costs. The savings available through prevention of snow-water damages may be more indirect and difficult to foresee; however, coupled with the fuel savings, *"maximum" insulation can be doubly rewarding!* This may mean that as much as 10 to 12 inches of insulation should be applied in attic areas; *AND it must be intelligently and carefully installed.*

Building code specifications usually specify the insulation requirements in terms of "*Minimum* Standards." Cost-conscious builders, under the pressure of price competition, tend to supply the bare minimum. Of course, additional insulation may be purchased, and there are many high-grade installations by quality builders; proof of these "quality jobs" can be observed in deep and winter-enduring roof-snow that melts and converts very, very slowly into ice-blocked eaves. Unwittingly the quality conscious builder or owner *rarely matches the adequate insulation with a comparable maximum of ventilation.*

Much contemporary residential construction barely meets minimum building codes; in many cases, houses are deficient not only in "thickness" or R-factor (that is, resistance to heat loss), but in addition, ceilings are prone to numerous voids and discontinuities in the protective insulation.

Builders often pay too little attention to the care with which insulation is installed. Continuity of the protective ceiling blanket is frequently interrupted by bridging, wires, recessed ceiling fixtures, scuttles, stair-entrances, and a variety of occupant actions such as careless attic storage, TV, Hi-Fi, and associated attic uses and misuses.

Extremely conscientious ceiling insulation application and care must not be considered trivial; cold-roofs for eave-ice prevention cannot tolerate the flaws, voids, thin spots, and discontinuities that are frequently associated with present construction practices.

Seldom considered and rarely examined attic spaces are too easily disregarded, ignored, forgotten, and neglected! *At least biennial inspections of ceiling insulation should be a maintenance routine;* sagged, compressed, wetted, blown-out, disturbed, or other diminished insulation should be located and corrected. Energy conservation programs may suggest your concern for insulation efficiencies; cold-roof ice-prevention *demands* it!

Disruptions in the ceiling insulation are costly not only in wasted room heat, but also because such flaws lead to roof hot spots that develop intensified localized roof-snow melting. Such hot spots are sometimes identified by spot shrinkages in the snow and related ice-dams downroof at the eave. An attic door or scuttle inadvertently left open may be revealed in the sagging snow-pattern on the roof.

Uninsulated chimneys (furnace flues), gas vents, bathroom, kitchen, clothes dryers, and any other warm exhaust equipment may contribute to a troublesome supply of attic heat; some pipes may have to be re-routed, and all of such attic heat contributors should be carefully wrapped with effective insulation. Eave-ice prevention technology requires *minimization of all attic heat sources*!

"Reading" roof-snow patterns to reveal insulation defenses of houses is as easy for the trained ice-prevention technologist as "reading a pass defense" is for an all-pro quarterback. The insulation picture blurs for ice-prevention analysts only when extremes of attic ventilation are encountered. Because almost all contemporary houses are vent-specified for *no more than minimums for vapor removal*, streetside observers will find properly vented houses extremely rare.

Ventilation Problems

Some building owners may find it necessary to dispel the notion that attic heat is an asset to be retained. Perhaps such misconceived reverence for attic heat was reasonable in early dwellings before insulation was available to slow escaping room heat to insignificant heat losses.

Anyone who has endured the discomforts of an uninsulated summer resort or motel cabin where sleep was delayed until early morning (when the overly abundant attic heat had dissipated) can appreciate the importance of ventilation in preventing attic heat buildup in summer; prevention of eave-ice with cold-roofs and cold attics demands a similar ventilation removal of attic heat in winter.

Builders and homeowners reluctant to accept ventilator-to-ceiling-area ratios of 1/900 to "as much as" 1/250 may be shocked with cold-roof designs demanding infinitely greater louvering—both in size, number, and distribution.

Standards in venting specifications, such as those quoted for attic vapor control, are both inadequate and meaningless in eave-ice prevention. Roof shapes, cornice styles, and appearance esthetics restrict many installations; available positions for louver distribution may limit ideal venting. Some cold-roof installations may have to settle for less than maximum ventilation; ice-prevention performances may, therefore, be proportionately less than fully effective.

Louver products of various types and shapes are presently available for venting both eaves and ridges.

Triangular gable louvers may be satisfactory for vapor dissipation if they are large enough and if they are strategically placed to utilize wind or air currents; seldom are gable louvers suitable for development of the eave-inlet ridge-outlet patterns of air-wash required for a properly functioning cold-roof.

Pitched bonnet-type roof vents, either powered or gravity exhausting, are seldom used in sufficient quantity and distribution to achieve the blanketing air-wash needed for complete cooling of the roof-surface.

Rectangular soffit louvers, circle-spot louvers, and alternated perforated soffit panels usually limit the eave inlets to less than desired inlet free-air venting and distribution. Generously perforated soffit materials may promote embarrassing wind-blasting of insulation. On house designs of little or no cornice projection, there is little or no room for soffit louvers; a proprietary dual-walled fascia (19) has been developed and effectively tested for such soffitless house designs, but the product is not commercially available.

Strip louvers on ridge and eaves are the simplest products for securing a good distribution of eave inlets and ridge outlets. Most ridge strip louvers are apparently designed to be unobtrusively small to avoid affronts to residential esthetics in styling. Such size limitations may leave much to be desired in deep-snow country where such ridge venting may become submerged and plugged by deep snow.

Because the naturally rising action of warming air generates eave-to-ridge movements for cooling roofs, the stack effect of cupolas and chimneys may find greater adaptations in future ice-prevention designs; a simple corbelling of the brick chimney as it passes through the attic can supply additional attic-venting flues and also contribute to the balanced appearance of a massive chimney emerging from the roof-ridge. Such stack actions, integrated with free distribution of eave inlets in soffit or fascias, effectively assure cool attics for summer comfort and cold-roofs for winter ice-prevention.

There appears to be a challenge for creative designers in the development of innovative devices for venting roofs for ice-prevention in harmony with the esthetics of acceptable appearance. Perhaps the multiple roof attractions of Chinese and Japanese structures may find a justifiable place in American home designs.

The watchword for venting attics, as with ceiling insulation, is simply: *"You can only install too little, never too much!"*

And yet, under certain *careless insulation practices,* it may, at first, appear that there is too much ventilation!

Too Much Venting

Where insulation is not carefully fastened or protected near the eaves, strong wind-blasts through freely vented eaves may blow insulation away to expose room ceiling areas to the attic's winterized temperatures. When the bared ceiling areas drop below the dew point or freezing point, those ceiling surfaces will develop wetted, stained, and frosted ceiling spots where room moisture has condensed on the cold ceiling areas.

When insulation has been blown backward from the plateline area, you commonly encounter a very infuriated homeowner, a bewildered builder, and a worker who has been ordered by the builder to "stuff the stuff along the platelines" to block recurrent wind-blasting of the plateline insulation. The builder blames the unhappy situation on the designer who "uses too much ventilation" and attempts to assure the hostile homeowner that "nothing like this will happen again!" No, it won't happen again when the venting is all but completely choked off, but it won't be long before the "ice-dam cometh" on the eave again, *again,* and *again!*

Is the blunder of insulation blow-outs due to "excessive" venting, or is it due to careless insulation application? If eave-ice prevention is an objective, there is no doubt that insulation must be more carefully positioned, and perhaps fastened, than has been customarily practiced by the building trades.

Restricted Plateline Gaps

The plateline region has been misunderstood, unrecognized, or ignored by many builders as a critically vulnerable region for heat-loss as well as a restricted passage that almost always limits and often inhibits air-flow from eave inlets to uproof outlets.

Exterior wall-to-ceiling corner joints are characterized by unavoidable cracks of: doubled plates; exterior sheathing edges; lath or wallboard joints at the interior corners; and usually *ceiling insulation that has been terminated at the platelines.* Wall insulation tends to sag beneath the plates and loses both its effectiveness as an insulator (and possibly its positioning) when moisture from condensate *or blocked snow-water* wets and soaks the sidewall insulation. Thermographic analyses disclose exterior platelines as major heat loss areas. The state of the art of residential design, construction, and research indicates gross neglect of this critical heat-loss region. Excessive heat losses at the exterior platelines also contribute to snow-water penetrations by supporting the liquid pool immediately over the plates. Quality construction calls for consideration, reduction, and preferably elimination of the plateline heat concentrations.

The exterior platelines require more and carefully positioned insulation; they must also retain spaces for a free flow of air past the usually restricted gaps above the plates. Such a dual requirement appears to produce an impasse for the majority of homes where the

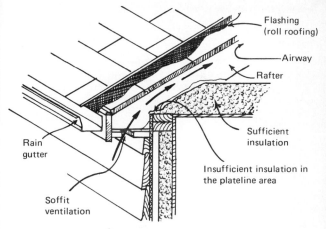

FIG. 9.21   The plate-gap (as the FPL report in Ref. 6 shows it).

roof-pitch, rafter-heel-cut, and sheathing allow less than 6 inches of clearance—often as little as 3½ inches.

The plateline gap difficulty is sketched in Fig. 9.21 as it appeared in a recent publication (of the Forest Products Laboratory, Madison, Wis.). Notice that free air-flows are indicated, but at an unfortunate sacrifice of insulation that was probably not intended. That sketch not only helps illustrate the restrictive nature of such exterior plate-lines, but it also demonstrates that the state of the art recognizes the *need* for adequate air flow into attics, but *has failed to explain HOW such essential airflow and insulation can be accomplished*!

The sketch of Fig. 9.22 illustrates a solution for compromising the restricted passages between *air-flow demands* and *insulation optimums*. The proprietary product (20) called "Air Passage Protector," is not commercially available.

Bridging the plateline gaps

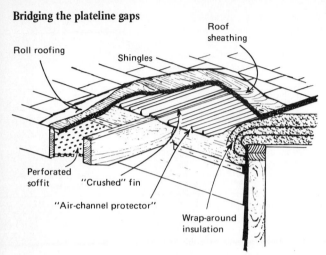

FIG. 9.22 Sketch of plate-gap with "air passage protector" in position between sheathing and wall plates.

New construction applications of this air passage protector add little to labor costs. Placement requires only positioning. The subsequently fastened sheathing holds the roll on to the rafters with the fins upward; the resilient insulation inserted beneath it supports and forces it upward against the sheathing, but is limited by the fins to assure air channels of fin depth between the insulation and the roof sheathing. Any fin on a rafter is simply crushed as the sheathing is nailed.

Positioned tightly between the sheathing and insulation, the spaced-fin roll not only creates and assures air channels at the restricted plateline gaps, but also serves as an interior flashing to help protect this leak-prone eave-zone from inadvertent roof-leaks above it.

With little additional labor cost, the roll-out-on-rafters procedures may be used on existing houses about to be reshingled. Often, especially where there has been a history of ice-dammed snow-water on the eaves, the weak and rotted roof sheathing should be replaced; the roll-out-on-rafters installation of the plateline air-passage protector (and flashing) material can be accomplished as simply as in new construction.

FIG. 9.23 Worker applying air passage protector along eaves.

Where reshingling is not needed on an existing dwelling, the insulation baffles may be prepared as finned *panels* sized for simple insertion *between* the rafters to bridge the plateline insulation. Such insertions may be made from the attic where that approach is feasible, or from ladder or scaffold along the exterior with the fascia removed for direct access to the plateline gaps.

9.15 PROVEN COLD-ROOF PERFORMERS

Existing houses which have been corrected for ice-dams offer convincing proof of the effectiveness and practicality of the materials and methods developed to convert roofs decked with thermal gradients into cold-roof performers. Proof of performances of these developments can be observed on a few houses in southwestern Wisconsin that were planned and constructed to perform as cold roofs. Most of the owners are unaware of the unique eave-ice protection built into their homes. ·

Figure 9.24 is a photo of a Colonial-type house constructed in 1933; it was generously insulated (even by today's standards). It has no vapor barriers; it had a 39-year history of blistered and peeling paint; it has also been decorated with eave ice-dams almost every winter until 1972.

In the summer of 1971, the house in Figure 9.24 was converted (with the materials and methods described in this publication) into a cold-roof house. Since the correction was made, the house has had no eave ice-dams.

Converting the house of Fig. 9.24 into a cold-roof performer not only eliminated the eave ice-dams, but altered the roof-snow melting pattern, as may be seen in Fig. 9.25. Notice the roof-snow edges at both rake and eave exhibit a ruffled concavity where wind actions have initiated roof-snow dissipation and removal. This house now sheds both rain *and snow-water* in an orderly drainage.

One of the most interesting "conversions" has been performed on a contemporary house distressingly troubled with ice-dammed snow-

FIG. 9.24 Well-insulated house built in 1933 with its almost annual eave ice-dam.

FIG. 9.25 The house in Fig. 9.24 after its modifications into a cold-deck performer.

FIG. 9.26 A house planned and constructed for cold-roof performance, but without the expense of the double-roof structure.

water. After corrections were made in late December, the snow-water penetrations stopped, *ice-dam growth ceased,* and by late February, the ruffled concave eave pattern (characteristic of a cold-roof) was evident and reassuring of the effectiveness of this rare, and not normally recommended, midwinter correction.

Detailed observations and comments of these and other "case history" performances in these developments have been logged in several audio-visual reports. The audio-visual presentations have been prepared for training installation personnel in the whys, wheres, whens, hows, and how-nots of the practical application of ice-prevention technology.

BIBLIOGRAPHY

1. "An Introduction To The IDP Rating." A survey and analysis of "Tough Winters" at twenty-five stations over five years — 1967-1971. By Howard L. Grange, R.P.E. An unpublished report of *Ice-Dam-Potential*.

2. "Audio-Visual Reports For Tape-Color Slide Review and Presentation" by Howard L. Grange, R.P.E. Personal papers and audio visuals prepared for consultation and teaching the methods and materials of Ice-Prevention Technology.

3. "Comparative Study Of The Effectiveness Of Fixed Ventilating Louvers" by H.S. Henrichs. A report for HC Products Co., Princeville, Ill. 61559.

4. "Condensation In Farm Buildings" by L.V. Teesdale, Engineer, Forest Product Laboratory. Information Reviewed and Reaffirmed January 1956 FPL No. 1186.

5. "Condensation Problems In Modern Buildings" by L.V. Teesdale, Engineer, Forest Products Laboratory Report for presentation before Conference on Air Conditioning, University of Illinois, March 8-9, 1939.

6. "Condensation Problems: Their Prevention And Solution" by L.O. Anderson, Engineer, Forest Products Laboratory, Forest Service, U.S. Department of Agriculture. Research Paper FPL 132, 1972.

7. "Home Builder Says Ice In Eaves May Cause Roof To Leak" by John Newhouse in Wisconsin State Journal, newspaper article of January 3, 1970.

8. "Insulation: Where And How Much" by Laurence Shuman. Technical Reprint Series No. 4 from HHFA Technical Bulletin No. 3 (March 1948) Housing and Home Finance Agency. From Supt. of Documents.

9. "Local Climatological Data." United States Department of Commerce, National Oceanic And Atmospheric Administration, Environmental Data Service, National Climatic Center, Federal Building, Asheville, N.C. 28801.

10. "Moisture Condensation" by Frank B. Rowley, Director, University of Minnesota Engineering Experiment Station; University of Illinois Bulletin, Vol. 44 No. 34, Jan. 27, 1947. Issued by The Small Homes Council Circular Series Index Number F6.2.

11. "Remedial Measures For Building Condensation Difficulties." Original reported dated September 1947, written by L.V. Teesdale, Engineer, Information reviewed and reaffirmed August 1962, Forest Products Laboratory Report N. 1710.

12. "Save Your Home From the MENACE OF MOISTURE." A pamphlet issued by the National Paint, Varnish, and Lacquer Association, Inc.

13. "Snow Country Design" by Ian Mckinlay, A.I.A. and W.F. Willis. R.P.E. procured from Mackinlay/Winnacker AIA & Associates, 5238 Claremont Ave., Oakland, Ca. 94618 for nominal fee.

14. "Some Construction Defects Which Permit Water To Menace Homes." Circular 763, issued October 1953 by the Scientific Section, National Paint, Varnish, and Lacquer Association, Inc.

15. "The Wald-Way System of Component House Construction" by Howard L. Grange, R.P.E. The book is a licensed private communication.

16. "Thermal Insulation Made Of Wood-Base Materials, Its Application And Use In Houses" by L.V. Teesdale, Engineer, Forest Product Laboratory, Revised October 1958, FPL No. 1740.

17. "Understanding The Mechanisms Of Deterioration Of House Paint" by F.L. Browne, Forest Products Laboratory. Reprint from November 1959, Forest Products Journal (Vol IX, No. 11).

18. United States Department of Commerce Weather Bureau Map and Data "Mean Monthly, and Annual Total Snowfall (inches." U.S. Dept. of Commerce, National Oceanic and Atmospheric Administration, Environmental Data Service, National Climatic Center, Federal Building, Asheville, NC 28801 *Taken from the Climatic Atlas of the U.S.*"

19. "U.S. Patent #2,797,180 'Ventilated Roof Construction'" Howard L. Grange, Inventor. Canadian Patent #949718.

20. "U.S. Patent #3,683,785 'Roof Construction Providing Air Flow From Eave to Ridge'" Howard L. Grange, Inventor. Canadian #995,869.

21. "Weather And The Hand Of Man." Environmental Science Services Administration Pamphlet of the Department of Commerce.

22. "What To Do About Condensation" by E.R. Queer and E.R. McLaughlin, Engineering Professors of Penn State College, a technical reprint from HHFA Technical Bulletin No. 4 (May 1948) of Housing And Home Finance Agency, Division Of Housing Research. From Supt. of Documents.

23. "Wood Decay In Houses and How To Control It." Home and Garden Bulletin No. 73, U.S. Department of Agriculture. From Supt. of Documents.

24. "Wood Handbook." Handbook No. 72. Forest Products Laboratory, Forest Service, United States Department of Agriculture. From Supt. of Documents.

25. "Wood-Frame Construction" by L.O. Anderson and O.C. Heyer, Engineers, Forest Products Laboratory, Forest Service, U.S. Department of Agriculture. Handbook No. 73; from Supt. of Documents.

NOTES

NOTES

Section **10**

ATTIC
VENTILATION[1]

---

[1]This section was extracted with permission from *Principles of Attic Ventilation,* 4th ed., by Clarke M. Wolfert, P.E., © 1985 by Air Vent, Inc., Peoria, IL.

Until recently there was limited interest in attic ventilation among builders, architects and engineers—to say nothing of the general homeowning public.

Recognition of the importance of effective attic ventilation to solve and prevent problems has grown rapidly, however, and today it is common practice to provide some sort of ventilation openings in attic or ceiling spaces. Vents of one type or another are found on single-family and multi-unit dwellings, manufactured homes, mobile homes and on farm and commercial buildings.

Passive solar housing designs, for example, invariably make use of ample and effective attic or ceiling space ventilation, recognition of its role in lowering energy costs, protecting roof materials, maintaining insulation R-value, and eliminating potential ice-damming problems.

Another significant development affecting attic ventilation requirements is the increasing use of whole house fans to exhaust warm air from living quarters into the attic space. Fan performance—and cooling results— are directly dependent on the amount of net free area provided by attic vents.

Also, the remodeling roofing industry is increasingly aware of the importance of proper ventilation to assure roof shingle durability and performance.

10.1   WHY VENTILATE?

The two fundamental benefits of an effective attic ventilation system are:

1. A cooler attic in summer
2. A dryer attic in winter

Both benefits result in energy saving, greater homeowner comfort and higher structural integrity of the dwelling.

SUMMER

How Heat Gets In

Summer sunshine causes a buildup of heat in the attic space, created by radiant heat from the sun increasing the roof temperature so that it re-radiates heat into the attic. During the night, the cooler outside air and absence of sunshine permit the attic to re-radiate or conduct stored heat to the atmosphere. However, an unventilated—or poorly ventilated— attic is not able to lose all of its stored heat in this manner. This results in a heat buildup over a period of days, with ever increasing attic temperatures.

How Much Gets In

If there is no attic ventilation, a 90° day with full sunshine can heat the roof sheath to 170° or more. Heat radiating from the sheathing down to the attic floor (or ceiling insulation) can raise its temperature to as much as 140°. (See Fig. 10.1.)

Homes today are being built with heavier insulation than ever before. This increases the need for effective attic ventilation. Without it, today's heavier insulation absorbs and holds more of the heat buildup in the attic during the day, making it less likely that all of the heat of the attic will be removed into the cooler nighttime air. Residual stored heat can build up over a period of hot days, maintaining the insulation at a higher temperature and increasing the heat radiation down to the rooms below.

What It Does

Overheated ceiling insulation conducts heat through to the ceiling, and this heat is then radiated downward to persons and objects in the rooms below (Fig. 10.2). There are two consequences of this:

1. The home's air conditioning system must operate for longer periods in order to remove this heat.

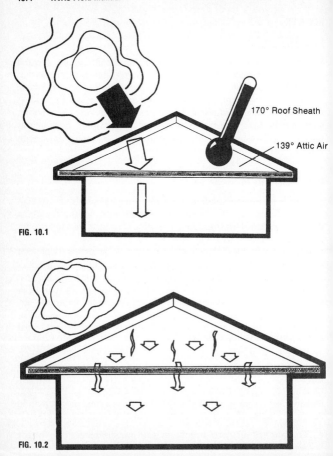

170° Roof Sheath

139° Attic Air

FIG. 10.1

FIG. 10.2

2. Because it is *radiant* heat, individuals in the home may feel warmer than the air temperature indicates, and their usual reaction is to put the thermostat at a lower setting, further burdening the system and causing higher energy usage. An overheated attic, combined with moisture, can also cause roof shingles to distort and deteriorate.

How to Prevent It

The solution to the problem is attic ventilation. A sufficient volume of ventilation air must be moved through the attic space—under varying conditions of wind force and direction.

Also the ventilation air must move in a uniform pattern along the entire underside of the roof, avoiding isolated non-ventilated "hot spot" areas. The best system is one that:

1. Does not *itself* use electrical power and thus nullify energy savings.
2. Completely ventilates and cools the underside of the roof sheathing, which is the source of heat radiation downward to the attic floor or top of the ceiling insulation.

WINTER

How Moisture Gets In

Winter conditions bring a different type of problem. During the cold months of the year, the air inside the home is warmer and carries more water vapor than the colder, dryer air in the attic. Cooking, laundry, showers, humidifiers and other activities using water contribute to this condition. There is a strong natural force, termed "vapor pressure," that causes water vapor to migrate from high-humidity air or materials to low-humidity air. This migration of water vapor passes through ceilings, insulation and wood—and even successfully circumvents a vapor barrier. It moves into the attic space where it can readily condense into liquid water on the cooler structural members—rafters, trusses, and especially the cold roof sheathing. (See Fig. 10.3.)

Another factor increasing the likelihood of a moisture condensation problem in many homes today is the use of an electrical heating system, more prevalent with the nationwide shortage of gas and oil. Because of the higher cost of heating with electricity, these homes are being built "tighter" and with greater insulation. This helps retain moisture in the home, which is transferred to the attic by the strong driving force—vapor pressure. The moisture problem is heightened by the fact that, in an electrically heated home, outside air is not drawn into the home to replace furnace combustion air. A minimum of dry outside air enters the home in winter; air within the home acquires a higher moisture content and becomes a more likely source of problems.

What It Does

The water vapor moving into the cold attic space in winter can be present in such a volume that it condenses on cold structural parts

FIG. 10.3

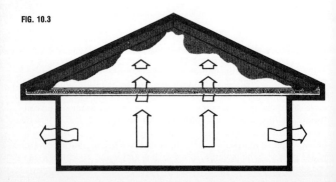

and drips down to soak ceiling insulation, compressing its volume and reducing its insulating effectiveness. As a consequence, heat loss through the ceiling increases and more energy is required to maintain the residence at a comfortable temperature.

Figure 10.4 shows how condensing moisture within an attic or ceiling space can dampen and compress insulation. Even small amounts of such condensation can have a substantial effect on R-value. Research has shown that, as moisture is added to fiberglass insulation, the R-value is reduced as shown below.

R-19 + 1½% moisture = R-12 (36% loss of R-value)
R-30 + 1½% moisture = R-19 (36% loss of R-value)
R-36 + 1½% moisture = R-23 (36% loss of R-value)

This loss of R-value takes place even without compaction of insulation, which can occur when moisture actually drips onto the insulating material, reducing R-value additionally and proportionately. (10% compaction reduces R-value by 10%, 20% compaction reduces it 20%, etc.)

In addition to its destructive effect upon attic insulation, moisture condensation can lead to many other problems: wood rot, paint peel-

FIG. 10.4

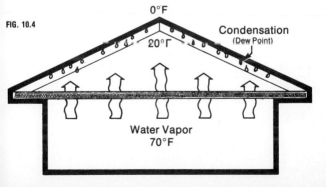

ing and deterioration of ceiling and roofing materials. It will, in fact, penetrate through the sheathing and will freeze, raise and rot out the shingles, sheathing and roofing materials.

How to Take It Out

As with heat buildup in summer, wintertime attic moisture is removed by ventilation air. (See Fig. 10.5.) The most effective system is one that moves the greatest volume of air through the attic space—under varying conditions of wind force and direction—*and* in a pattern of uniform flow across the underside of the roof sheathing. This is because the roof sheathing is the coldest surface in the attic and the prime location for moisture condensation.

Ice-Dams—A Special Problem

In the northern two-thirds of the U.S. (see Figs. 10.6 and 10.7) there can be a serious problem caused by snow melting on a warm roof surface, usually near the ridge, the water running down to the colder roof overhang and then refreezing into ice. Subsequent melting of

FIG. 10.5

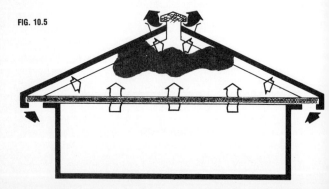

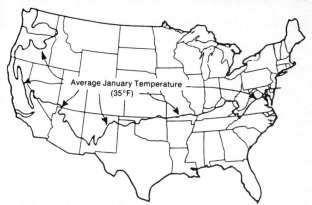

FIG. 10.6 Map of the United States associating moisture damage susceptibility with temperature.

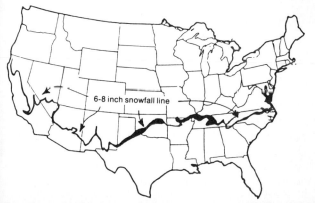

FIG. 10.7 U.S. Weather Bureau map and data of mean annual total snowfall with 6- to 8-in line delineated.

roof-snow can then accumulate as a pocket of snow-water that eventually backs up under the shingles to cause major damage in the plateline area.

This damage can appear in the form of soaked and inefficient insulation; stained, cracked, and spalled plaster or sheetrock; damp,

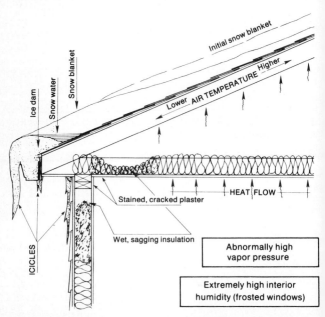

FIG. 10.8 This sketch of the ice-dam problem identifies both the ice dam and its damage. Of course, all the damage illustrated may (or may not) occur at any one instance. The damages illustrated here are far more common and costly than is generally acknowledged.

odorous, and rotting wall cavities; and stained, blistered, and peeling wall paint, both inside and outside the home.

In some cases, the buildup of ice weight can cause structural damage to the roof. Other times there is a snow or ice fall from the roof, damaging gutters, facia and shrubbery, and creating other problems. (See Fig. 10.8.)

A Growing Problem

Ice-dam problems have become more frequent in recent years because of present construction methods and standards along with concern over energy conservation. Typical American houses built before 1930 were constructed with relatively steeply pitched roofs, with "open" (spaced) sheathing, and mostly with wood shingles. These shingles had gaps between them which served to ventilate attics and to cool the roofdeck. In addition, most pre-'30's houses had no insulation in either ceilings or walls because there was little interest in conserving fuel.

On these older homes, the excessive attic heat supported rapid flows of snow-water downroof to the cooler eave projection and beyond. When air temperatures were less than extremely cold and attic heat was in plentiful supply, rapid melting of roof-snow quickly cleared the roof of snow—with only a gutterful or eave-line of ice and icicles. These houses flowed comparatively rapidly and copiously with snow-water; their roofs cleared fairly rapidly; snow-water penetrations were of comparatively short duration; and massive eave-ice usually was all that remained of the roof-snow until the next snowfall.

Because of concern for energy conservation, modern homes and pre-'30's homes which have been "modernized" now must meet certain standards, codes, and recommendations for insulation and ventilation. However, these standards frequently represent minimum requirements, and as a result, warm, if not hot, attics are an undesirable characteristic of millions of American homes. Roof-snow blanketing on modern houses, in contrast with older homes, melts slowly;

beneath the cover water pools may threaten penetration in the house for periods of weeks or months.

Solution—A Cold Roof

Though a number of factors enter into the formation of ice-dams, the fundamental problem is undesirable attic heat which results in a warm roof surface. The solution is to maintain a cold roof.

The effectiveness of cold-surface snow behavior to avoid massive ice formations is exhibited everywhere in winterized snow country. There are no massive ice formations on felled timber, rock outcrops, woodpiles, picnic tables, sheds, and unheated buildings. In homes, continuous removal of attic air to maintain the roofdeck at atmospheric air temperatures requires minimizing the amount of heat entering the attic, together with free movement of air from eave inlets to uproof (ridge) outlets. In short, it means insulating room ceilings far more than customary to minimize heat losses and attic temperatures, and ventilating profusely at all eaves and ridge for a natural flow of air to sweep out the warmed attic air.

It is difficult to stipulate the amount of insulation which will be sufficient to adequately contain waste heat because of the innumerable variables in residential designs and requirements. In general, though, a home is not adequately protected until insulation approaches 10 to 12 inches or an R-value of 38.

It is important to emphasize that adequate insulation alone will not prevent ice-dams; it must be done in conjunction with adequate ventilation.

Ventilation Is Important

When it comes to preventing ice-dams, it is impossible to have too much ventilation. Ventilation should be equally distributed between inlet and outlet areas. Overall, the goal is to have a sweep of cooler outside air entering through the soffit which moves along the ridge rafters and exits through a ridge vent near the peak of the house.

There are other approaches to ventilation, including individual soffit vents, individual roof louvers, and power vents. In general, however, these do not perform as well as the continuous ventilation method.

Eliminating a Building Code Requirement

The problems caused by ice damming have been recognized in the universal building code, which requires an additional course of roofing on homes built in regions where ice-dams may form.

The requirement is as follows: When the roof has a slope of $4/12$ or more and an overhang of less than 36 inches, a course of either 90-lb mineral surfaced roll roofing or 50-lb smooth roll roofing must be installed over the single underlay, extending from the edge of the roof to a minimum of 12 inches beyond the interior face of the wall. For slopes less than $4/12$ and with any amount of overhang, the extra course must extend 24 inches beyond the interior wall face and shall be double and cemented.

This is a substantial additional expense that can be avoided with a cold roof system of ventilation. Builders can obtain a code variance eliminating the requirement, based upon their use of ridge and soffit venting to assure cold-roof performance throughout the winter months, preventing the ice-dam formation.

10.2   HOW VENTILATION WORKS

What the Wind Does

There are two natural forces that can move air through an attic. They are the pressure caused by wind striking against the building and the thermal effect of warm air rising within the attic.

There is also an artificial force that can move the air—a ventilating fan powered by electricity.

Of the three forces, *the wind is the most constant and is often the strongest.* As it moves against and around the residential structure it creates areas of positive and negative pressure. When these are known and understood, intake vents can be placed in areas of positive pressure and exhaust vents placed in areas of negative pressure in order to assure a continuous air flow.

Wind Force

The force of the wind blowing against the side or end of a house creates a positive pressure area at the point of contact. It then flares out, jumping in a vertical or horizontal direction—or both—depending on the configuration of the building at the point of impact. Within this "jump" area a negative pressure is created. (See Fig. 10.9.)

Wind Direction

While the wind appears to maintain a general direction, such as easterly or westerly, it does in fact shift direction frequently, often by as much as 20 or 30 degrees. This shift in direction can change a positive pressure area on a residential structure to a negative pressure area, or vice versa. (See Fig. 10.10.) A vent that was an intake can, a few minutes later, become an exhaust vent. An exhaust vent can become an intake vent, with the undesirable effect of permitting rain and snow infiltration—or reducing ventilation to only a small portion of the attic.

Locating exhaust venting at that portion of the attic structure that is in a negative pressure area *regardless of wind direction* is the first objective for a ventilation system that will be continuously effective.

Thermal Effect

In addition to wind movement, there is another natural force which affects attic ventilation. It is the thermal effect, caused by the well known principle that "warm air rises." Air in the attic is usually

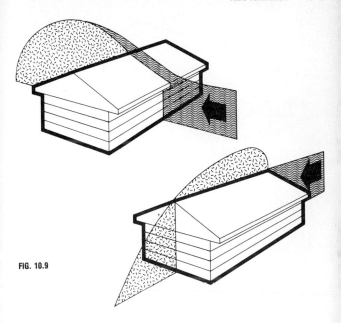

FIG. 10.9

warmer than the outside air—during daytime hours—and tends to rise and exhaust at the ridge if there is venting there to allow it. (See Fig. 10.11.)

Also, the air within the attic circulates by convection, warm air rising to the top and cooler air sinking to the attic floor. Exhaust venting at the ridge and intake venting at the soffit assists in this natural movement.

Thermal effect is not a major force in a natural ventilation system, however. The wind effect is far greater. Even a small breeze provides

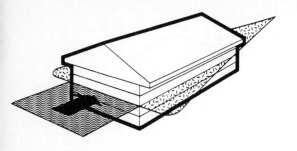

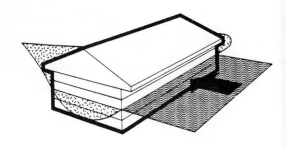

OUTSIDE WIND DIRECTION

INSIDE AIR FLOW

FIG. 10.10

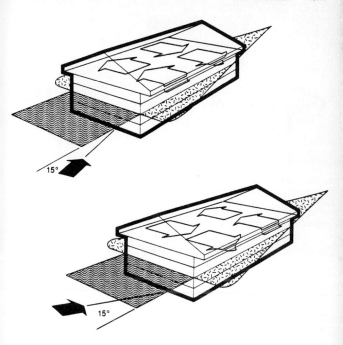

15°

15°

   POSITIVE AIR PRESSURE

   NEGATIVE AIR PRESSURE

   DEAD AIR SPACE

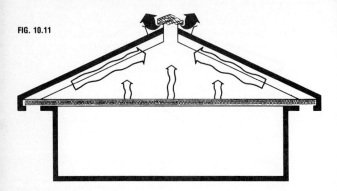

FIG. 10.11

more air movement through a correctly vented attic than thermal effect.

In winter, especially, today's heavier ceiling insulation greatly diminishes heat loss to the attic, resulting in attic air and outside air being nearly the same temperature, for minimal thermal effect.

The Inertia of Air

An extremely important factor in maintaining good attic ventilation is the inertia of moving air. Air in motion acquires a momentum, similar to moving water or any moving object, which tends to keep it moving in the same direction.

A correctly designed attic ventilation system keeps air moving in the same direction. This makes the inertia of air work to maintain a constant air flow. In effect, it provides more ventilation with less wind.

Conversely, air not in motion tends to remain so and requires more force to begin movement than it does to maintain movement once it has started.

Worse, if vents are placed so that, as the wind changes direction, the air flow through the attic is reversed, very little ventilation will take place. If every few minutes an intake vent becomes an exhaust vent, and vice versa, there is no effective and continuing air flow but only a back-and-forth pulsating movement with little air entering or leaving the attic.

The inertia of air is best utilized when intake vents remain intake vents and exhaust vents remain exhaust vents, regardless of wind force or direction.

High and Low Balance

Specifications for an attic ventilation system often include the requirement that 50% of the vent area be located high in the attic and 50% low. This is intended for balance, so that there will be equal vent areas for air flow into and out of the attic.

It is better to specify that the net free area of vents located in a positive wind pressure area be approximately equal to the net free area of vents located in a negative pressure area. (See Fig. 10.12.) Only in a ridge vent system are these positions always "low" for intake and "high" for exhaust.

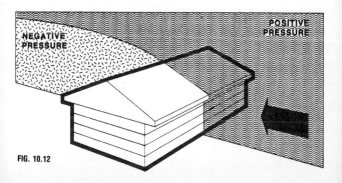

FIG. 10.12

If a ventilation system is not balanced in this manner, the effective ventilation rate is reduced to the air flow through the smaller of the two vent areas.

Underside of Roof Sheathing

While an attic ventilation system may move air throughout all parts of the attic, there is actually only one area that needs to be bathed with moving air—the underside of the roof sheathing. (See Fig. 10.13.)

Since heat buildup is caused by a radiation effect from the roof sheathing downward to the attic floor, or to the top of the insulation, cooling the roof sheathing effectively cancels this effect and maintains a desirable attic floor temperature.

The winter problem of condensation is also solved by ventilating the underside of the roof sheathing. The roof is the coldest structure in the attic, being in contact with outside air. Water vapor readily condenses there, and by ventilating all parts of the undersurface of the roof, vapor combines with the cold, dry air and is carried out of the attic before it can condense and cause problems.

A ventilation system that brings outside air across the attic floor,

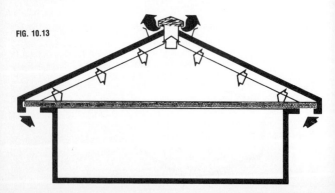

FIG. 10.13

or top of ceiling insulation, rather than across the roof sheathing, not only cannot prevent condensation but removes heat from the insulation, adding to heat loss from the residence. It is far better to let the insulation remain as warm as possible and let the roof sheathing be cold, but well ventilated.

How Rain and Snow Get In

Another sometimes troublesome source of moisture in the attic is the infiltration of rain and snow through vents. It occurs when intake air enters through the vent, and if wind conditions cause air to enter at high velocity, rain or snow may enter with it in spite of louvers designed to prevent infiltration.

The quantity of this infiltration can be sufficient, at times, to dampen ceiling insulation and even leak through to the ceiling below.

The problem is eliminated or at least minimized when the ventilation system is designed so that air intake is confined to those vents located where rain or snow infiltration is unlikely—such as in the soffit. Also, such vents should be placed adjacent to—or even outboard of—the facia. This provides for any moisture infiltration to either drain back out through the vent or be confined to the soffit, away from insulation, where it can evaporate without causing damage.

Figure 10.14 shows accepted application of continuous soffit vent in conventional construction. Figures 10.15 and 10.16 illustrate a method of combining a drip edge and intake venting on either a soffitless or soffited roof. Figure 10.17 shows placement of regularly spaced eave vents in the soffit.

10.3   TYPES OF ATTIC VENTILATION

Soffit Vents

A common method of ventilating attic space is with soffit vents, usually but not always in combination with other vents in the roof or

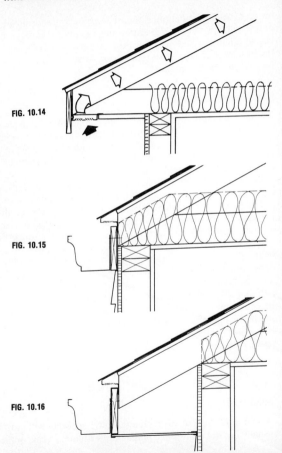

FIG. 10.14

FIG. 10.15

FIG. 10.16

FIG. 10.17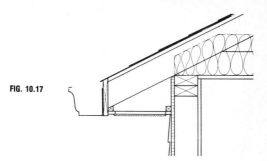

gable ends. The soffit is an excellent location for venting, particularly for intake venting, because it is less exposed to rain and snow. Also, the wind is parallel to the vent regardless of direction. With soffit venting on both sides of the structure, there is always an equal amount of venting in a positive pressure area as there is in a negative pressure area, regardless of wind direction. Thus a soffit-vent-only system is always in balance.

The problem with a soffit-vent-only system is that ventilation is confined to the attic floor. There is no air movement into the upper part of the attic, and consequently very little moisture removal. There is no air movement due to thermal effect. There is virtually no air movement across the underside of the roof sheathing. (See Fig. 10.18.)

Roof Vents—Alone and Combined with Soffit Vents

Roof vents, while they can be installed near the ridge and thus provide an escape route for overheated air, have a number of disadvantages. If used with no other type of venting, the pattern of air circulation is very small, confined to the space immediately surrounding the vent.

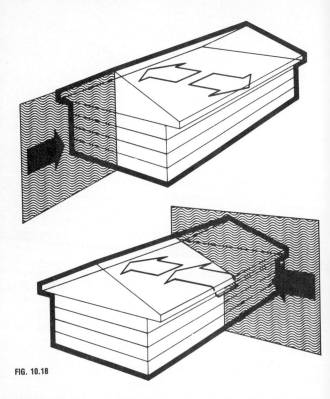

FIG. 10.18

(See Fig. 10.19.) Air will enter through some vents and exhaust through others, depending on wind pressure areas on the roof. Weather infiltration can be a problem under these circumstances.

The usual installation combines roof vents with soffit vents. The

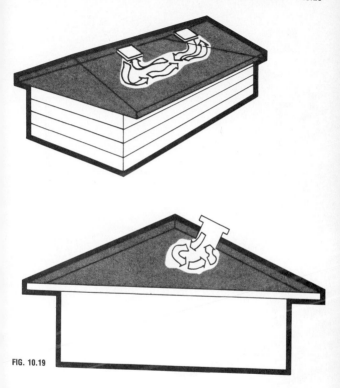

FIG. 10.19

air flow pattern is not uniform (see Fig. 10.20), with large portions of the roof sheathing unventilated. Also, since roof vents provide only 40 to 80 in² of net free area apiece, it is virtually impossible to have enough roof vents to balance the soffit vents. Air usually enters

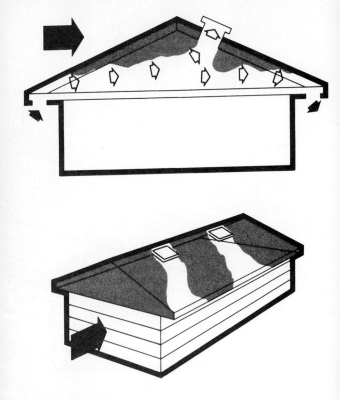

FIG. 10.20

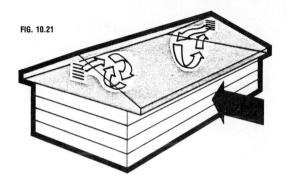

FIG. 10.21

one soffit and flows out the other soffit, with some finding its way upward to the roof vents.

Gable and End Wall Louvers—Alone and Combined with Soffit Vents

The gable or end wall louver, whether triangular or rectangular in shape, offers a ventilation air pattern as limited as the roof vent. When there are louvers in each gable—and no soffit venting—the ventilated area depends upon wind direction. Figure 10.21 shows what happens when the wind is perpendicular to the ridge. Each louver acts as both intake and exhaust, and only the ends of the attic receive any air circulation. Figure 10.22 shows the flow of air when the wind is parallel to the ridge—in at one louver, dipping down to the attic floor in the middle, then up and out the other louver. The roof sheathing is almost entirely unventilated.

With soffit vents added, the air flow is changed in that there is movement across the attic floor—similarly to the pattern of soffit vents used alone. As Figs. 10.23 and 10.24 indicate, gable end louvers

FIG. 10.22

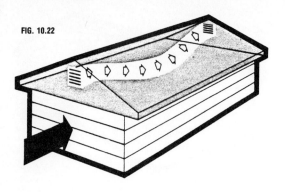

FIG. 10.23

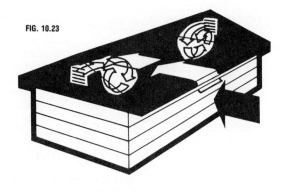

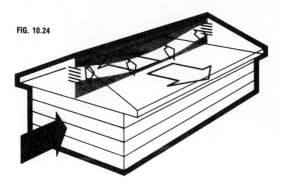

FIG. 10.24

and soffit vents largely operate independently of each other instead of combining into an effective system.

Turbine Vents

A roof vent with a turbine wheel mounted on top of it continues to act as a roof vent, with the same pattern of air flow as shown in Figs. 10.19 and 10.20. At low wind speeds, when the movement of ventilation air into and out of the attic is most needed, the turbine vent is no better or worse than an ordinary roof vent. It has been shown in tests that air may actually enter through the turbine vent, depending on wind pressure conditions. Air, rain, and snow may enter, even with the turbine wheel turning, just as it would with a regular roof louver.

Power Fans

Vents in the roof or gable ends employing a powered fan to move air out of the attic are of questionable value. In the first place, they

FIG. 10.25

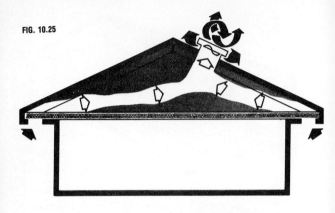

FIG. 10.26

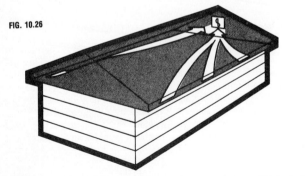

cannot save energy. The power required to operate the fan exceeds the power saved in air conditioning the residence, compared to any fixed louver ventilation system meeting FHA Minimum Property Standards. Depending on their location, air moves in a direct path or shaft from the intake vent—such as in the soffits—to the fan, creating a narrow, restricted pattern. (See Figs. 10.25 and 10.26.)

If the power fan is controlled by a thermostat switch, there is no fan operation during the winter months when air circulation over the roof sheathing is needed in order to prevent moisture condensation. Thus a humidistat must be added to the installation in order to effect winter ventilation, which in any event occurs in an ineffective pattern. The need for electrical installation adds to the high initial cost of a powered vent, with maintenance and repair costs also a factor.

Ridge Vent Combined with Soffitt Vents

There is one attic ventilation system that effectively uses the two natural forces of wind pressure and thermal effect to continuously and uniformly ventilate the entire underside of the roof sheathing. It is the combination of a continuous ridge vent and an equal net free area of soffit venting, half of it on each side of the structure.

Figure 10.27 shows the air flow pattern of a ridge vent system. Intake is at the soffit. Air moves along the roof sheathing and exhausts at the ridge. *This occurs regardless of wind directions*. Since air flow

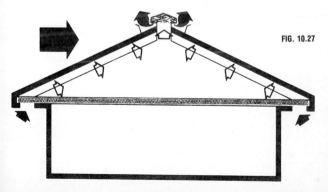

FIG. 10.27

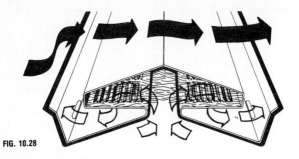

FIG. 10.28

is always in the same direction, it creates an inertia of constant air movement that is a principal reason a ridge vent system is so efficient. Figure 10.28 shows the importance of the baffle on AV Ridge Filtervent.™ When wind is perpendicular to the ridge, or coming at an angle, it strikes the baffle and jumps over the top of the ridge. A venturi action is created, causing both the windward and lee sides of the ridge vent to be in a negative pressure area. This pulls air out of the attic space.

As illustrated in Fig. 10.9, if the wind is coming against the end of the house it jumps upward and over the ridge, causing a negative pressure area. Thus, regardless of wind direction, ridge vent is always in negative pressure and always acts as an exhaust vent.

When there is little or no wind, thermal effect acts to maintain air circulation across the underside of the roof sheathing. Warm air rises to the ridge and exhausts through the ridge vent and is replaced by cooler air entering at the soffit. No other attic ventilation method utilizes thermal effect as well as a ridge vent system.

A ridge vent system brings a uniform flow of air across the underside of the roof sheathing for the entire length of the attic, leaving no unventilated areas of the sheathing. *No other system provides this type of air flow pattern.*

Contemporary Roof Designs

Certain residential designs create roofs without a traditional ridge, the top of the roof meeting a vertical wall either at its peak or at some point prior to the peak. (See Fig. 10.29.) Products are available to provide continuous ventilation along the top of the roof in these cases and consist of one-half of a ridge vent in a configuration to fit the roof.

How Vent Systems Compare in Effective Air Flow

The preceding descriptions cover the general air flow *patterns* of the main types of attic vents. The other important consideration in judging a ventilation system is the *quantity* of air the vent system moves through the attic space under given wind conditions.

Comprehensive tests have been conducted by university labora-

FIG. 10.29

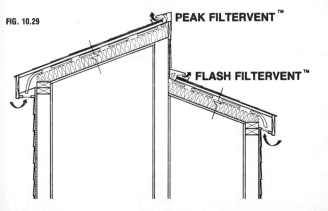

PEAK FILTERVENT™

FLASH FILTERVENT™

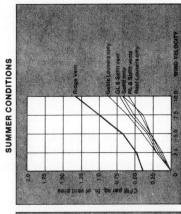

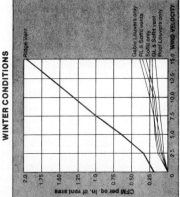

FIG. 10.30

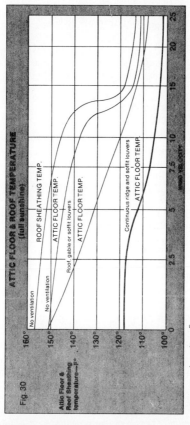

ATTIC FLOOR & ROOF TEMPERATURE
(full sunshine)

Fig. 30

No ventilation — ROOF SHEATHING TEMP.

No ventilation — ATTIC FLOOR TEMP.

Roof gable or soffit louvers — ATTIC FLOOR TEMP.

Continuous ridge and soffit louvers — ATTIC FLOOR TEMP.

Attic floor & Roof Sheathing temperature—°F

WIND VEL. M.P.H.

FIG. 10.30 (*continued*)

tories and private industry measuring the cubic feet of air per minute (CFM) moving into and out of an attic for each of the vent systems shown in the three graphs in Fig. 10.30. Test results are expressed as CFM per square inch of vent area and are shown for varying wind velocities. Both winter and summer conditions are shown.

It is readily seen that a ridge vent system far exceeds other systems in moving more effective ventilating air through the attic space.

Attic Floor Temperature Reduction with Various Vent Systems

Because there is a wide variance in the pattern of air flow and quantity of flow between a ridge vent system and all other systems, it is logical to expect that there will be a corresponding difference in the reduction of attic floor temperatures.

Test results bear this out, as shown in Fig. 10.30. Note that with wind velocity at zero, *only the ridge vent system reduces* attic floor temperature from the 150°+ condition when there is no wind. This is due to thermal effect, the warm air rising along the underside of the roof sheath and exhausting at the ridge, being replaced by lower temperature air at the soffit.

Also, at all wind velocities *the ridge vent system maintains a lower attic floor temperature than other systems,* even when wind exceeds 25 mph.

10.4   HOW MUCH ATTIC VENT IS REQUIRED?

In past years there was little knowledge available to help builders determine how much vent area to provide in a residential attic. Builders tended to select one or another type of vent and installed it without aiming for a specific amount of moisture removal or a specific temperature reduction. They were just providing "ventilation."

FHA Minimum Property Standards

So limited has been the information on and concern for proper vent area that the Federal Housing Administration's Minimum Property Standards for one and two living units states:

> Attics and spaces between roof and top floor ceiling may have a free ventilating area of ⅟₃₀₀ of the horizontal area when (a) a vapor barrier having a transmission rate not exceeding one perm is installed on the warm side of the ceiling or (b) at least 50 per cent of the required ventilating area is provided with fixed louvers located in the upper portion of the space to be ventilated (at least 3 ft above eave or cornice vents) with the remainder of the required ventilation provided by eave or cornice vents.

This ratio of ⅟₃₀₀ translates into slightly less than one half square inch of vent area for each square foot of attic floor. It is barely adequate and presumes that the vent system is properly balanced, with equal net free area in the high vents and the low vents. Frequently, however, this is not the case. There are many homes being built today with no consideration for this balance, and, in fact, even less total vent net free area than the standards require.

With today's greater insulation requirements and a continuing need for energy conservation, both the total amount and placement of attic venting is critically important.

Recommended Rate of Air Flow

Before determining how much vent area is required for a given attic space it is necessary to have a specific rate of air flow as an objective. Tests at the University of Illinois revealed that attic temperature is reduced by 44.5% when air flow is 1.5 ft³/min per square foot of attic floor area. This will provide a floor temperature reduction of 24°F. This reduction is true regardless of ambient air temperature and was true on all of the tests, with ambient air temperature ranging from 85 to 110°F. The university reported that some further reduction of temperature occurs as air flow is increased to 2.0 ft³/min per square foot of floor area, but it is clear from Fig. 10.31 that most of the removable heat is gone once the ventilation rate reaches 1.5 ft³/min per square foot.

A ridge vent system (ridge vent plus soffit vents) provides this rate of air flow when there is 1.5 in² of net free area per square foot of attic floor. No other vent system can do this.

Summer Requirement vs. Winter Requirement

While it is true that the summer requirement for a ventilation air flow sufficient to prevent heat buildup is greater than the winter requirement to remove moisture, there is an increasing need for higher air flow rates in winter months. The electric utility industry is recommending ceiling insulation factors as high as R-36, and with heavier insulation the attic air is kept at a lower temperature. This lower temperature air absorbs less water vapor, so that there is a greater tendency for moisture to condense on the roof sheathing, rafters and other structural parts.

In fact, attic temperature may even be low enough to cause escaping residential moisture to freeze in the upper portion of ceiling insulation, thawing at a later time to dampen the insulation.

For these reasons, a good vapor barrier is recommended in order to minimize moisture migration to the insulation and the attic space, along with effective attic ventilation to remove water vapor contained in the air. Rate of water vapor removed is shown in Fig. 10.32.

Builders should use a vent system which will remove at least 3.0 to 4.0 grains of moisture per hour, when using R-19 or heavier ceiling insulation. That requires a ventilation rate of 0.5 ft³/min per square foot of attic floor area. This rate can be accomplished with a ridge vent system that provides 1 square inch of net free area per square foot of attic floor. (This is double the FHA Minimum Property Standard.)

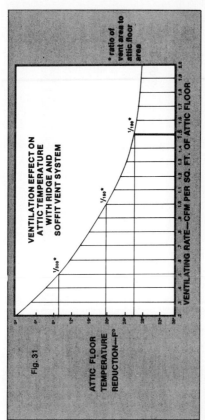

Fig. 31

VENTILATION EFFECT ON
ATTIC TEMPERATURE
WITH RIDGE AND
SOFFIT VENT SYSTEM

ATTIC FLOOR
TEMPERATURE
REDUCTION—F°

VENTILATING RATE—CFM PER SQ. FT. OF ATTIC FLOOR

* ratio of
vent area to
attic floor
area

FIG. 10.31

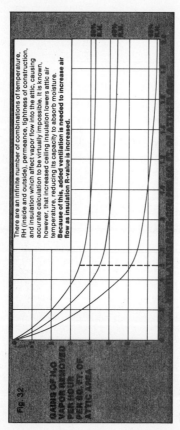

There are an infinite number of combinations of temperature, RH (inside and outside), permeance, tightness of construction, and insulation which affect vapor flow into the attic, causing accurate calculation to be virtually impossible. It is known, however, that increased ceiling insulation lowers attic air temperature, reducing its capacity to absorb moisture. **Because of this, added ventilation is needed to increase air flow as insulation R-value is increased.**

Fig. 32

GARS OF H₂O
VAPOR REMOVED
PER HOUR
PER SQ. FT. OF
ATTIC AREA

FIG. 10.32

However, if good summer heat removal as well as good winter moisture removal is desired, a ridge vent system providing 1.5 in^2 of net free area per square foot of attic floor should be used and will furnish 1.5 ft^3/min of air flow—ample for both summer and winter conditions.

Required Vent Area for Various Systems

It was illustrated in graphs earlier that there is a difference in the rate of air flow through various vent systems. Using this data, Table 10.1 has been prepared showing the amount of net free area each vent system must have in order to provide both summer and winter air flow. Note that less is required with a ridge vent system. In some cases, other systems require so much net free area that installation of that much venting is virtually impossible.

Ridge-and-Soffit System Meets All Needs

The clear conclusion drawn from the foregoing information is that the most cost-effective method of meeting code requirements is with a ridge vent and soffit vent system. At the same time it provides the ventilation system that removes the most heat from the attic space in summer and the most moisture in winter.

10.5 VENTILATING FOR THE WHOLE HOUSE FAN

Formerly a popular cooling device before central air conditioning was commonplace, the large ceiling-mounted fan (Fig. 10.33) is again becoming popular as a viable alternative to air conditioning, particularly on summer days when heat and humidity are less than maximum. The fan reduces energy cost, providing a cooling method that

TABLE 10.1   Net Free Vent Area Required for Summer and Winter Air Flow

VENT AREA REQUIRED	SUMMER 1.5 CFM per sq. ft. attic area Sq. in. of net free vent area per sq. ft. of attic floor area	WINTER 0.5 CFM per sq. ft. attic area Sq. in. of net free vent area per sq. ft. of attic floor area
VENT SYSTEM		
Roof Louvers	3.3	11.1
Soffit Louvers	2.9	6.5
Gable Louvers	2.4	5.1
Roof Louvers and Soffit Vents	3.1	6.0
Gable Louvers and Soffit Vents	2.7	6.0
Ridge Flitervent™ and Soffit Vents	1.5	1.0

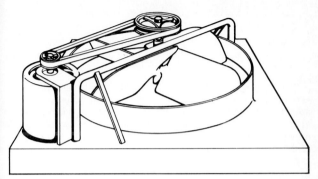

FIG. 10.33

uses approximately ¹⁄₁₀ the power requirement of a 3-ton central air conditioning system.

How It Cools

A whole house fan discharges air from the living space into the attic, cooling the home two ways: (1) When outside air is cooler than inside air, the fan draws the cooler air into the house to provide immediate relief to occupants and reduce inside air temperature. (2) The flow of air removes heat from the walls and objects in the home.

These cooling opportunities usually occur at night and in the early morning hours, often reducing temperatures enough so that daytime air conditioning is not required.

Two Essential Considerations

In designing a whole house fan installation, there are two essential considerations:

1. Fan size; i.e., how much air must be circulated to cool the house.
2. Exhaust ventilation; i.e., how much and what type of attic ventilation must be present to handle the exhaust capacity of the fan.

Item No. 1 above depends upon climate and user preference. In very hot and humid conditions, a net change of air once a minute appears to be satisfactory. In less extreme conditions, such as in more northern states, an air change every two minutes provides ample comfort. (It should be noted that some experts believe that only half of this air movement is sufficient.)

Item No. 2 relates directly to achieving the rated air movement of the fan and will be covered later in this section.

Determining Correct Fan Size

The following example illustrates how to determine correct fan size.

Assuming a house with a floor area of 1250 ft^2 and standard 8-ft. ceiling heights, which provides a total volume of 10,000 ft^3. To produce an air flow equivalent to one complete air change every two minutes, the requirement would be a fan of 5,000-ft^3/min capacity. The formula is:

$$\frac{\text{Volume of home in ft}^3}{\text{No. of air changes desired per min}} = \text{fan capacity (ft}^3\text{/min)}$$

Two factors which may influence the fan capacity decision are: (a) fan noise level, and (b) allowance for unavoidable dead air space in the home. Either or both of these factors could reduce the CFM requirement for the fan.

Whole house fans are available from a large number of manufacturers in a range of sizes and capacities. Table 10.2 lists some typical house sizes and the correct fan size for each, to achieve a net change of air of once a minute. In this table fan capacity is shown in cubic

TABLE 10.2 Fan Sizing for Houses

House area, ft²	House volume*, ft³	Fan size, ft³/min	Blade size (dia.), in	NFAVA**, ft²
750	6000	3000	20	4.00
1000	8000	4000	24	5.33
1250	10,000	5000	24	6.66
1500	12,000	6000	30	8.00
1750	14,000	7000	30	9.33
2000	16,000	8000	36	10.66

*8-ft ceiling height assumed.
**Net Free Attic Ventilation Area required to accommodate whole house fan exhaust.

feet of air flow per minute against an assumed static pressure of 0.1 inch of water pressure. This is the industry standard and represents the normal resistance encountered when there is unrestricted air flow out of the attic space. The final column of the table shows the net free area of attic ventilation required if the rated fan capacity is to be achieved.

Why Sufficient Attic Ventilation Is Critical

Often overlooked in whole house fan installations is the provision of sufficient net free area of attic ventilation to permit the fan to operate at its rated capacity. The three- or four-bladed whole house fan is designed for moving a high volume of air against a very low back pressure. As stated earlier, the standard is 0.1 inch of static water pressure, and as back pressure increases beyond this amount, the fan volume capacity drops sharply.

In the curve shown in Fig. 10.34 the fan has a rated capacity of 5375 ft³/min at 0.1 inches of water pressure. (0.1 inch of water is the equivalent of x lbs/in², which is a very small force.) If the back pressure of this particular fan is increased to 0.2 inches of water (i.e.,

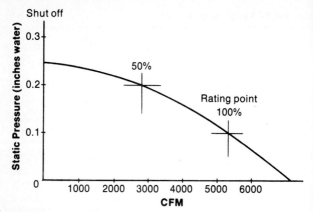

FIG. 10.34 Typical fan curve.

$2 \times$ lbs/in^2), the fan capacity is reduced 50%; and if the back pressure is 0.25 inches of water (i.e., $2.5 \times$ lbs/in^2), the fan ceases to move air.

Whole house fan energy saving calculations are based on having the fan operate at rated capacity. From the above it is easy to see that even a small amount of back pressure eliminates the effectiveness of a whole house fan. Back pressure comes principally from inadequate exhaust venting—inadequate in both amount and placement.

The static pressure of 0.1 inch of water pressure is equivalent to the resistance caused by attic vents and the turning of the air stream as the air flows through the attic. *Forcing a fan to operate against a pressure higher than its rated capacity not only reduces the volume of air flow but can create excessive fan noise and rapid deterioration of the fan motor due to overheating.* Operation can be interrupted due to circuit breaker tripping, or the motor heater controls will cause erratic operation.

Determining NFAVA

To determine the necessary net free attic ventilation area required for a whole house fan installation, the simple and recommended method is to use the industry standard rule of allowing approximately 1.33 ft^2 of net free attic vent area for every 1,000 ft^3/min of fan capacity. Following this rule, the required NFAVA for various fan sizes is shown in Table 10.2.

Which Type of Attic Ventilation is Best?

Very few homes have enough net free area of attic ventilation to satisfy the requirements for a whole house fan. In almost all cases, attic venting must be added. For this, the *type* of attic vents selected is an important consideration.

Because their design dictates their location in the attic, vents are categorized into two general types: high and low vents. The best ventilation for all purposes is always a combination of high and low vents, preferably balanced so that the net free area of the high vents is equal to the net free area of the low vents. It should be recognized, however, that in many existing homes and some new ones, this balance is not possible and the required net free area must be placed in the best alternative locations.

Another consideration is the desirability of having multiple exhaust vents to minimize the effect of the wind blowing against the vent and increasing static pressure. This can be a significant factor, for a *15 mph wind creates a pressure of 0.1 inch* (*water pressure*).

If this is being applied against the *only* attic vent, it effectively doubles the resistance that the fan must work against. Whenever possible, vents should be placed so that they enhance, rather than inhibit, the operation of the fan.

Important also is the selection of attic vents that will not admit snow and rain. This occurs when the vent becomes an intake opening for outside air under certain wind conditions. An example would be gable end louvers at each end of the attic. Often air enters at one

TABLE.10.3   Attic Ventilators

Type of vent	Net free attic vent area, ft^2
High Vents	
Ridge Filtervent™*	
One 8-ft piece	1.00
Typical Installations:	
24 ft (3 pieces)	3.00
32 ft (4 pieces)	4.00
40 ft (5 pieces)	5.00
Roof Louvers	
One roof louver	0.35
Typical Installations:	
2 roof louvers	0.70
3 roof louvers	1.05
4 roof louvers	1.40
5 roof louvers	1.75
Turbine Vent	
One turbine vent	0.78
Typical Installation	
2 turbine vents	1.56
Rectangular Gable Vents	
Size	
14 in × 24 in	1.09
18 in × 24 in	1.42
24 in × 30 in	2.42
Triangular Gable Vents	
Base Length	
30 in	0.57
48 in	1.00
72 in	1.09
96 in	1.91
120 in	2.89

*Ridge Filtervent is a patented product of Air Vent Inc.

TABLE 10.3 Attic Ventilators (continued)

Type of vent	Net free attic vent area, ft²
Low Vents	
Unscreened AVI Eave Vent 16 in × 8 in	0.5
Screened Eave Vents 16 in × 8 in 16 in × 4 in	0.36 0.15
Strip Vent and Drip Edge Vent One 8-ft piece	0.5
Perforated Aluminum Soffit 1 ft²	0.1
Lanced Aluminum Soffit 1 ft³	0.025–0.05

louver and exits at the other. Another example would be a residence without soffit venting and with several high vents in the gables or roof. Air will exhaust through one or more high vents and enter through other high vents, depending on vent placement.

Table 10.3 lists the types of attic ventilators currently being marketed, along with their net free vent areas.

Table 10.3 clearly shows that the most practical and effective method of providing increased net free attic ventilation area is with Ridge Filtervent™. Not only does Ridge Filtervent™ provide the necessary net opening at the lowest cost, but it is the only vent system that *always* aids attic exhaust by utilizing the principle of warm air rising and the lowering of air pressure immediately above the ridge as air flow is lifted by the baffles, creating a venturi effect. Thus Ridge Filtervent™ works with the whole house fan, not against it.

Four Points to Remember

1. Whole house fans represent a cost-effective alternative to air conditioning, providing comfort at a fraction of air conditioning's operating cost.
2. The correct size fan can be determined, based on house size and climate conditions.
3. Adequate net free area of attic ventilation must be provided to assure that the fan will move its rated volume of air.
4. Lack of sufficient attic ventilation will prevent adequate cooling, increase fan noise, shorten fan life, and cancel the cost efficiency of a whole house fan system.

NOTES

NOTES

AIR DISTRIBUTION

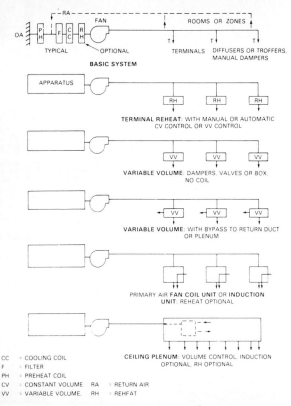

CC = COOLING COIL
F = FILTER
PH = PREHEAT COIL
CV = CONSTANT VOLUME   RA = RETURN AIR
VV = VARIABLE VOLUME.   RH = REHEAT

FIG. 11.1 Single path air systems. (*From HVAC Duct Construction Standards—Metal and Flexible, 1st ed., © 1985 by the Sheet-Metal and Air Conditioning Contractor's National Association. Used with permission of the copyright holder.*)

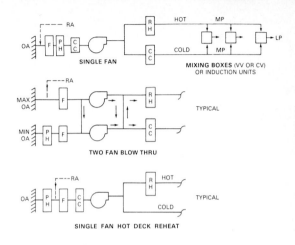

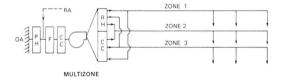

F	= FILTER	VV	= VARIABLE VOLUME
CC	= COOLING COIL	CV	= CONSTANT VOLUME
PH	= PREHEAT COIL	LP	= LOW PRESSURE DUCT
OA	= OUTSIDE AIR	MP	= MEDIUM PRESSURE OR
RH	= REHEAT COIL		HIGH PRESSURE DUCT
		RA	= RETURN AIR

FIG. 11.2 Dual-path air systems. (*From HVAC Duct Construction Standards— Metal and Flexible, 1st ed.,* © 1985 by the Sheet-Metal and Air Conditioning Contractor's National Association. Used with permission of the copyright holder.)

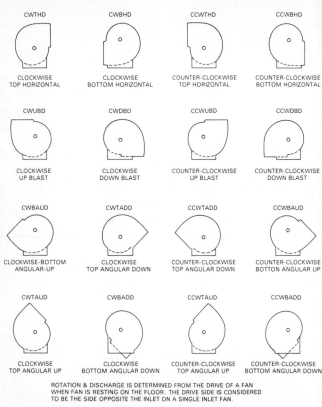

ROTATION & DISCHARGE IS DETERMINED FROM THE DRIVE OF A FAN
WHEN FAN IS RESTING ON THE FLOOR. THE DRIVE SIDE IS CONSIDERED
TO BE THE SIDE OPPOSITE THE INLET ON A SINGLE INLET FAN.

FIG. 11.3 Fan rotation and discharge positions. (*From HVAC Duct Construction Standards—Metal and Flexible, 1st ed., © 1985 by the Sheet-Metal and Air Conditioning Contractor's National Association. Used with permission of the copyright holder.*)

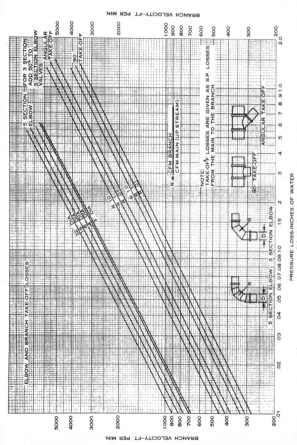

FIG. 11.4 Elbow and branch takeoff losses. (*Reproduced with permission of The Trane Company, LaCrosse, WI.*)

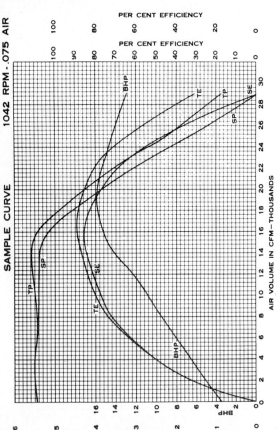

FIG. 11.5 Performance of an airfoil blade fan. (Reproduced by permission of The Trane Company, LaCrosse, WI.)

TABLE 11.1 Sound Analysis

ITEM	OCTAVE BAND CENTER FREQUENCIES						
	106	212	425	850	1700	3400	6800
1. FAN SOUND POWER LEVEL FOR 3 1/2 IN. FAN STATIC AND 13,000 CFM AT 1030 RPM, AIRFOIL TYPE FAN	99	101	95	93	94	87	81
2. REDUCTION IN FAN SOUND POWER LEVEL DUE TO THE BRANCH TAKEOFF	11	11	11	11	11	11	11
3. OCTAVE BAND POWER LEVEL AT OUTLET 1 (LINE 1 MINUS LINE 2)	88	90	84	82	83	76	70
4. NATURAL ATTENUATION							
A. 42 IN. ELBOW WITHOUT TURNING VANES	4	8	5	3	3	3	3
B. 22 FT OF 28 X 42 IN. DUCT	2	2	1	1	1	1	1
C. 11 IN. ELBOW WITHOUT TURNING VANES	0	0	4	8	5	3	3
D. 30 FT OF 11 X 11 IN. DUCT	6	4	3	3	3	3	0
E. END REFLECTION ATTENUATION	9	5	1	0	0	0	0
5. TOTAL NATURAL ATTENUATION	23	19	14	15	12	10	10
6. NET FAN SOUND POWER LEVEL AT OUTLET 1 WITHOUT SOUND TREATMENT	65	71	70	67	71	66	60
7. N.C. 40 PERMISSIBLE SOUND LEVEL	59	52	46	42	40	38	37
8. MINIMUM REQUIRED ATTENUATION	6	19	24	25	31	28	23

SOURCE: The Trane Company, LaCrosse, WI. Reproduced by permission.

TABLE 11.2   Ventilation Requirements for Various Functional Areas

Functional Areas	Ventilation Requirement
Auditoriums, churches, dance halls	4–30 air changes* per hour 10–65 cfm/occupant (5–30 lps/occupant) 1.5–2 cfm/ft² (7.6–10.2 lps/m²)
Barbershops and cafes	7.5 air changes per hour
Bedrooms	1 air change per hour
Billiard halls and bowling alleys	6–20 air changes per hour
Classrooms	
Colleges	25–40 cfm/occupant (12–19 lps/occupant)
Schools	30–40 cfm/occupant (14–19 lps/occupant) 2 cfm/ft² (10.2 lps/m²)
Corridors	4 air changes per hour ½ cfm/ft² (2.5 lps/m²)
Dining rooms	4–40 air changes per hour 1.5 cfm/ft² (7.6 lps/m²)
Garages	6–12 air changes per hour
Guest rooms	3–5 air changes per hour
Gymnasiums	12 air changes per hour 1.5 cfm/ft² (7.6 lps/m²)
Halls (residence)	1–3 air changes per hour
Kitchens	4–60 air changes per hour 2–4 cfm/ft² (10.2–20.3 lps/m²)
Laboratories	6–20 air changes per hour
Living rooms (residence)	1–2 air changes per hour
Lobbies	3–4 air changes per hour
Locker rooms	2–10 air changes per hour 2 cfm/ft² (10.2 lps/m²)
Lounges	6 air changes per hour
Mechanical rooms	3–12 air changes per hour
Operating rooms	50 cfm/occupant (24 lps/occupant)
Projection booths	30 air changes per hour 1.5 cfm/ft² (7.6 lps/m²)
Reading rooms	3–5 air changes per hour
Stores (retail)	6–12 air changes per hour 4 cfm/ft² (20.3 lps/m²)
Toilets	
Private	1–5 air changes per hour
Public	10–30 air changes per hour 2 cfm/ft² (10.2 lps/m²)
Waiting rooms	4–6 air changes per hour

*An *air change* is the infiltration of the air in a room by, or the ventilation of a room with, a volume of air equivalent to the volume of the room.

SOURCE: A. M. Khashab, *Heating, Ventilating, and Air Conditioning Systems Estimating Manual*, 2d ed., McGraw-Hill, New York, © 1984. Used with permission of the publisher.

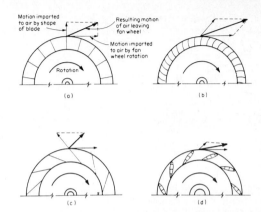

FIG. 11.6 Types of centrifugal fan blades: (*a*) radial; (*b*) forward-curved; (*c*) backward-inclined; (*d*) airfoil (forward-inclined). (*From A. M. Khashab, Heating, Ventilating, and Air Conditioning Systems Estimating Manual, 2d ed., McGraw-Hill, New York, © 1984. Used with permission of the publisher.*)

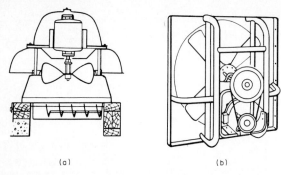

(a)   (b)

FIG. 11.7 Propeller fans: mainly for moving air from one non-air-conditioned space to another, they are normally used with ductwork; there are both roof and wall-mounted types; shown here are (a) roof type and (b) V-belt-driven. (*From A. M. Khashab, Heating, Ventilating, and Air Conditioning Systems Estimating Manual, 2d ed., McGraw-Hill, New York, © 1984. Used with permission of the publisher.*)

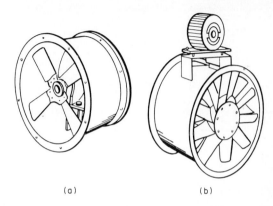

(a)   (b)

FIG. 11.8 Axial fans. (*a*) Tube-axial (direct-driven): a heavy-duty propeller fan, normally used without inlet or outlet vanes. The air is discharged in helical motion. These fans are constructed with up to eight wheel blades, which are wider than those of propeller fans. (*b*) Vane-axial: normally has a surrounding cylinder, similar to those of tube-axial design, but with air guide vanes to straighten the helical air-discharge pattern. (*From A. M. Khashab, Heating, Ventilating, and Air Conditioning Systems Estimating Manual, 2d ed., McGraw-Hill, New York, © 1984. Used with permission of the publisher.*)

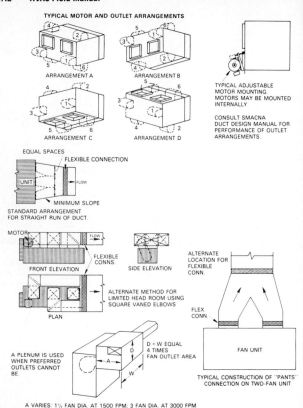

FIG. 11.9 Typical HVAC unit connections. (*From HVAC Duct Construction Standards—Metal and Flexible, 1st ed.,* © 1985 by the Sheet-Metal and Air Conditioning Contractor's National Association. Used with permission of the copyright holder.)

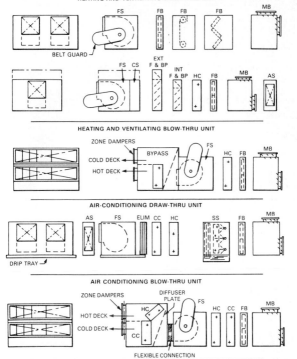

HEATING AND VENTILATING DRAW-THRU UNIT

HEATING AND VENTILATING BLOW-THRU UNIT

AIR-CONDITIONING DRAW-THRU UNIT

AIR CONDITIONING BLOW-THRU UNIT

AS	ACCESS SECTION	EXT F & BP	EXTERNAL FACE AND BYPASS DAMPER	FS	FAN SECTION
CS	COIL SECTION	INT F & BP	INTERNAL FACE AND BYPASS DAMPER	FB	FILTER BOX
CC	COOLING COIL	ELIM	ELIMINATORS	MB	MIXING BOX
HC	HEATING COIL			SS	SPRAY SECTION

FIG. 11.10 Terminology for central station apparatus. (*From HVAC Duct Construction Standards—Metal and Flexible, 1st ed., © 1985 by the Sheet-Metal and Air Conditioning Contractor's National Association. Used with permission of the copyright holder.*)

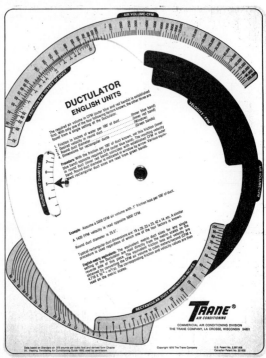

FIG. 11.11 Ductulator—USCS (English) units. The Ductulator is an accurate, easy-to-use calculating device that aids in the layout of air-handling systems, sizing of ducts, and checking of existing duct systems. With a single setting, the answers to any duct sizing problem may be read directly from the face of the Ductulator. Conversion of units, USCS to metric, or vice versa, is accomplished by setting up a problem on the scales of one unit system, reversing sides of the Ductulator, and reading the converted values from the scales of the other unit system. (*Reproduced by permission of The Trane Company, LaCrosse, WI.*)

FIG. 11.12 Ductulator—SI metric units. The reverse side of the Ductulator shown in Fig. 11.11. Its purpose and how to convert from metric to USCS units are the same as given for Fig. 11.11. (*Reproduced by permission of The Trane Company, LaCrosse, WI.*)

NOTES

AIR DUCT
DESIGN

FIG. 12.1 Ductulator. A mechanical device for calculating the sizes of ducts. For conversion from USCS units to metric and vice versa, see Fig. 11.11. (*Reproduced by permission of The Trane Company, LaCrosse, WI.*)

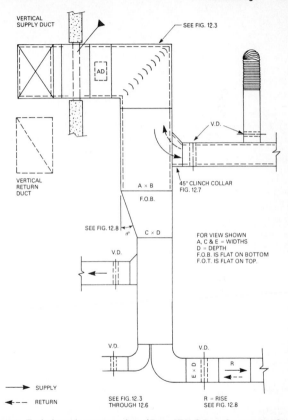

FIG. 12.2 Typical supply or return duct. (*From HVAC Duct Construction Standards—Metal and Flexible, 1st ed., © 1985 by the Sheet-Metal and Air Conditioning Contractor's National Association. Used with permission of the copyright holder.*)

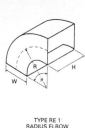

TYPE RE 1
RADIUS ELBOW
(CENTERLINE R $= \frac{3W}{2} =$ STD RADIUS)

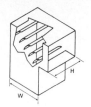

TYPE RE 2
SQUARE THROAT ELBOW
WITH VANES

TYPE RE 3
RADIUS ELBOW
WITH VANES

TYPE RE 4
SQUARE THROAT ELBOW
WITHOUT VANES
(1000 FPM MAXIMUM VELOCITY)

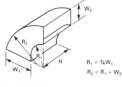

$R_1 = \frac{3}{4}W_1$
$R_2 = R_1 + W_2$

TYPE RE 5
DUAL RADIUS ELBOW

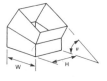

MITERED ELBOW

BEAD, CROSSBREAK AND REINFORCE FLAT SURFACES AS IN STRAIGHT DUCT

FIG. 12.3 Rectangular elbows. (*From HVAC Duct Construction Standards—Metal and Flexible, 1st ed.,* © 1985 by the Sheet-Metal and Air Conditioning Contractor's National Association. Used with permission of the copyright holder.)

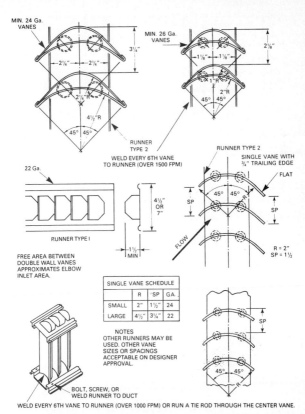

FIG. 12.4 Vanes and vane runners. (*From HVAC Duct Construction Standards— Metal and Flexible, 1st ed., © 1985 by the Sheet-Metal and Air Conditioning Contractor's National Association. Used with permission of the copyright holder.*)

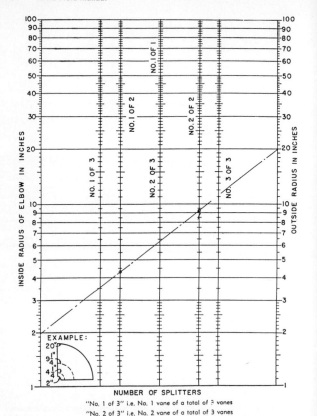

FIG. 12.5 Number of short-radius vanes. (*From HVAC Duct Construction Standards—Metal and Flexible, 1st ed., © 1985 by the Sheet-Metal and Air Conditioning Contractor's National Association. Used with permission of the copyright holder.*)

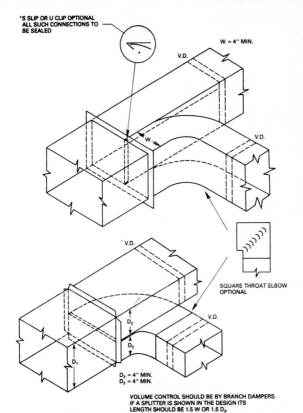

FIG. 12.6 Parallel-flow branches. (*From HVAC Duct Construction Standards—Metal and Flexible, 1st ed.,* © 1985 by the Sheet-Metal and Air Conditioning Contractor's National Association. Used with permission of the copyright holder.*)

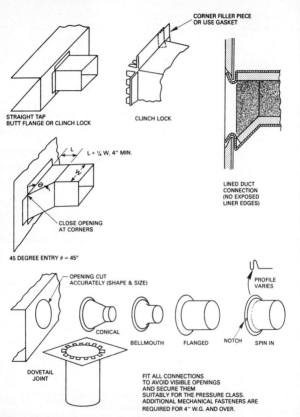

FIG. 12.7 Branch connections. (*From HVAC Duct Construction Standards—Metal and Flexible, 1st ed.,* © 1985 by the Sheet-Metal and Air Conditioning Contractor's National Association. Used with permission of the copyright holder.)

OFFSETS 2 AND 3 AND TRANSITIONS MAY HAVE EQUAL OR UNEQUAL INLET
AND OUTLET AREAS. TRANSITIONS MAY CONVERT DUCT PROFILES TO
ANY COMBINATION FOR RECTANGULAR, ROUND OR FLAT OVAL SHAPES.

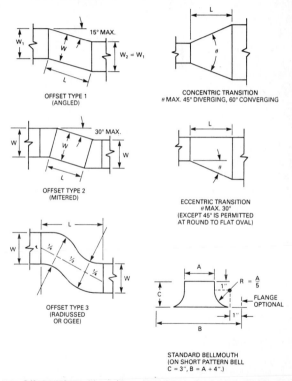

OFFSET TYPE 1
(ANGLED)

CONCENTRIC TRANSITION
θ MAX. 45° DIVERGING, 60° CONVERGING

OFFSET TYPE 2
(MITERED)

ECCENTRIC TRANSITION
θ MAX. 30°
(EXCEPT 45° IS PERMITTED
AT ROUND TO FLAT OVAL)

OFFSET TYPE 3
(RADIUSSED
OR OGEE)

STANDARD BELLMOUTH
(ON SHORT PATTERN BELL
C = 3″, B = A + 4″.)

$R = \dfrac{A}{5}$

FLANGE
OPTIONAL

FIG. 12.8 Offsets and transitions. (*From HVAC Duct Construction Standards—
Metal and Flexible, 1st ed., © 1985 by the Sheet-Metal and Air Conditioning
Contractor's National Association. Used with permission of the copyright
holder.*)

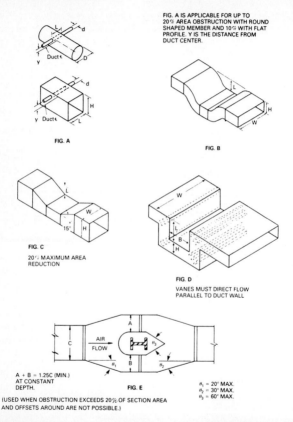

FIG. A IS APPLICABLE FOR UP TO 20% AREA OBSTRUCTION WITH ROUND SHAPED MEMBER AND 10% WITH FLAT PROFILE. Y IS THE DISTANCE FROM DUCT CENTER.

FIG. A

FIG. B

FIG. C

20% MAXIMUM AREA REDUCTION

FIG. D

VANES MUST DIRECT FLOW PARALLEL TO DUCT WALL

A + B = 1.25C (MIN.) AT CONSTANT DEPTH.

AIR FLOW

FIG. E

θ_1 = 20° MAX.
θ_2 = 30° MAX.
θ_3 = 60° MAX.

(USED WHEN OBSTRUCTION EXCEEDS 20% OF SECTION AREA AND OFFSETS AROUND ARE NOT POSSIBLE.)

FIG. 12.9 Obstructions. (*From HVAC Duct Construction Standards—Metal and Flexible, 1st ed.,* © 1985 by the Sheet-Metal and Air Conditioning Contractor's National Association. Used with permission of the copyright holder.)

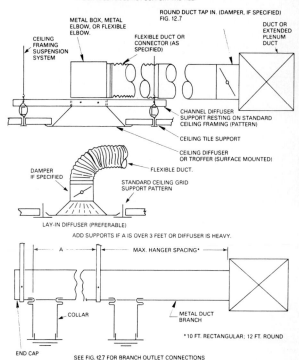

THE CEILING SUPPORT SYSTEM MUST SUPPORT DIFFUSER
WEIGHT WHEN FLEXIBLE CONNECTIONS ARE USED!
A PROPERLY SIZED HOLE IS PROVIDED IN THE CEILING TILE.
THE DIFFUSER DOES NOT SUPPORT THE TILE.

ROUND DUCT TAP IN. (DAMPER, IF SPECIFIED)
FIG. 12.7

METAL BOX, METAL
ELBOW, OR FLEXIBLE
ELBOW.

DUCT OR
EXTENDED
PLENUM
DUCT

CEILING
FRAMING
SUSPENSION
SYSTEM

FLEXIBLE DUCT OR
CONNECTOR (AS
SPECIFIED)

CHANNEL DIFFUSER
SUPPORT RESTING ON STANDARD
CEILING FRAMING (PATTERN)

CEILING TILE SUPPORT

CEILING DIFFUSER
OR TROFFER (SURFACE MOUNTED)

DAMPER
IF SPECIFIED

FLEXIBLE DUCT.

STANDARD CEILING GRID
SUPPORT PATTERN

LAY-IN DIFFUSER (PREFERABLE)

ADD SUPPORTS IF A IS OVER 3 FEET OR DIFFUSER IS HEAVY.

A

MAX. HANGER SPACING*

COLLAR

METAL DUCT
BRANCH

*10 FT. RECTANGULAR; 12 FT. ROUND

END CAP

SEE FIG. 12.7 FOR BRANCH OUTLET CONNECTIONS

FIG. 12.10 Ceiling diffuser branch ducts. (*From HVAC Duct Construction Standards—Metal and Flexible, 1st ed., © 1985 by the Sheet-Metal and Air Conditioning Contractor's National Association. Used with permission of the copyright holder.*)

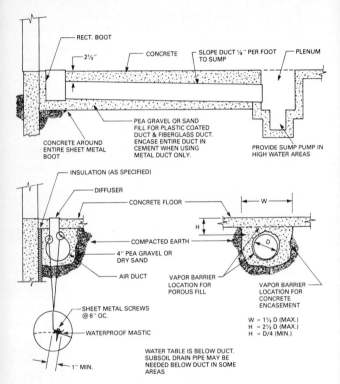

FIG. 12.11 Typical underslab duct. (*From HVAC Duct Construction Standards— Metal and Flexible, 1st ed.,* © 1985 by the Sheet-Metal and Air Conditioning Contractor's National Association. Used with permission of the copyright holder.*)

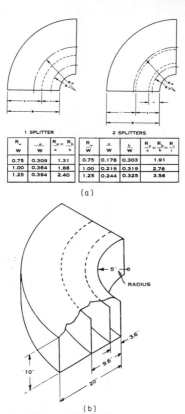

FIG. 12.12 (*a*) Correct location of splitters; (*b*) improving aspect ratio with splitters. (*Reproduced by permission of The Trane Company, LaCrosse, WI.*)

TABLE 12.1   Pressure-Velocity Classification for Ductwork

VELOCITY LEVEL (MAX.)	2000 FPM		2500 FPM			4000 FPM				PRE-CONTRACT DETERMINATION BY DESIGNER	
STATIC PRESSURE CLASS (INCHES W.G.)	+1/2"	-1/2"	+1"	-1"	+2"	-2"	+3"	-3"	+4"	+6"	+10"
Rectangular Style	A	A	STD	STD	STV	A	A	A	A	A	A
Round Style	A	A	STD	STD	STV	A	A	A	A	A	A
Flat Oval Style	A		STD		STV		A		A	A	
Flexible Style	A	A	STD	STD	STV	A	A		A	A	A

NOTES:

1. "STD" denotes standard (non variable volume) air duct construction requirements (regardless of actual velocity level) for compliance with this document—for all cases in which the designer does not predesignate the pressure classification specifically applicable for the duct system independent of fan static rating. STV denotes standard construction classification for variable volume ducts for compliance with this document when the designer does not predesignate a class for such application. See S1.4 on Page 8.

2. "A" denotes other pressure classes for which construction or installation details are given in this document and are *available* for designation in contract documents prepared by designers.

3. See Section S1.8 for sealing requirements related to duct pressure class.

4. The pressure classification number herein denotes construction suitable for a maximum level not less than the maximum operating pressure in the portion of the system receiving the classification from the designer.

5. The velocity classification indicates the normal range of service. The designer's selection of pressure class for the purpose of construction is acknowledgement of acceptable design velocity level and pressure level.

6. The designation of a pressure class pertains to straight duct and duct fittings exclusive of equipment and special components inserted in the ductwork system; such other items are governed by separate specifications in the contract documents.

7. No negative pressure construction for 4", 6" or 10" w.g. is provided herein. Designers should consult other references, including the SMACNA Round and Rectangular Industrial Duct Construction Standards.

8. Other section numbers and titles, figures, tables, and page numbers mentioned in these notes refer to the original source document.

SOURCE: *HVAC Duct Construction Standards—Metal and Flexible,* 1st ed., © 1985 by the Sheet-Metal and Air Conditioning Contractor's National Association. Used with permission of the copyright holder.

TABLE 12.2   Rectangular Duct Hangers Minimum Sizes

MAXIMUM HALF OF DUCT PERIMETER	Pair at 10 ft. Spacing		Pair at 8 ft. Spacing		Pair at 5 ft. Spacing		Pair at 4 ft. Spacing	
	STRAP	WIRE/ROD	STRAP	WIRE/ROD	STRAP	WIRE/ROD	STRAP	WIRE/ROD
$\frac{P}{2}$ = 30"	1" × 22 ga.	10 ga. (.135")	1" × 22 ga.	10 ga. (.135")	1" × 22 ga.	12 ga. (.106")	1" × 22 ga.	12 ga. (.106")
$\frac{P}{2}$ = 72"	1" × 18 ga.	3/8"	1" × 20 ga.	1/4"	1" × 22 ga.	1/4"	1" × 22 ga.	1/4"
$\frac{P}{2}$ = 96"	1" × 16 ga.	3/8"	1" × 18 ga.	3/8"	1" × 20 ga.	3/8"	1" × 22 ga.	1/4"
$\frac{P}{2}$ = 120"	1½" × 16 ga.	1/2"	1" × 16 ga.	3/8"	1" × 18 ga.	3/8"	1" × 20 ga.	1/4"
$\frac{P}{2}$ = 168"	1½" × 16 ga.	1/2"	1½" × 16 ga.	1/2"	1" × 16 ga.	3/8"	1" × 18 ga.	3/8"
$\frac{P}{2}$ = 192"	—	1/2"	1½" × 16 ga.	1/2"	1" × 16 ga.	3/8"	1" × 16 ga.	3/8"
$\frac{P}{2}$ = 193" up	SPECIAL ANALYSIS REQUIRED							

WHEN STRAPS ARE LAP JOINED USE THESE MINIMUM FASTENERS:	SINGLE HANGER MAXIMUM ALLOWABLE LOAD	
	STRAP	WIRE OR ROD (Dia.)
1″ × 18, 20, 22 ga. — two #10 or one 1/4″ bolt 1″ × 16 ga. — two 1/4″ dia. 1″ × 16 ga. — two 3/8″ dia. Place fasteners in series, not side by side.	1″ × 22 ga. — 260 lbs. 1″ × 20 ga. — 320 lbs. 1″ × 18 ga. — 420 lbs. 1″ × 16 ga. — 700 lbs. 1½″ × 16 ga. —1100 lbs.	0.106″ — 80 lbs. 0.135″ — 120 lbs. 0.162″ — 160 lbs. 1/4″ — 270 lbs. 3/8″ — 680 lbs. 1/2″ —1250 lbs. 5/8″ —2000 lbs. 3/4″ —3000 lbs.

NOTES:
1. Dimensions other than guage are in inches.
2. Tables allow for duct weight, 1 lb./sf insulation weight and normal reinforcement and trapeze weight, but no external loads!
3. For custom design of hangers, designers may consult SMACNA's rectangular industrial duct standards, the AISI Cold Formed Steel Design Manual and the AISC Steel Construction Manual.
4. Straps are galvanized steel; other materials are uncoated steel.
5. Allowable loads for P/2 assume that ducts are 16 ga. maximum, except that when maximum duct dimension (w) is over 60″ then P/2 maximum is 1.25 w.
6. For upper attachments see Fig. 4-2.
7. For lower attachments see Fig. 4-4.
8. For trapeze sizes see Table 4-3 and Fig. 4-5.
9. 12, 10 or 8 ga. wire is steel of black annealed, bright basic or galvanized type.
10. Other figure and table numbers mentioned in these notes refer to original source document.

SOURCE: *HVAC Duct Construction Standards—Metal and Flexible*, 1st ed., © 1985 by the Sheet-Metal and Air Conditioning Contractor's National Association. Used with permission of the copyright holder.

TABLE 12.3   Minimum Hanger Sizes for Round Ducts

Dia.	Maximum Spacing	Wire Dia.	Rod	Strap
10" dn	12'	One 12 ga.	1/4"	1" × 22 ga.
11-18"	12'	Two 12 ga. or One 8 ga.	1/4"	1" × 22 ga.
19-24"	12'	Two 10 ga.	1/4"	1" × 22 ga.
25-36"	12'	Two 8 ga.	3/8"	1" × 20 ga.
37-50"	12'	↑	Two 3/8"	Two 1" × 20 ga.
51-60"	12'	↑	Two 3/8"	Two 1" × 18 ga.
61-84"	12'	↑	Two 3/8"	Two 1" × 16 ga.

NOTES:

1. Straps are galvanized steel; rods are uncoated or galvanized steel; wire is black annealed, bright basic or galvanized steel. All are alternatives.

2. See Figure 4-4 for lower supports.

3. See Figures 4-2 and 4-3 for upper attachments.

4. Table allows for conventional wall thickness, and joint systems plus one lb/sf of insulation weight. If heavier ducts are to be installed, adjust hanger sizes

to be within their load limits; see allowable loads with Table 4-1.

5. Designers: For industrial grade supports, including saddles, single point trapeze loads, longer spans and flanged joint loads, see SMACNA's Round Industrial Duct Construction Standards.

6. See Figures 3-9 and 3-10 for flexible duct supports.

7. Other figure and table numbers mentioned in these notes refer to original source document.

SOURCE: *HVAC Duct Construction Standards—Metal and Flexible*, 1st ed., © 1985 by the Sheet-Metal and Air Conditioning Contractor's National Association. Used with permission of the copyright holder.

TABLE 12.4   Allowable Loads for Trapeze Angles

Length	1 x 1 x 16 ga	1 x 1 x 1/8"	1-1/2 x 1-1/2 x 16 ga.	1-1/2 x 1-1/2 x 1/8"	1-1/2 x 1-1/2 x 3/16"	1-1/2 x 1-1/2 x 1/4" or 2 x 2 x 1/8"	2 x 2 x 3/16"	2 x 2 x 1/4"	2-1/2 x 2-1/2 x 3/16"	2-1/2 x 2-1/2 x 1/4"
18"	80	150	180	350	510	650	940	1230	1500	1960
24"	75	150	180	350	510	650	940	1230	1500	1960
30"	70	150	180	350	510	650	940	1230	1500	1960
36"	60	130	160	340	500	620	920	1200	1480	1940
42"	40	110	140	320	480	610	900	1190	1470	1930
48"	—	80	110	290	450	580	870	1160	1440	1900
54"	—	40	70	250	400	540	840	1120	1400	1860
60"	—	—	—	190	350	490	780	1060	1340	1800
66"	—	—	—	100	270	400	700	980	1260	1720
72"	—	—	—	—	190	320	620	900	1180	1640
78"	—	—	—	—	80	210	500	790	1070	1530
84"	—	—	—	—	—	80	380	660	940	1400
96"	—	—	—	—	—	—	—	320	600	1060
108"	—	—	—	—	—	—	—	—	150	610

SECTION PROPERTIES										
I_x	.012	.022	.041	.078	.110	.139/ .190	.272	.348	.547	.703
A	.12	.234	.180	.359	.527	.688/ .484/ .130	.715	.938	.902	1.19
Z	.016	.031	.037	.072	.104	.130/ .130	.190	.247	.303	.394
LB/LF	.44	.80	.66	1.23	1.80	2.34/ 1.65	2.44	3.19	3.07	4.10

IT IS ASSUMED THAT STEEL MATERIAL WITH A YIELD STRENGTH OF 25,000 PSI OR GREATER IS USED.

LOADS ABOVE ASSUME THAT A HANGER ROD IS 6" MAX. DISTANCE FROM DUCT SIDE.

IF THE ROD IS 2" AWAY FROM THE DUCT THE ALLOWABLE LOAD INCREASES SIGNIFICANTLY.

See Fig. 4-5 in original document for load calculation method.

See Table 4-I in original document for rod and strap load limits.

SOURCE: *HVAC Duct Construction Standards—Metal and Flexible*, 1st ed., © 1985 by the Sheet-Metal and Air Conditioning Contractor's National Association. Used with permission of the copyright holder.

TABLE 12.5 Galvanized Sheet Thickness Tolerances

Gage	Thickness In Inches			Weight				Thickness In Millimeters		
	Min.	Max.	Nom.	Min. lb/sf	Nom. lb/sf	Max. lb/sf	Nom. kg/m²	Nom.	Min.	Max.
33	.0060	.0120	.0090	.2409	.376	.486		.2286	.1524	.3048
32	.0104	.0164	.0134	.4204	.563	.665		.3404	.2642	.4166
31	.0112	.0172	.0142	.4531	.594	.698		.3607	.2845	.4369
30	.0127	.0187	.0157	.5143	.656	.759	3.20	.3988	.3188	.4788
29	.0142	.0202	.0172	.5755	.719	.820		.4369	.3569	.5169
28	.0157	.0217	.0187	.6367	.781	.881	3.81	.4750	.3950	.5550
27	.0172	.0232	.0202	.6979	.844	.943		.5131	.4331	.5931
26	.0187	.0247	.0217	.7591	.906	1.004	4.42	.5512	.4712	.6312
25	.0207	.0287	.0247	.8407		1.167		.6274	.5274	.7274
24	.0236	.0316	.0276	.9590	1.156	1.285	5.64	.7010	.6010	.8010
23	.0266	.0346	.0306	1.0814		1.408		.7772	.6772	.8772
22	.0296	.0376	.0336	1.2038	1.406	1.530	6.86	.8534	.7534	.9534
21	.0326	.0406	.0366	1.3263		1.653		.9296	.8296	1.0296
20	.0356	.0436	.0396	1.4486	1.656	1.775	8.08	1.006	.906	1.106
19	.0406	.0506	.0456	1.6526		2.061		1.158	1.028	1.288
18	.0466	.0566	.0516	1.8974	2.156	2.305	10.52	1.311	1.181	1.441
17	.0525	.0625	.0575	2.1381		2.546		1.461	1.331	1.591
16	.0575	.0695	.0635	2.342	2.656	2.832	12.96	1.613	1.463	1.763
15	.0650	.0770	.0710	2.6481		3.138		1.803	1.653	1.953
14	.0705	.0865	.0785	2.8725	3.281	3.525	16.01	1.994	1.784	2.204
13	.0854	.1014	.0934	3.4804		4.133		2.372	2.162	2.582
12	.0994	.1174	.1084	4.0516	4.531	4.786	22.21	2.753	2.523	2.983
11	.1143	.1323	.1233	4.6595		5.394		3.132	2.902	3.362
10	.1292	.1472	.1382	5.2675	5.781	6.002	28.21	3.510	3.280	3.740
9	.1442	.1622	.1532	5.8795		6.614		3.891	3.661	4.121
8	.1591	.1771	.1681	6.4874	6.875	7.222		4.270	4.040	4.500

NOTES:

1. Based on ASTM A525 (Hot Dip Galvanized sheet) and Manufacturers' Standards.
2. Tolerances are valid for 48'' and 60'' wide coil and cut length stock—other dimensions apply to other sheet widths and to strip.
3. The lock forming grade of steel will conform to ASTM A527.
4. The steel producing industry recommends that steel be ordered by decimal thickness only. Thickness and zinc coating class can be stenciled on the sheet. The gage designation is retained for residual familiarity reference only.
5. Minimum weight in this table is based on the following computation:

   Minimum sheet thickness minus 0.001'' of G60 coating times 40.8 lb per s.f. per inch plus 0.0369 lb/sf of zinc.

   G90 stock would be comparably calculated from:

   (t − .00153'') 40.8 + 0.0564 = minimum weight.

6. However, scale weight may run 2% (or more) greater than theoretical weight. Actual weight may be near 40.82 lb. per s.f. per inch.

   G60 coating, per ASTM A525 and ASTM A90, has 0.60 oz/sf (triple spot test) total for two sides. 0.59 oz/sf of zinc equals 0.001''. 1 oz is 0.0017''.

   G90 coating is 0.90 oz/sf (triple spot), or 0.00153''. Magnetic gage measurement of zinc coating may have 15% error.

TABLE 12.6 Manufacturers' Standard Gage Thicknesses— Uncoated Steel

M.S. Gage	Weight lb/sf (kg/m²)	Thickness Nominal	Hot Rolled Min	Hot Rolled Max	Cold Rolled Min	Cold Rolled Max	ANSI STANDARD B32.3 Preferred in Thickness Millimeters First	Second
28	.625 (3.051)	.0149 in .378 mm			.0129 in .328 mm	.0169 in .429 mm	.30	.35
26	.750 (3.661)	.0179 in .455 mm			.0159 in .404 mm	.0199 in .505 mm	.40	.45
24	1.000 (4.882)	.0239 in .607 mm			.0209 in .531 mm	.0269 in .683 mm	.50	.55
22	1.250 (6.102)	.0299 in .759 mm			.0269 in .683 mm	.0329 in .836 mm	.60	.65 .70
20	1.500 (7.323)	.0359 in .912 mm			.0329 in .836 mm	.0389 in .988 mm	.80	.90
18	2.000 (9.764)	.0478 in 1.214 mm	.0428 in 1.087 mm	.0528 in 1.341 mm	.0438 in 1.113 mm	.0518 in 1.316 mm	1.00 1.2	1.10 1.4
16	2.500 (12.205)	.0598 in 1.519 mm	.0538 in 1.367 mm	.0658 in 1.671 mm	.0548 in 1.392 mm	.0648 in 1.649 mm	1.6	1.8

14	3.125 (15.256)	.0747 in 1.897 mm	.0677 in 1.720 mm	.0817 in 2.075 mm	.0697 in 1.770 mm	.0797 in 2.024 mm	2.0	2.2
12	4.375 (21.359)	.1046 in 2.657 mm	.0966 in 2.454 mm	.1126 in 2.860 mm	.0986 in 2.504 mm	.1106 in 2.809 mm	2.5	2.8
10	5.625 (27.461)	.1345 in 3.416 mm	.1265 in 3.213 mm	.1425 in 3.619 mm	.1285 in 3.264 mm	.1405 in 3.569 mm	3.0	3.2
8	6.875 (33.564)	.1644 in 4.176 mm	.1564 in 3.973 mm	.1724 in 4.379 mm			3.5	3.8
							4.0	

Notes:

Manufacturers Standard Gage is based on a theoretical steel density of 489.6 lb/cf, or 40.80 lb/sf per inch of thickness plus 2.5% normally experienced increase in delivery weight. Thus, the weight basis associated with thickness specifications is 41.82 lb/sf per inch.

U.S. Standard Gage, the legal gage since 1893, although based on the density of wrought iron 480 #/cf, uses 40.00 lb/sf/in for both iron and steel. Thus, U.S. gage thicknesses are derived from weights 2% lighter than steel.

The table is based on 48" width coil and sheet stock. 60" stock has the same tolerance for gages listed except for 16 gage which has ±0.007" in hot rolled sheet.

Thickness and weight in customary units are based on data in the AISI Carbon Sheet Steel Products Manual. Metric conversions listed here are straight multiplications for comparison purposes. Individual manufacturers may quote other tolerances.

ANSI is the American National Standards Institute. Standard B-32.3 actually covers a wider range of thickness than listed here.

SOURCE: *HVAC Duct Construction Standards—Metal and Flexible,* 1st ed., © 1985 by the Sheet-Metal and Air Conditioning Contractor's National Association. Used with permission of the copyright holder.

TABLE 12.7 Stainless Steel Thicknesses

| Gage | Thickness in Inches | | | | Weight | | | | Thickness in Millimeters | | |
| | Min | Max | Tolerance | Nom | lb/Sf | | Kg/m² | | Nom | Min | Max |
					300	400	300	400			
31	.0089	.0129	.002	.0109	.459	.451	2.239	2.200	.2769	.2269	.3269
30	.0105	.0145	.002	.0125	.525	.515	2.561	2.512	.3175	.2675	.3675
29	.0121	.0161	.002	.0141	.591	.579	2.883	2.825	.3581	.3081	.4081
28	.0136	.0176	.002	.0156	.656	.644	3.200	3.142	.3962	.3462	.4462
27	.0142	.0202	.003	.0172	.722	.708	3.522	3.454	.4369	.3569	.5169
26	.0158	.0218	.003	.0188	.788	.773	3.844	3.771	.4775	.3975	.5575
25	.0189	.0249	.003	.0219	.919	.901	4.483	4.395	.5562	.4762	.6362
24	.0220	.0280	.003	.0250	1.050	1.030	5.122	5.025	.6350	.5550	.7150
23	.0241	.0321	.004	.0281	1.181	1.159	5.761	5.654	.7137	.6137	.8137
22	.0273	.0353	.004	.0313	1.313	1.288	6.405	6.283	.7950	.6950	.8950
21	.0304	.0384	.004	.0344	1.444	1.416	7.044	6.908	.8738	.7738	.9738
20	.0335	.0415	.004	.0375	1.575	1.545	7.683	7.537	.9525	.8525	1.0525
19	.0388	.0488	.005	.0438	1.838	1.803	8.966	8.796	1.1125	.9835	1.2425
18	.0450	.0550	.005	.0500	2.100	2.060	10.245	10.050	1.2700	1.1400	1.4000
17	.0513	.0613	.005	.0563	2.363	2.318	11.528	11.308	1.4300	1.3000	1.5600
16	.0565	.0685	.006	.0625	2.625	2.575	12.806	12.562	1.5875	1.4375	1.7375
15	.0643	.0763	.006	.0703	2.953	2.897	14.406	14.133	1.2856	1.6356	1.9356

14	.0711	.0851	.007	.0781	3.281	3.219	16.006	15.704	1.9837	1.8037	2.1637
13	.0858	.1018	.008	.0938	3.938	3.863	19.211	18.845	2.3825	2.1825	2.5825
12	.0100	.1184	.009	.1094	4.594	4.506	22.411	21.982	2.7788	2.5488	2.9788
11	.1150	.1350	.010	.1250	5.250	5.150	25.612	25.124	3.1750	2.9250	3.4250
10	.1286	.1526	.012	.1406	5.906	5.794	28.812	28.265	3.5712	3.2712	3.8712
9	.1423	.1703	.014	.1563	6.563	6.438	32.017	31.407	3.9700	3.6100	4.3300
8	.1579	.1859	.014	.1719	7.219	7.081	35.217	34.544	4.3663	4.0063	4.7263

ASTM-A167—"Stainless and Heat-Resisting Chromium-Nickel Steel Plate, Sheet, and Strip" (Properties of the 300 series)

ASTM-A480—"Standard Specification for General Requirements for Flat-Rolled Stainless and Heat-Resisting Steel Plate, Sheet, and Strip"

Finishes: *No. 1 Finish*—Hot-rolled, annealed, and descaled.
 No. 2 D Finish—Cold-rolled, dull finish.
 No. 2 B Finish—Cold-rolled, bright finish.
 Bright Annealed Finish—A bright cold-rolled finish retained by final annealing in a controlled atmosphere furnace.
 No. 3 Finish—Intermediate polished finish, one or both sides.
 No. 4 Finish—General purpose polished finish, one or both sides.
 No. 6 Finish—Dull satin finish, Tampico brushed, one or both sides.
 No. 7 Finish—High luster finish.
 No. 8 Finish—Mirror finish.

The 300 series weight is based on 0.2916 lb per square foot per inch of thickness (or 504 lb/cf).

The 400 series weight is based on 0.2861 lb per square foot per inch (or 494 lb/cf).

ASTM-A666 covers the structural grade of stainless steel (not used for ducts). For design criteria, generally, consult the AISI Stainless Steel Cold-Formed Structural Design Manual. For general application and corrosion data consult the AISI Design Guidelines for the Selection and Use of Stainless Steels.

SOURCE: *HVAC Duct Construction Standards—Metal and Flexible*, 1st ed., © 1985 by the Sheet-Metal and Air Conditioning Contractor's National Association. Used with permission of the copyright holder.

TABLE 12.8 Aluminum Sheet Thicknesses—Alloy 3003-H14

	Thickness in Inches			Weight		Thickness in Millimeters		
Nom.	Tolerance 48" & (60") Width	Min.	Max.	lb/ft²	Kg/m²	Nom.	Min.	Max.
.016	.002	.014	.018	.228	1.114	.4064	.3556	.4572
.020	.0025 (.003)	.0175	.0225	.285	1.393	.508	.4445	.5715
.024	.0025 (.003)	.0215	.0265	.342	1.671	.6096	.5461	.6731
.025	.0025 (.003)	.0225	.0275	.356	1.7398	.635	.5715	.6985
.032	.0025 (.0035)	.0295	.0345	.456	2.228	.8128	.7493	.8763
.040	.003 (.004)	.037	.043	.570	2.786	1.016	.9398	1.0922
.050	.004 (.005)	.046	.054	.713	3.484	1.27	1.1684	1.3716
.063	.004 (.005)	.059	.067	.898	4.389	1.600	1.4986	1.7018
.080	.004 (.006)	.076	.084	1.140	5.571	2.032	1.9340	2.1336
.090	.004	.086	.094	1.283	6.270	2.286	2.1844	2.3876

		.095	.105	1.426	6.969	2.54	2.413	2.667
.100	.005 (.007)							
.125	.005 (.007)	.12	.13	1.782	8.709	3.175	3.048	3.302

Weight is based on 14.256 lb per square foot per inch of thickness (or 171.1 lb/ft3). Alloy 1100 is of slightly lower density.

Specification references: ASTM B209 Standard Specification for Aluminum Alloy Sheet and Plate which references ANSI Standard H-35.2 Dimensional Tolerances for Aluminum Mill Products.

Other useful references are published by the Aluminum Association: Specifications for Aluminum Structures; Engineering Data for Aluminum Structures; Aluminum Standards and Data.

SOURCE: *HVAC Duct Construction Standards—Metal and Flexible*, 1st ed., © 1985 by the Sheet-Metal and Air Conditioning Contractor's National Association. Used with permission of the copyright holder.

TABLE 12.9 Duct Surface Area in Square Feet per Linear Foot

DUCT DIMENSION (WIDTH)

	6"	8"	10"	12"	14"	16"	18"	20"	22"	24"	26"	28"	30"	36"	42"	48"	54"	60"	66"	72"	84"	96"	108"
6"	1.	2.33	2.67	3.00	3.33	3.67	4.00	4.33	4.67	5.00	5.33	5.67	6.00	7.00	8.00	9.00	10.00	11.00	12.00	13.00	15.00	17.00	19.00
8"		2.67	3.00	3.33	3.67	4.00	4.33	4.67	5.00	5.33	5.67	6.00	6.33	7.33	8.33	9.33	10.33	11.33	12.33	13.33	15.33	17.33	19.33
10"			3.33	3.67	4.00	4.33	4.67	5.00	5.33	5.67	6.00	6.33	6.67	7.67	8.67	9.67	10.67	11.67	12.67	13.67	15.67	17.67	19.67
12"				4.00	4.33	4.67	5.00	5.33	5.67	6.00	6.33	6.67	7.00	8.00	9.00	10.00	11.00	12.00	13.00	14.00	16.00	18.00	20.00
14"					4.67	5.00	5.33	5.67	6.00	6.33	6.67	7.00	7.33	8.33	9.33	10.33	11.33	12.33	13.33	14.33	16.33	18.33	20.33
16"						5.33	5.67	6.00	6.33	6.67	7.00	7.33	7.67	8.67	9.67	10.67	11.67	12.67	13.67	14.67	16.67	18.67	20.67
18"							6.00	6.33	6.67	7.00	7.33	7.67	8.00	9.00	10.00	11.00	12.00	13.00	14.00	15.00	17.00	19.00	21.00
20"								6.67	7.00	7.33	7.67	8.00	8.33	9.33	10.33	11.33	12.33	13.33	14.33	15.33	17.33	19.33	21.33
22"									7.33	7.67	8.00	8.33	8.67	9.67	10.67	11.67	12.67	13.67	14.67	15.67	17.67	19.67	21.67
24"										8.00	8.33	8.67	9.00	10.00	11.00	12.00	13.00	14.00	15.00	16.00	18.00	20.00	22.00
26"											8.67	9.00	9.33	10.33	11.33	12.33	13.33	14.33	15.33	16.33	18.33	20.33	22.33
28"												9.33	9.67	10.67	11.67	12.67	13.67	14.67	15.67	16.67	18.67	20.67	22.67
30"													10.00	11.00	12.00	13.00	14.00	15.00	16.00	17.00	19.00	21.00	23.00
36"														12.00	13.00	14.00	15.00	16.00	17.00	18.00	20.00	22.00	24.00
42"															14.00	15.00	16.00	17.00	18.00	19.00	21.00	23.00	25.00
48"																16.00	17.00	18.00	19.00	20.00	22.00	24.00	26.00
54"																	18.00	19.00	20.00	21.00	23.00	25.00	27.00
60"																		20.00	21.00	22.00	24.00	26.00	28.00
66"																			22.00	23.00	25.00	27.00	29.00
72"																				24.00	26.00	28.00	30.00
84"																					28.00	30.00	32.00
96"																						32.00	34.00
108"																							36.00

DUCT DIMENSION (DEPTH)

SOURCE: *HVAC Duct Construction Standards—Metal and Flexible*, 1st ed., © 1985 by the Sheet-Metal and Air Conditioning Contractor's National Association. Used with permission of the copyright holder.

TABLE 12.10   Surface Areas of Rectangular Ducts

Width + Depth, in*	Area,† ft²/ft	Width + Depth, in*	Area,† ft²/ft	Width + Depth, in*	Area,† ft²/ft
10	1.67	47	7.83	84	14.00
11	1.83	48	8.00	85	14.17
12	2.00	49	8.17	86	14.34
13	2.17	50	8.34	87	14.50
14	2.34	51	8.50	88	14.67
15	2.50	52	8.67	89	14.83
16	2.67	53	8.83	90	15.00
17	2.83	54	9.00	91	15.17
18	3.00	55	9.17	92	15.34
19	3.17	56	9.34	93	15.50
20	3.34	57	9.50	94	15.67
21	3.50	58	9.67	95	15.83
22	3.67	59	9.83	96	16.00
23	3.83	60	10.00	97	16.17
24	4.00	61	10.17	98	16.34
25	4.17	62	10.34	99	16.50
26	4.34	63	10.50	100	16.67
27	4.50	64	10.67	101	16.83
28	4.67	65	10.83	102	17.00
29	4.83	66	11.00	103	17.17
30	5.00	67	11.17	104	17.34
31	5.17	68	11.34	105	17.50
32	5.34	69	11.50	106	17.67
33	5.50	70	11.67	107	17.83
34	5.67	71	11.83	108	18.00
35	5.83	72	12.00	109	18.17
36	6.00	73	12.17	110	18.34
37	6.17	74	12.34	111	18.50
38	6.34	75	12.50	112	18.67
39	6.50	76	12.67	113	18.83
40	6.67	77	12.83	114	19.00
41	6.83	78	13.00	115	19.17
42	7.00	79	13.17	116	19.34
43	7.17	80	13.34	117	19.50
44	7.34	81	13.50	118	19.67
45	7.50	82	13.67	119	19.83
46	7.67	83	13.83	120	20.00

*1 in = 25.4 mm, 1 mm = 0.0394 in.
†Per foot of duct length. 1 ft²/ft = 0.305 m²/m, 1 m²/m = 3.28 ft²/ft.

SOURCE: A. M. Khashab, *Heating, Ventilating, and Air Conditioning Systems Estimating Manual,* 2d ed., McGraw-Hill, New York, © 1984. Used with permission of the publisher.

TABLE 12.11 Angle, Bar, and Channel Properties

Size	Description	Weight lbs/ft	Rated Z	Rated I_x
1 x 1 x 1/8	Angle	.80	.031	.022
1¼ x 1¼ x 1/8	Angle	1.02	.049	.044
1½ x 1/8	Bar		.035	.035
1 x ¼	Bar	.85	.042	.021
1½ x 1½ x 1/8	Angle	1.23	.072	.078
1½ x 1½ x ¼	Angle	2.34	.134	.139
2 x 2 x 1/8	Angle	1.65	.131	.190
2 x ¼	Bar	1.70	.167	.167
2 x 2 x 3/16	Angle	2.44	.190	.272
2 x 2 x 5/16	Angle	3.92	.30	.416
3 x ¼	Bar	2.55	.375	.563
2½ x 2½ x ¼	Angle	4.1	.394	.703
3 x 5/16	Bar	3.19	.469	.703
2½ x 2½ x 3/8	Angle	5.9	.566	.984
3 x 3 x ¼	Angle	4.9	.577	1.24
3 x 2 x 3/8	Angle	5.9	.781	1.53

4 x 4 x ¼	Angle	6.6	1.05	3.04
4 x 4 x 5/16 4 x ½ C4	Angle Bar Channel	8.2 6.8 5.4	1.29 1.334 1.93	3.71 2.667 3.85
4 x 3½ x ½ C5	Angle Channel	11.9 6.7	1.94 3.00	5.32 7.49
C5	Channel	9	3.56	8.90

Z is section modulus in in.3

Stress is M/Z where M is bending moment.

I$_x$ is moment of inertia in in.4

For steel the rigidity index EI is $290 \times 10^5 \times I_x$

SOURCE: *HVAC Duct Construction Standards—Metal and Flexible*, 1st ed., © 1985 by the Sheet-Metal and Air Conditioning Contractor's National Association. Used with permission of the copyright holder.

TABLE 12.12   Surface Areas of Round Ducts

Diameter, in*	Area per Unit Duct Length, ft²/ft	Diameter, in*	Area per Unit Duct Length, ft²/ft	Diameter, in*	Area per Unit Duct Length, ft²/ft
4	1.05	16	4.19	28	7.33
5	1.31	17	4.45	29	7.59
6	1.57	18	4.72	30	7.85
7	1.83	19	4.98	31	8.11
8	2.09	20	5.24	32	8.38
9	2.36	21	5.50	33	8.65
10	2.62	22	5.76	34	8.91
11	2.88	23	6.02	35	9.17
12	3.15	24	6.28	36	9.43
13	3.40	25	6.54	37	9.69
14	3.67	26	6.80	38	9.95
15	3.93	27	7.07	39	10.21
				40	10.47

*1 in = 25.4 mm; 1 mm = 0.0394 in.
†1 ft²/ft = 0.305 m²/m; 1 m²/m = 3.28 ft²/ft.

SOURCE: A. M. Khashab, *Heating, Ventilating, and Air Conditioning Systems Estimating Manual*, 2d ed., McGraw-Hill, New York, © 1984. Used with permission of the publisher.

NOTES

NOTES

AIR CONDITIONING

TABLE 13.1 Air Conditioning Formulas and Data

Formula	Legend	Formula	Legend
$T_1 = \dfrac{T_{cl}}{12,000}$	T_1 = load expressed in tons of refrigeration T_{cl} = total cooling load, expressed in Btu/h 1 ton of refrigeration = amount of cooling that can be done by 1 ton of ice melting in a 24-h period	$q_{ef} = \text{watts} \times 3.42 \times 1.25$	q_{ef} = gain from fluorescent lighting (25% increase for ballast), expressed in Btu's
$Q = \dfrac{q_s}{1.08(T_a - T_d)}$	Q = outside air entering conditioned room (make-up air), expressed in cfm (ft³/min) q_s = sensible load due to outside ventilation air, expressed in Btu/h T_a = indoor temperature (dry bulb)	$q_{fan} = R \times Q_s$	q_{fan} = gain from fan motor, expressed in Btu's R = 0.05 for small systems 0.35 for large systems Q_s = total sensible load, expressed in Btu/h
		$q_{os} = \text{number of persons} \times H_s$	q_{os} = sensible heat load from occupants, expressed in Btu/h H_s = sensible heat per person—see ASHRAE tables

T_d = dry-bulb temperature of air discharged from air-handling unit

$q_s = 1.08Q(T_o - T_i)$

T_o = design dry-bulb temperature of outside air

T_i = design dry-bulb temperature of conditioned air

q_1 = latent load due to outside ventilation air, expressed in Btu/h

$q_1 = 0.67Q(G_o - G_i)$

G_o = moisture content of outside air, expressed in g/lb of air

G_i = moisture content of inside air, expressed in g/lb of air

q_e = gain from electric light fixtures, expression in Btu's

q_e = watts × 3.42

q_{o1} = number of persons × H_1

q_{o1} = latent heat load from occupants, expressed in Btu/h

H_1 = latent heat per person—see ASHRAE tables

$Q = \dfrac{qs}{1.08\,\Delta T}$

ΔT = temperature difference between room and air leaving coil (usually 18°F)

$V = \dfrac{Q}{A_c}$

V = air velocity, ft/min

A_c = coil face area (ft²)

$Q_g = \dfrac{24 \times \text{tons of refrigeration}}{\Delta T_1}$

Q_g = amount of chilled water in gal/min to be circulated

ΔT_1 = temperature rise (°F) of water on passing through cooling coil (usually 8°F or 10°F)

TABLE 13.2   Weight of Dry Air at Various Pressures and Temperatures at Sea Level

Temp of air °F	Temp of air °C	Gauge Pressure, Pounds — Weight in Pounds per Cubic Foot																					
		0	5	10	20	30	40	50	60	70	80	90	100	110	120	130	140	150	175	200	225	250	300
-20	-28.9	.0900	.1205	.1515	.2125	.2744	.3360	.3970	.4580	.5190	.5800	.6410	.7020	.7635	.8250	.8860	.9480	1.010	1.165	1.318	1.465	1.625	1.930
-10	-23.3	.0882	.1184	.1485	.2090	.2685	.3283	.3880	.4478	.5076	.5674	.6272	.6870	.7470	.8070	.8680	.9280	.9890	1.139	1.288	1.438	1.588	1.890
0	-17.8	.0864	.1160	.1455	.2040	.2630	.3215	.3800	.4385	.4970	.5555	.6140	.6725	.7310	.7900	.8490	.9080	.9680	1.114	1.260	1.406	1.553	1.850
10	-12.2	.0846	.1136	.1425	.1995	.2568	.3145	.3720	.4292	.4863	.5433	.6006	.6580	.7160	.7740	.8320	.8890	.9470	1.090	1.233	1.376	1.520	1.810
20	-6.7	.0828	.1113	.1395	.1950	.2516	.3071	.3645	.4205	.4770	.5330	.5890	.6450	.7010	.7570	.8130	.8690	.9270	1.067	1.208	1.348	1.489	1.770
30	1.1	.0811	.1088	.1366	.1916	.2465	.3015	.3570	.4121	.4672	.5221	.5771	.6320	.6870	.7420	.7970	.8520	.9080	1.046	1.184	1.322	1.460	1.735
40	4.5	.0795	.1067	.1338	.1876	.2415	.2954	.3503	.4038	.4576	.5114	.5652	.6190	.6730	.7270	.7810	.8350	.8900	1.025	1.161	1.296	1.431	1.701
50	10.0	.0780	.1045	.1310	.1839	.2367	.2905	.3432	.3960	.4487	.5014	.5541	.6070	.6600	.7130	.7660	.8190	.8730	1.006	1.139	1.271	1.403	1.668
60	15.6	.0765	.1025	.1283	.1803	.2323	.2842	.3362	.3882	.4402	.4921	.5441	.5960	.6490	.7040	.7520	.8040	.8560	.9880	1.116	1.245	1.376	1.636
70	21.1	.0750	.1005	.1260	.1770	.2280	.2791	.3302	.3808	.4316	.4824	.5332	.5840	.6350	.6860	.7370	.7890	.8390	.9670	1.095	1.223	1.350	1.604
80	26.7	.0736	.0988	.1239	.1738	.2237	.2739	.3242	.3726	.4229	.4729	.5224	.5720	.6220	.6730	.7240	.7740	.8240	.9490	1.073	1.199	1.323	1.573
90	32.2	.0723	.0970	.1218	.1707	.2195	.2688	.3182	.3670	.4154	.4639	.5122	.5610	.6110	.6600	.7090	.7590	.8090	.9320	1.054	1.177	1.300	1.544
100	37.8	.0710	.0953	.1196	.1678	.2155	.2635	.3113	.3602	.4079	.4561	.5033	.5520	.6010	.6480	.6960	.7450	.7940	.9140	1.035	1.155	1.276	1.517
110	43.3	.0698	.0937	.1176	.1650	.2115	.2593	.3070	.3478	.4011	.4481	.4950	.5420	.5890	.6360	.6850	.7340	.7820	.9010	1.021	1.135	1.255	1.491
120	48.9	.0686	.0921	.1155	.1618	.2080	.2548	.3018	.3473	.3944	.4403	.4866	.5330	.5790	.6260	.6730	.7200	.7670	.8840	1.001	1.118	1.234	1.465
130	54.4	.0674	.0905	.1135	.1590	.2045	.2505	.2966	.3446	.3924	.4296	.4770	.5210	.5700	.6160	.6620	.7090	.7540	.8690	.984	1.099	1.214	1.440
140	60.0	.0663	.0889	.1115	.1565	.2015	.2465	.2915	.3366	.3813	.4193	.4711	.5130	.5560	.6050	.6460	.6960	.7410	.8170	.974	1.092	1.197	1.416
150	65.6	.0652	.0875	.1096	.1540	.1985	.2425	.2870	.3312	.3760	.4193	.4636	.5070	.5500	.5960	.6370	.6850	.7300	.8080	.953	1.064	1.175	1.392
175	79.4	.0626	.0840	.1054	.1482	.1910	.2335	.2755	.3181	.3607	.4033	.4450	.4870	.5310	.5730	.6160	.6580	.7010	.8080	.914	1.021	1.128	1.337
200	93.3	.0609	.0809	.1014	.1430	.1840	.2248	.2660	.3054	.3473	.3882	.4291	.4700	.5110	.5520	.5920	.6330	.6740	.7760	.876	.982	1.084	1.287
225	107	.0581	.0779	.0976	.1373	.1770	.2163	.2555	.2949	.3344	.3738	.4120	.4520	.4910	.5310	.5700	.6090	.6490	.7470	.846	.944	1.043	1.240
250	121	.0560	.0751	.0941	.1323	.1705	.2085	.2466	.2854	.3223	.3602	.3969	.4360	.4740	.5130	.5510	.5890	.6270	.7220	.817	.912	1.007	1.197
275	135	.0541	.0726	.0910	.1278	.1645	.2011	.2378	.2745	.3111	.3478	.3844	.4210	.4580	.4940	.5310	.5680	.6050	.6970	.789	.881	.972	1.155
300	149	.0523	.0702	.0881	.1230	.1592	.1945	.2300	.2658	.3020	.3362	.3716	.4070	.4440	.4780	.5140	.5490	.5850	.6730	.763	.852	.940	1.118
350	177	.0489	.0656	.0825	.1160	.1495	.1825	.2160	.2492	.2824	.3156	.3488	.3820	.4150	.4490	.4820	.5160	.5490	.6330	.715	.799	.883	1.048
400	204	.0463	.0621	.0779	.1090	.1405	.1720	.2035	.2349	.2661	.2974	.3287	.3600	.3930	.4230	.4540	.4860	.5170	.5960	.674	.753	.831	.987
450	232	.0437	.0586	.0735	.1033	.1330	.1628	.1925	.2220	.2515	.2810	.3105	.3400	.3690	.3990	.4290	.4580	.4880	.5620	.637	.711	.786	.934
500	260	.0414	.0555	.0696	.0981	.1263	.1543	.1824	.2105	.2385	.2662	.2940	.3230	.3510	.3790	.4070	.4360	.4640	.5340	.604	.675	.746	.885
550	288	.0394	.0528	.0661	.0934	.1198	.1464	.1730	.1998	.2262	.2528	.2794	.3060	.3330	.3590	.3860	.4130	.4400	.5040	.571	.638	.705	.841
600	316	.0376	.0504	.0631	.0885	.1140	.1395	.1650	.1904	.2158	.2412	.2668	.2920	.3170	.3430	.3680	.3930	.4190	.4830	.547	.611	.675	.801

Based on perfect gas laws and air weight of .08071 lbs. per cu. ft at 32°F and barometric pressure of 14.696 lbs. per sq. in.
Based on perfect gas laws and air weight of 1.29285 kg/m³ at 0°C and barometric pressure of 101.33 kPa A
Pressure—psiG X 6.895 = pressure—kPa G
Lb/ft³ X 16.018 = Kg/m³

SOURCE: Courtesy Ingersoll-Rand Co.

TABLE 13.3 Properties of Hydrocarbon and Special Refrigerant Vapors

Gas	Chemical Formula	Alternate Designation	Molecular Weight	Boiling Point °F at 14.696 PSIA	Values at 14.696 PSIA & 60°F — Specific Gravity (Air = 1.00)	Density lb/cu ft	Specific Volume cu ft/lb	Specific Heat of Constant Pressure at 14.696 PSIA Btu/lb/°F — Minus 40°F	60°F	150°F	300°F	Ratios of Specific Heats K = Cp ÷ Cv at 14.696 PSIA — Minus 40°F	60°F	150°F	300°F	Molar Heat Capacity Cp at 14.696 PSIA and 130°F Btu/°/Mole	Critical Temperature Rankine	Critical Pressure PSIA
Methane	CH_4	C_1	16.04	−259	0.555	0.0424	23.61	.506	.527		.624	1.33	1.31		1.25	8.95	344	673
Acetylene	C_2H_2		26.04	−119	0.899p	0.0686p	14.58p	.353	.397	.427	.449	1.31	1.26	1.24	1.21	11.12	557	905
Ethylene	C_2H_4	Ethene	28.05	−155	0.969p	0.0739p	13.53p	.312	.362	.406	.478	1.29	1.24	1.23	1.17	11.39	510	742
Ethane	C_2H_6		30.07	−128	1.047	0.0799	12.52	.365	.410	.458	.543	1.22	1.19	1.17	1.14	13.77	550	708
Propylene	C_3H_6	Propene	42.08	−54	1.453p	.1109p	9.021p	.303	.354	.399	.473	1.18	1.15	1.14	1.11	16.79	657	667
Propane	C_3H_8	C_3	44.09	−44	1.547	.1180	8.471	.333	.389	.443	.534	1.16	1.13	1.13	1.09	19.33	666	617
Butadiene 1,2	C_4H_6		54.09	+51	1.867p	.1425p	7.018p		.346	.387	.451	1.12	1.12	1.10	1.09	20.93	799	653
Butadiene 1,3	C_4H_6		54.09	+24	1.867p	.1425p	7.018p		.341	.392	.468	1.12	1.12	1.11	1.09	21.26	766	628
Isobutylene	C_4H_8	i-Butene	56.10	+20	1.937p	.1478p	6.766p		.370	.419	.493	1.11	1.11	1.10	1.08	21.28	753	580
Butylene	C_4H_8		56.10	+21	1.937p	.1481p	6.746p		.355	.406	.484	1.11	1.10	1.10	1.08	22.51	756	583
Isobutane	C_4H_{10}	i-C_4	58.12	+11	2.068	.1578	6.339		.387	.443	.535	1.10	1.10	1.09	1.07	22.78	735	529
n-Butane	C_4H_{10}	n-C_4	58.12	+31	2.071	.1581	6.337		.391	.444	.532	1.11	1.09	1.08	1.07	23.75	765	551
Isopentane	C_5H_{12}	i-C_5	72.15	+82	2.491p	.190p	5.262p	.401b	.439		.529	1.076	1.07	1.06	1.04	31.67	830	483
n-Pentane	C_5H_{12}	n-C_5	72.15	+97	2.491p	.190p	5.263p	.401b	.441		.528	1.076	1.07	1.05	1.04	31.82	846	489
Benzene	C_6H_6		78.11	+176	2.697p	.206p	4.860p		.301b	.360	.526	1.096	1.06	1.06	1.03	23.51	1012	714
n-Hexane	C_6H_{14}	n-C_6	86.17	+156	2.975p	.227p	4.406p		.443b		.525	1.04b	1.04b			38.17	915	440
n-Heptane	C_7H_{16}	n-C_7	100.20	+209	3.459p	.264p	3.789p		.446b		.524	1.04b	1.04b			47.49	973	397
n-Octane	C_8H_{18}	n-C_8	114.22	+258	3.943p	.301p	3.324p		.446b			1.03	1.03			57.00	1025	362
Refrigerant 11 ***	CCl_3F		137.38	+75	4.736	.3635b	2.739b		.134b	.141	.156		1.14b	1.13	1.10	19.37	848	635
Refrigerant 12 ***	CCl_2F_2		120.93	−22	4.27	.326	3.067		.145g	.141	.164	1.17	1.14g	1.15		17.33	694	597
Refrigerant 13 ***	$CClF_3$		104.47	−115	3.62	.276	3.624		.150							17.13	544	561
Refrigerant 21 ***	$CHCl_2F$		102.93	+48	3.63	.277	3.608	.133	.136	.148	.169	1.18	1.18	1.15	1.12	13.23	813	750
Refrigerant 22 ***	$CHClF_2$		86.48	−41	3.05	.233	4.299		.149	.161	.182	1.20	1.20	1.17	1.14	13.92	665	716
Refrigerant 113 ***	CCl_2FCClF_2		187.39	+118	6.48	.464	2.155		.157	.172	.179				1.07	13.93	877	495
Refrigerant 114 ***	$CClF_2CClF_2$		170.93	+38	6.08	.444	2.155		.157	.168	.188				1.07	28.72	754	474

SOURCE: Courtesy Ingersoll-Rand Co.

TABLE 13.4   Properties of Miscellaneous Gases

Gas	Chemical Formula	Alternate Designation	Molecular Weight	Boiling Point °F	Specific Gravity (Air=1.00) Values at 14.696 PSIA & 60°F	Density lb/cu ft at 14.696 PSIA	Specific Volume cu ft/lb at 14.696 PSIA (See Notes)	Sp. Heat Cp Btu/lb°F −40°F	60°F	150°F	300°F	Ratio K=Cp/Cv −40°F	60°F	150°F	300°F	Molar Heat Capacity Cp at 150°F Btu/p/mole and 14.696 PSIA	Critical Temp °Rankine	Critical Pressure PSIA
Air (dry) **			28.97	−318	1.000	.0763	13.106	.240	.240	.241	.243	1.40	1.40	1.40	1.40	6.98	239	547
Ammonia	NH$_3$		17.03	−28	0.594	.0454	22.05		.506	.525	.556		1.30	1.30	1.30	8.94	730	1639
Argon	A		39.94	−303	1.380	.1053	9.497	.125	.125	.125	.124	1.67	1.67	1.67	1.67	4.99	272	705
Carbon Dioxide	CO$_2$		44.01		1.528	.1166	8.576	.189	.201	.213	.254	1.30	1.30	1.28	1.25	9.37	548	1073
Carbon Monoxide	CO		28.01	−312	0.967	.0738	13.55	.248	.248	.249	.252	1.40	1.40	1.40	1.40	6.97	242	507
Chlorine	Cl$_2$		70.91	−30	2.48	.1886	5.30	.115	.115			1.34	1.35			8.15d	751	1119
Ethylene Oxide	H$_2$C$_2$O		44.05	+51	1.52	.116	8.62	.225n	.264n	.302n	.355n	1.25n	1.21n	1.19n		14.10	844	1043
Helium	He		4.003	−451	0.138	.0105	91.66	1.25	1.25	1.25		1.66	1.66	1.66		5.00	*e	*e
Hydrogen	H$_2$		2.016	−423	0.0696	.00531	188.32		3.409	3.442	3.462					6.94	*e	*e
Hydrogen Chloride	HCl		36.47	−121	1.271	.0970	10.31	.194	.194			1.41	1.41	1.41		7.06d	585	1200
Hydrogen Sulphide	H$_2$S		34.08	−77	1.175	.0897	11.15	.233	.238	.243	.251	1.34	1.34	1.32	1.30	8.28	673	1306
Methyl Chloride	CH$_3$Cl		50.49	−11	1.777	.1356	7.372		.1991					1.29f		10.05f	749	969
Neon	Ne		20.19	−411	0.697	.0532	18.81	.246	.246	.246	.246	1.66	1.66	1.66	1.66	4.97	80	385
Nitric Oxide	NO		30.01	−240	1.038	.0792	12.62	.239	.238	.238	.239	1.38	1.39	1.39	1.38	7.14	323	956
Nitrogen	N$_2$		28.02	−320	0.967	.0738	13.55	.249	.249	.249	.250	1.40	1.40	1.40	1.40	6.98	227	492
Nitrous Oxide	N$_2$O		44.02	−127	1.531	.1168	8.56	.218	.21	.22	.226	1.40	1.30			9.2d	558	1054
Oxygen	O$_2$		32.00	−297	1.105	.0843	11.86	.123	.219	.219	.158	1.19	1.17	1.16	1.14	7.07	278	732
Phosgene	COCl$_2$		98.92	+46	3.41	.262	3.82		.136	.146						14.44	820	823
Sulphur Dioxide	SO$_2$		64.06	+14	2.254	.1720	5.814	.147				1.25	1.25	1.25		9.42d	775	1142
	CH$_3$C$_6$H$_5$		92.13	+231	3.181p	.243p	4.121p		.3468	.379		1.078			1.04	31.87	1069	611
Water Vapor	H$_2$O	Steam	18.02	+212	0.622b	.0373b	26.80b		.4946	.55c		1.31c	1.326	1.32b	1.31c	8.94	1165	3187

NOTES TO TABLES

a - An average for 0°F to 300°F.

b - At the boiling point.

c - Approximate average for 212°F to 600°F and 14.7 psia to 200 psia.

d - At 60°F.

e - These are Effective Values to be used only for Generalized Compressibility Charts and gas mixtures. Actual values are

	T_c °R	P_c PSIA
Helium	9.7°R	33.2 psia
Hydrogen	59.9°R	188 psia

f - At 77°F.

g - As a perfect gas.

h - Within plus or minus 5 percent.

p - As a perfect gas.

** Normal Atmospheric Air contains some moisture. For convenience it is common to consider that, at 68°F and 14.696 psia, the air is at 36 percent relative humidity, weighs 0.075 lb/cu ft, and has a k value of 1.395. (Based on ASME Test Code for Displacement Compressors).

*** This group of refrigerants is known by trade names such as Freon, Genetron, etc.

SOURCE: Ingersoll-Rand Co.

TABLE 13.5   Latent Heat of Vaporization versus Boiling Point

No.	Refrigerant Name	Normal Boiling Pt °F	Latent Heat at Boiling Point λ Btu/lb·mol	Trouton Constant λ/°R*
630	Methyl Amine[a]	23.0	11141	23.08
717	Ammonia	−28.0	10036	23.25
764	Sulfur Dioxide	13.6	10705	22.62
631	Ethyl Amine	68.0	11645	22.07
611	Methyl Formate[a]	100.0	12094	21.61
504		−71.0	8282	21.31
23	Trifluoromethane	−115.7	7325	21.29
21	Dichlorofluoromethane	47.8	10557	20.80
30	Methylene Chloride[a]	120.0	11598	19.66
C318	Octafluorocyclobutane	21.5	10017	20.81
22	Chlorodifluoromethane	−41.4	8687	20.76
40	Methyl Chloride	−10.8	9305	20.73
506		9.9	9644	21.44
113	Trichlorotrifluoroethane	117.6	11828	20.49
152a	Difluoroethane	−13.0	9045	20.25
502		−49.9	8280	20.21
114	Dichlorotetrafluoroethane	38.8	10005	20.07
216	Dichlorohexafluoropropane	96.2	11154	20.07
11	Trichlorofluoromethane	74.9	10648	19.92
505	—	−21.8	8735	19.95
500	—	−28.3	8588	19.91
290	Propane	−43.7	8026	19.29
14	Tetrafluoromethane	−198.3	5146	19.69
600	Butane	31.1	9641	19.64
13B1	Bromotrifluoromethane	−72.0	7607	19.62
12	Dichlorodifluoromethane	−21.6	8591	19.61
142b	Chlorodifluoroethane	14.4	9297	19.61
115	Chloropentafluoroethane	−38.4	8245	19.57
503		−126.1	6483	19.43
1270	Propylene	−53.9	7931	19.55
600a	Isobutane	10.9	9103	19.34
13	Chlorotrifluoromethane	−114.6	6670	19.33
1150	Ethylene	−154.7	5793	19.00
170	Ethane	−127.9	6296	18.98
50	Methane	−258.7	3521	17.52

[a]Not at normal atmospheric pressure.   [b]Normal boiling temperatures.

SOURCE: Reprinted by permission from *ASHRAE Handbook—1985 Fundamentals* (Inch-Pound Edition).

TABLE 13.6   Estimated Equivalent Rated Full-Load Hours of Operation for Properly Sized Equipment during Normal Cooling Season

City	Hours	City	Hours
Albuquerque, NM	800–2200	Indianapolis, IN	600–1600
Atlantic City, NJ	500–800	Little Rock, AR	1400–2400
Birmingham, AL	1200–2200	Minneapolis, MN	400–800
Boston, MA	400–1200	New Orleans, LA	1400–2800
Burlington, VT	200–600	New York, NY	500–1000
Charlotte, NC	700–1100	Newark, NJ	400–900
Chicago, IL	500–1000	Oklahoma City, OK	1100–2000
Cleveland, OH	400–800	Pittsburgh, PA	900–1200
Cincinnati, OH	1000–1500	Rapid City, SD	800–1000
Columbia, SC	1200–1400	St. Joseph, MO	1000–1600
Corpus Christi, TX	2000–2500	St. Petersburg, FL	1500–2700
Dallas, TX	1200–1600	San Diego, CA	800–1700
Denver, CO	400–800	Savannah, GA	1200–1400
Des Moines, IA	600–1000	Seattle, WA	400–1200
Detroit, MI	700–1000	Syracuse, NY	200–1000
Duluth, MN	300–500	Trenton, NJ	800–1000
El Paso, TX	1000–1400	Tulsa, OK	1500–2200
Honolulu, HI	1500–3500	Washington, DC	700–1200

SOURCE: Reprinted by permission from *ASHRAE Handbook—1981 Fundamentals.*

TABLE 13.7   Over-All Heat-Transfer Coefficients for Glass and Glass Blocks

DESCRIPTION	OUTDOOR EXPOSURE
SINGLE-GLASS WINDOWS	1.06
DOUBLE-GLASS WINDOWS	0.64
TRIPLE-GLASS WINDOWS	0.34
GLASS BLOCK	0.56
GLASS BLOCK	0.48

SOURCE: Reproduced by permission of The Trane Company, LaCrosse, WI.

TABLE 13.8   Outside Design Temperatures and Latitudes for Various Cities in the United States

STATE	CITY	DESIGN TEMP. DB	DESIGN TEMP. WB	NORTH LATITUDE, DEGREES
ALABAMA	ANNISTON	95	75	33
	BIRMINGHAM	95	78	33
	MOBILE	95	80	31
ALASKA	JUNEAU	65	52	58
ARIZONA	PHOENIX	105	76	33
	TUCSON	105	72	32
ARKANSAS	LITTLE ROCK	95	78	35
CALIFORNIA	FRESNO	105	74	37
	LOS ANGELES	90	70	34
	SACRAMENTO	100	72	38
	SAN FRANCISCO	85	65	38
COLORADO	DENVER	95	64	40
	PUEBLO	95	65	38
CONNECTICUT	HARTFORD	93	75	42
DELAWARE	WILMINGTON	95	78	40
DISTRICT OF COLUMBIA	WASHINGTON	95	78	39
FLORIDA	JACKSONVILLE	95	78	30
	MIAMI	91	79	26
GEORGIA	ATLANTA	95	76	34
	SAVANNAH	95	78	32
HAWAII	HONOLULU	83	73	21
IDAHO	BOISE	95	65	44
	POCATELLO	95	65	43
ILLINOIS	CHICAGO	96	75	43
	PEORIA	98	77	42
INDIANA	SPRINGFIELD	95	78	38
	EVANSVILLE	95	78	41
	FORT WAYNE	95	75	40
	INDIANAPOLIS	95	76	42
IOWA	DES MOINES	95	78	42
	DUBUQUE	95	78	42
	SIOUX CITY	95	76	39
KANSAS	KANSAS CITY	100	76	38
	WICHITA	100	75	38
KENTUCKY	ASHLAND	95	76	38
	LOUISVILLE	95	78	38

State	City			
LOUISIANA	NEW ORLEANS	95	80	30
	SHREVEPORT	100	78	32
MAINE	PORTLAND	90	73	44
MARYLAND	BALTIMORE	95	78	39
MASSACHUSETTS	BOSTON	92	75	42
	HOLYOKE	93	75	42
MICHIGAN	DETROIT	95	75	42
	GRAND RAPIDS	95	75	43
MINNESOTA	DULUTH	93	73	47
	MINNEAPOLIS	95	75	45
MISSISSIPPI	JACKSON	95	78	32
MISSOURI	KANSAS CITY	100	76	39
	SPRINGFIELD	100	75	37
	ST. LOUIS	95	78	39
MONTANA	BILLINGS	90	66	46
	HELENA	95	67	46
NEBRASKA	LINCOLN	95	78	41
	OMAHA	95	78	41
NEVADA	RENO	95	65	39
NEW HAMPSHIRE	CONCORD	90	73	43
NEW JERSEY	TRENTON	95	78	40
NEW MEXICO	ALBUQUERQUE	95	70	35
	SANTA FE	93	65	36
NEW YORK	ALBANY	93	75	43
	BUFFALO	93	73	43
	NEW YORK	95	75	41
NORTH CAROLINA	ASHEVILLE	93	75	36
	GREENSBORO	95	78	36
NORTH DAKOTA	BISMARCK	95	73	47
OHIO	CINCINNATI	95	78	39
	CLEVELAND	95	75	42
OKLAHOMA	OKLAHOMA CITY	101	77	35
	TULSA	101	77	36
OREGON	PORTLAND	90	68	45
PENNSYLVANIA	PHILADELPHIA	95	78	40
	PITTSBURGH	95	75	40
RHODE ISLAND	PROVIDENCE	93	75	42
SOUTH CAROLINA	CHARLESTON	95	78	33
	GREENVILLE	95	75	35
SOUTH DAKOTA	RAPID CITY	95	70	44
TENNESSEE	CHATTANOOGA	95	76	35
	MEMPHIS	95	78	35

(continued)

TABLE 13.8 *(continued)*

STATE	CITY	DESIGN TEMP.		NORTH LATITUDE, DEGREES
		DB	WB	
TEXAS	DALLAS	100	78	33
	EL PASO	100	69	32
	GALVESTON	95	80	29
	HOUSTON	95	80	30
	SAN ANTONIO	100	78	29
UTAH	SALT LAKE CITY	95	64	41
VIRGINIA	NORFOLK	95	78	37
	RICHMOND	95	78	38
	ROANOKE	95	78	37
WASHINGTON	SEATTLE	85	65	48
	SPOKANE	95	65	48
WEST VIRGINIA	CHARLESTON	95	75	42
WISCONSIN	EAU CLAIRE	95	75	45
	MADISON	95	75	43
	MILWAUKEE	95	75	43
WYOMING	CHEYENNE	95	65	41

SOURCE: Reproduced by permission of The Trane Company, LaCrosse, WI.

TABLE 13.9 Design Temperature Differences

ITEM NO.	ITEM	TEMPERATURE DIFFERENCE* F
1	WALLS, EXTERIOR	17
2	GLASS IN EXTERIOR WALLS	17
3	GLASS IN PARTITIONS	10
4	STORE SHOW WINDOWS HAVING A LARGE LIGHTING LOAD	30
5	PARTITIONS	10
6	PARTITIONS, OR GLASS IN PARTITIONS, ADJACENT TO LAUNDRIES, KITCHENS, OR BOILER ROOMS	25
7	FLOORS ABOVE UNCONDITIONED ROOMS	10
8	FLOORS ON GROUND	0
9	FLOORS ABOVE BASEMENTS	0
10	FLOORS ABOVE ROOMS OR BASEMENTS USED AS LAUN-DRIES, KITCHENS, OR BOILER ROOMS	35
11	FLOORS ABOVE VENTED SPACES	17
12	FLOORS ABOVE UNVENTED SPACES	0
13	CEILINGS WITH UNCONDITIONED ROOMS ABOVE	10
14	CEILINGS WITH ROOMS ABOVE USED AS LAUNDRIES, KITCHENS, ETC.	20
15	CEILING WITH ROOF DIRECTLY ABOVE (NO ATTIC)	17
16	CEILING WITH TOTALLY ENCLOSED ATTIC ABOVE	17
17	CEILING WITH CROSS-VENTILATED ATTIC ABOVE	17

*These temperature differences are based on the assumption that the air conditioning system is being designed to maintain an inside temperature 17 F lower than the outdoor temperature. For air conditioning systems designed to maintain a greater temperature difference than 17 F between the inside and outside, add to the values in the above table, the difference between the assumed design temperature difference and 17 F.

SOURCE: Reproduced by permission of The Trane Company, LaCrosse, WI.

TABLE 13.10   Equivalent Full-Load Hours of Operation per Year

	ATLANTA	BALTI-MORE	BOSTON	CHICAGO	DALLAS	DENVER	DETROIT	LOS ANGELES	MIAMI	MILWAU-KEE
RESTAURANTS	1750	1620	1050	1250	2240	1050	1250	1150	2020	1050
DRUG STORES	1700	1580	1030	1220	2170	1030	1220	1120	1950	1030
CAFETERIAS	1370	1270	825	990	1750	825	990	910	1580	825
JEWELRY STORES	1020	950	620	750	1300	620	750	700	1170	620
BARBER SHOPS	1020	950	620	750	1300	620	750	700	1170	620
NIGHT CLUBS	1010	940	610	730	1280	610	730	675	1150	610
THEATERS	650-1000	600-1000	400-650	500-800	850-1400	400-650	500-800	475-750	800-1300	400-650
DRESS SHOPS	940	870	565	675	1200	565	675	630	1080	565
LARGE OFFICES	915	850	550	660	1180	550	660	610	1060	550
DEPARTMENT STORES	850	790	515	650	1100	515	650	600	1000	515
SPECIALTY SHOPS (5 & 10)	840	780	510	640	1080	510	640	590	975	510
RESIDENCES	810	750	490	600	1050	490	600	550	950	490
SHOE STORES	650	600	400	500	850	400	500	475	775	400
BEAUTY SHOPS	625	580	380	450	800	380	450	425	750	380
SMALL OFFICES	540	500	425	450	700	425	450	410	650	425
RECREATION SPACES	520	480	450	400	675	450	400	380	650	450
FUNERAL PARLORS	460	425	350	375	600	350	375	350	575	350

SOURCE: Reproduced by permission of The Trane Company, LaCrosse, WI.

MINNE-APOLIS	NEW ORLEANS	OKLA. CITY	PHILA-DELPHIA	PHOE-NIX	PORTLAND, ORE.	SAN FRAN.	SAINT LOUIS	WASH., D.C.	NEW YORK
1050	2020	2240	1480	2240	1050	450	2020	1620	1430
1030	1950	2170	1440	2170	1030	400	1950	1580	1400
825	1580	1750	1160	1750	825	350	1580	1270	1120
620	1170	1300	875	1300	620	250	1170	950	850
620	1170	1300	875	1300	620	250	1170	950	850
610	1150	1280	860	1280	610	240	1150	940	840
400-650	800-1300	850-1400	550-920	850-1400	400-650	200-400	800-1300	600-1000	550-900
565	1080	1200	800	1200	565	225	1080	870	780
550	1060	1180	775	1180	550	200	1060	850	750
515	1000	1100	725	1100	515	175	1000	790	700
510	975	1080	710	1080	510	175	975	780	700
490	950	1050	690	1050	490	170	950	750	670
400	775	850	575	850	400	150	775	600	550
380	750	800	540	800	380	150	750	580	525
425	650	700	490	700	425	125	650	500	475
450	650	675	470	675	450	125	650	480	450
350	575	600	410	600	350	100	575	425	400

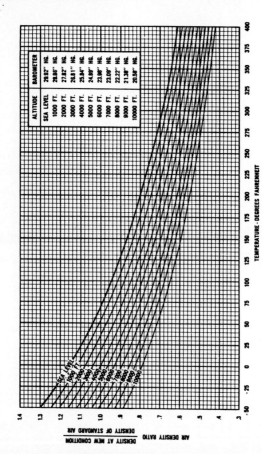

FIG. 13.1 Graph for determining air density at temperatures and altitudes other than standard. (*Reproduced by permission of The Trane Company, LaCrosse, WI.*)

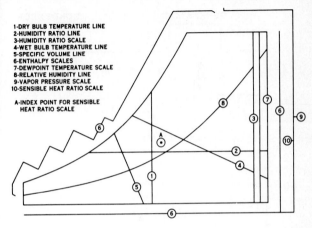

1-DRY BULB TEMPERATURE LINE
2-HUMIDITY RATIO LINE
3-HUMIDITY RATIO SCALE
4-WET BULB TEMPERATURE LINE
5-SPECIFIC VOLUME LINE
6-ENTHALPY SCALES
7-DEWPOINT TEMPERATURE SCALE
8-RELATIVE HUMIDITY LINE
9-VAPOR PRESSURE SCALE
10-SENSIBLE HEAT RATIO SCALE

A-INDEX POINT FOR SENSIBLE
  HEAT RATIO SCALE

FIG. 13.2 Lines and scales on the Trane Psychrometic Chart. (*Reproduced by permission of The Trane Company, LaCrosse, WI.*)

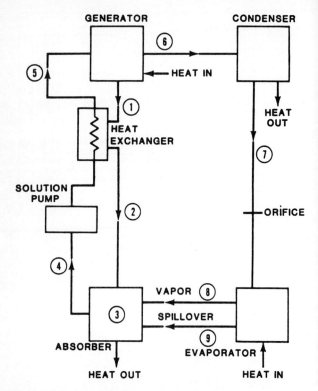

FIG. 13.3 Lithium bromide-water single-stage absorption refrigeration cycle. (*Reproduced by permission from ASHRAE Handbook– 1981 Fundamentals.*)

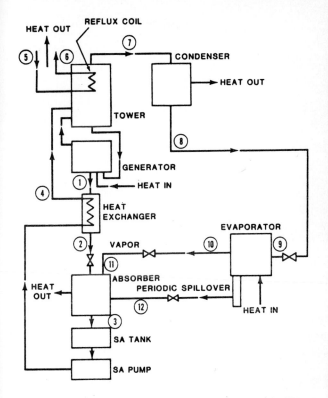

FIG. 13.4 Ammonia-water single-stage absorption refrigeration cycle. (*Reproduced by permission from ASHRAE Handbook—1981 Fundamentals.*)

NOTES

ENERGY CONSERVATION

TABLE 14.1   Potential Energy Conservation Checklist

BUILDINGS

1 Add insulation to attic spaces

2 Insulate top of basement walls

3 Weatherstrip around doors and openable windows

4 Insulate backplate of convectors, radiators, electric baseboard with foil faced rigid fiberglass

5 Caulk around exterior doors and windows

6 Close up unused doors during heating season where possible

7 Insert rigid insulation in unused windows during heating season

8 Insulate outside walls of pipe tunnels to at least 2 feet below grade

9 Install automatic closers on exterior doors

10 Install astragals to close opening between double doors

11 Install vestibules, especially at exterior doors used frequently

12 Install storm doors on infrequently-used outside doors

13 Install storm windows where single-glazing now exists

14 Install awnings over windows in air-conditioned spaces

15 Install drapes and/or venetian blinds in windows

16 Install two-inch styrofoam insulation on top of flat roofs

17 Apply epoxy resin over porous exterior masonry walls

18 Add insulation to the inside of exterior masonry walls

19 Add insulation to the outside of exterior masonry walls and cover with stucco, siding, etc.

20 Install insulating window coverings

21 Excavate around building perimeter and insulate exterior surface of basement walls

22 Close up and insulate unneeded windows permanently

HEATING SYSTEMS

1 Install automatic night temperature setback controls

2 Provide locking covers on thermostats

3 Change range of thermostats to a wider band (dead band between heat and cool)

4 Install self-contained thermostat control valves on uncontrolled radiation

5 Replace existing thermostats with new dual temperature thermostats controlled by a 7-day time clock

6 Improve boiler or furnace efficiency (test for air/fuel mix-oxygen test)

7 Install automatic stack dampers on gas-fired furnaces and domestic hot water heaters

8 Insulate hot air supply ducts and outside air ducts up to the damper in furnace rooms

9 Add insulation to surface of boilers

10 Insulate bare domestic hot water piping and hot water heaters

(continued)

TABLE 14.1   Potential Energy Conservation Checklist (*continued*)

11 Insulate bare steam condensate return lines

12 Add insulation to existing steam lines

13 Add resilient edge seals to outside air dampers

14 Seal air leaks around boiler access doors to combustion chamber

15 Install automatic indoor-outdoor control to maintain lowest hot water boiler temperature necessary for space heating needs

16 Install an automatic combustion air damper

17 Install turbulators in firetube boilers after obtaining the boiler manufacturers approval

18 Install replacement burners that burn either natural gas or fuel oil

19 Replace gas pilots with electric or electronic pilots

20 Install a "shell-head" adapter at the end of the gun on oil burners firing No. 2 fuel oil

21 Install glass doors on fireplace openings

VENTILATING SYSTEMS

1 Disconnect interlock between ventilation supply fan and exhaust fan, and install a manual timer switch to control the exhaust fan

2 Provide a manual positioning switch (or potentiometer) for outside air dampers on ventilation supply air

3 Consider installing destratification fans in the high ceiling area of the building

4 Reduce ventilation rate to code minimum

5 Seal with duct tape or caulk all joints in ductwork systems

6 Reduce toilet exhaust to code minimum

7 Provide light switch control of the toilet exhaust fan

8 Install timers to automatically turn off exhaust fans

9 Install smoke removal units in meeting rooms

10 Set the minimum position on unit ventilator outside air dampers to fully closed

11 Replace gravity backdraft dampers in exhaust fan or relief system with insulated powered dampers

12 Replace existing outside air dampers with insulating dampers

13 Install a baffle in the kitchen hood

PLUMBING SYSTEMS

1 Install water dams in toilet water closets

2 Reduce urinal flushing

3 Install 7-day time clock to control domestic hot water circulating pump

4 Install 7-day time clock and controls to reduce the temperature of domestic hot water heater that is not required

TABLE 14.1   Potential Energy Conservation Checklist (*continued*)

5 Install separate domestic hot water heater for summer use

6 Relocate electric domestic hot water heater closer to use area when possible

ELECTRICAL SYSTEMS

1 Replace selected incandescent bulbs with fluorescent circular adapter units

2 Replace incandescent light fixtures with fluorescent light fixtures

3 Install photocells to automatically shut off outdoor or vestibule security lighting when adequate daylight is available

4 Install additional switches to control lighting in unused areas

5 Install dimmers to reduce lighting level when possible

6 Install manual wall timers in selected rooms to automatically turn off lights

7 Insall a time clock to shut off refrigerated drinking fountains during unoccupied periods

8 Install a timer and disconnect lights on refrigerated vending machines

9 Replace outdoor security lighting with sodium vapor fixtures

10 Replace incandescent gymnasium, banquet hall, meeting room lighting with mercury vapor, metal halide, or sodium vapor fixtures

SIZE, GROSS SQ. FT. _____

AREA COOLED _____ AREA HEATED _____

TYPE(S) OF OCCUPANCY: (% OR SQ. FT.)

Office _____ (Other) _____

Warehouse _____ (Other) _____

Manufacturing _____ (Other) _____

Retail _____

Lobbies & Mall _____
(Enclosed)

BUILDING USE AND OCCUPANCY

Fully Occupied: (50% or more of normal)

Weekdays (Hours) _____ to _____

Weekends (Hours) _____ to _____

_____ to _____ Sunday

_____ to _____ Holidays

Remarks: Describe below if occupancy differs for different floors, areas, buildings: _____

_____

_____

LIGHTING SURVEY

1. Interior Lighting Type _____ Watts/Ft² Offices _____

   _____ Other _____

   Total Install KW _____

   On-Off from Breaker Panel? _____

   Wall Switches? _____ Control Switching? _____

   Operating Schedule _____

2. Exterior Lighting Type _____ Sq. Ft. Served: _____

   Total KW _____ Foot-Candles: _____

Remarks: _____

   _____

Operating Schedule: _____

   _____

   _____

   _____

FIG. 14.1 Space conditioning equipment and schedules. *(From Energy Conservation with Comfort, 2d ed., © 1979 by Honeywell. Used with permission of the copyright holder.)*

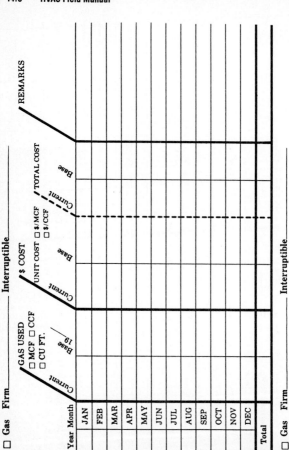

FIG. 14.2 Gas consumption tabulation. (*From Energy Conservation with Comfort, 2d ed., © 1979 by Honeywell. Used with permission of the copyright holder.*)

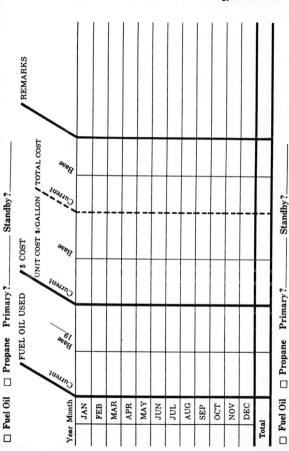

FIG. 14.3 Fuel oil and propane consumption tabulation. (*From Energy Conservation with Comfort, 2d ed. © 1979 by Honeywell. Used with permission of the copyright holder.*)

☐ **Electricity**

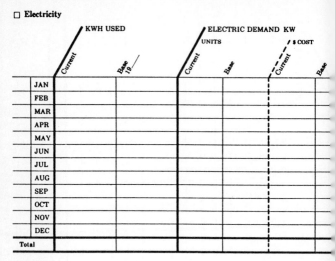

☐ **Electricity**

FIG. 14.4 Electricity consumption tabulation. (*From Energy Conservation with Comfort, 2d ed.,* © 1979 by Honeywell. Used with permission of the copyright holder.*)

| $ COST | | | | REMARKS |
| UNIT COST | | TOTAL COST | | |
Current	Base	Current	Base	

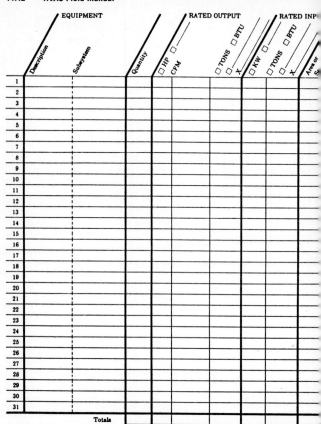

FIG. 14.5 Building survey—equipment list tabulation. (*From Energy Conservation with Comfort, 2d ed., © 1979 by Honeywell. Used with permission of the copyright holder.*)

System:_____ Date:_____ Building/Location:_____

Weekdays	OPERATING SCHEDULE			OUTSIDE AIR CFM	HOURS PER WK SPACE IS OCCUPIED	NIGHT SET BACK?	REMARKS

TABLE 14.2   Energy Equivalents

One	Is Equal to	One	Is Equal to
Btu (mean)	778.104 foot-pounds	Foot-pound	1.3557 joules
U.S. horsepower-hour	2544.65 Btu	Btu (mean)	1054.90 joules
Kilowatthour	3412.66 Btu	Watthour	3600.00 joules
Btu (mean)	252.0 calories	Calorie	4.186 joules

SOURCE: A. M. Khashab, *Heating, Ventilating, and Air Conditioning Systems Estimating Manual*, 2d ed., McGraw-Hill, New York, © 1984. Used with permission of the publisher.

TABLE 14.3 Power Equivalents

One	Is Equal to	One	Is Equal to
U.S. horsepower	550 foot-pounds per second	U.S. horsepower	745.70 watts
U.S. horsepower	0.70685 Btu per second	Metric horsepower	735.50 watts
U.S. horsepower	76.04 kilogram-meters per second	Boiler horsepower	9809.50 watts
Boiler horsepower	33,475 Btu per hour	Btu per hour	0.2929 watt
Ton of refrigeration	12,000 Btu per hour	Ton of refrigeration	3516.8 watts
Metric horsepower	75 kilogram-meters per second	Calorie per second	4.184 watts

SOURCE: A. M. Khashab, *Heating, Ventilating, and Air Conditioning Systems Estimating Manual,* 2d ed., McGraw-Hill, New York, © 1984. Used with permission of the publisher.

NOTES

CLIMATIC CONDITIONS

TABLE 15.1 Climatic Conditions for the United States [a]

Note: AP = airport; AFB = air force base; CO = office location within an urban area.

Col. 1	Col. 2		Col. 3		Col. 4	Col. 5		Col. 6			Col. 7	Col. 8		
	Latitude [b]		Longitude [c]		Elevation [c]	Winter [d] °F		Summer [e] °F						
						Design Dry-Bulb		Design Dry-Bulb and Mean Coincident Wet-Bulb			Mean Daily	Design Wet-Bulb		
State and Station	°	'	°	'	Ft	99%	97.5%	1%	2.5%	5%	Range	1%	2.5%	5%
ALABAMA														
Alexander City	33	0	86	0	660	18	22	96/77	93/76	91/76	21	79	78	78
Anniston AP	33	4	85	5	599	18	22	97/77	94/76	92/76	21	79	78	78
Auburn	32	4	85	3	730	18	22	96/77	93/76	91/76	21	79	78	78
Birmingham AP	33	3	86	5	610	17	21	96/74	94/75	92/74	21	78	77	76
Decatur	34	4	87	0	580	11	16	95/75	93/74	91/74	22	78	77	76
Dothan AP	31	1	85	2	321	23	27	94/76	92/76	91/76	20	80	79	78
Florence AP	34	5	87	4	528	17	21	97/74	94/74	92/74	22	78	77	76
Gadsden	34	0	86	0	570	16	20	96/75	94/75	92/74	22	78	77	76
Huntsville AP	34	4	86	4	619	11	16	95/75	93/74	91/74	23	78	77	76
Mobile AP	30	4	88	2	211	25	29	95/77	93/77	91/76	18	80	79	78
Mobile CO	30	4	88	1	119	25	29	95/77	93/77	91/76	16	80	79	78
Montgomery AP	32	2	86	2	195	22	25	96/76	95/76	93/76	21	79	79	78
Selma-Craig AFB	32	2	87	0	207	22	26	97/78	95/77	93/77	21	81	80	79
Talladega	33	2	86	1	565	18	22	97/77	94/76	92/76	21	79	78	78
Tuscaloosa AP	33	1	87	4	170r	20	23	98/75	96/76	94/76	22	79	78	77
ALASKA														
Anchorage AP	61	1	150	0	90	-23	-18	71/59	68/58	66/56	15	60	59	57
Barrow (S)	71	2	156	5	22	-45	-41	57/53	53/50	49/47	12	54	50	47
Fairbanks AP (S)	64	5	147	5	436	-51	-47	82/62	78/60	75/59	24	64	62	60
Juneau AP	58	3	134	4	17	-4	1	74/60	70/58	67/57	15	61	59	58
Kodiak	57	3	152	3	21	10	13	69/58	65/56	62/55	10	60	58	56
Nome AP	64	3	165	3	13	-31	-27	66/57	62/55	59/54	10	58	56	55
ARIZONA														
Douglas AP	31	3	109	3	4098	27	31	98/63	95/63	93/63	31	70	69	68
Flagstaff AP	35	1	111	4	6973	-2	4	84/55	82/55	80/54	31	61	60	59
Fort Huachuca AP (S)	31	3	110	2	4664	24	28	95/62	93/62	90/62	27	69	68	57

Station	Lat	Long	Elev	99%	97.5%	1%	2.5%	5%	DR	WB	WB	WB	WB
Kingman AP	35 2	114 0	3446	18	25	103/65	100/64	97/64	30	70	69	70	66
Nogales	31 2	111 0	3800	28	32	99/64	96/64	94/64	31	71	70	71	69
Phoenix AP(S)	33 3	112 3	1117	31	34	109/71	107/71	105/71	27	76	75	75	78
Prescott AP	34 4	112 0	5014	4	9	96/61	94/60	92/60	30	66	65	64	78
Tuscon AP (S)	32 1	111 0	2584	28	32	104/66	102/66	100/66	26	72	72	71	78
Winslow AP	35 0	110 4	4880	5	10	97/61	95/60	93/60	32	66	65	65	75
Yuma AP	32 4	114 4	199	36	39	111/72	109/72	107/71	27	79	78	78	78
ARKANSAS													
Blytheville AFB	36 0	90 0	264	10	15	96/78	94/77	91/76	21	81	80	80	78
Camden	33 1	92 5	116	18	23	98/76	96/76	94/76	21	80	79	79	78
El Dorado AP	33 1	92 5	252	18	23	98/76	96/76	94/76	21	80	79	79	78
Fayetteville AP	36 0	94 1	1253	7	12	97/72	94/73	92/73	23	77	76	76	75
Fort Smith AP	35 2	94 2	449	12	17	101/75	98/76	95/76	24	80	79	80	78
Hot Springs	34 3	93 1	535	17	23	101/77	97/77	94/77	22	80	80	80	78
Jonesboro	35 5	90 4	345	10	15	96/78	94/77	91/76	21	81	80	81	80
Little Rock AP (S)	34 4	92 1	257	15	20	99/76	96/77	94/77	22	80	80	80	78
Pine Bluff AP	34 2	92 0	204	16	22	100/78	97/77	95/78	22	81	80	81	80
Texarkana AP	33 3	94 0	361	18	23	98/76	96/77	93/76	21	80	79	80	79
CALIFORNIA													
Bakersfield AP	35 2	119 0	495	30	32	104/70	101/69	98/68	32	73	73	71	70
Barstow AP	34 5	116 5	2142	26	29	106/68	104/68	102/67	37	71	70	71	70
Blythe AP	33 4	114 3	390	30	33	112/71	110/71	110/71	28	75	75	75	74
Burbank AP	34 1	118 2	699	37	39	95/68	91/68	88/67	25	71	71	71	69
Chico	39 5	121 5	205	28	30	103/69	101/68	98/67	36	71	70	70	68

(continued)

aTable 15.1 was prepared by ASHRAE Technical Committee 4.2, Weather Data, from data compiled from official weather stations where hourly weather observations are made by trained observers.

bLatitude, for use in calculating solar loads, and longitude are given to the nearest 10 minutes. For example, the latitude and longitude for Anniston, Alabama, are given as 33°34 and 85°55, respectively, or 33°40 and 85°50.

cElevations are ground elevations for each station. Temperature readings are generally made at an elevation of 5 ft above ground, except for locations marked r, indicating roof exposure thermometer.

dPercentage of winter design data shows the percent of the 3-month period, December through February.

ePercentage of summer design data shows the percent of the 4-month period, June through September.

SOURCE: Reprinted by permission from *ASHRAE Handbook—1985 Fundamentals* (Inch-Pound Edition).

TABLE 15.1 Climatic Conditions for the United States (continued)

Col. 1	Col. 2 Lati-tude° '	Col. 3 Longi-tude° '	Col. 4 Eleva-tion Ft	Col. 5 Winter Design Dry-Bulb 99%	97.5%	Col. 6 Summer Design Dry-Bulb and Mean Coincident Wet-Bulb 1%	2.5%	5%	Col. 7 Mean Daily Range	Col. 8 Design Wet-Bulb 1%	2.5%	5%
State and Station												
Concord	38 0	122 0	195	24	27	100/69	97/68	94/67	32	71	70	68
Covina	34 5	117 5	575	32	35	98/69	95/68	92/67	31	73	71	70
Crescent City AP	41 5	124 0	50	31	33	68/60	65/59	63/58	18	62	60	59
Downey	34 0	118 0	116	37	40	93/70	89/70	86/69	22	72	71	70
El Cajon	32 4	117 0	525	42	44	83/69	80/69	78/68	30	71	70	68
El Centro AP (S)	32 5	115 0	−30	35	38	112/74	110/74	108/74	34	81	80	78
Escondido	33 0	117 0	660	39	41	89/68	85/68	82/68	30	71	70	69
Eureka/ Arcata AP	41 0	124 1	217	31	33	68/60	65/59	63/58	11	62	60	59
Fairfield- Travis AFB	38 2	122 0	72	29	32	99/68	95/67	91/66	34	70	68	67
Fresno AP (S)	36 5	119 4	326	28	30	102/70	100/69	97/68	34	72	71	70
Hamilton AFB	38 0	122 3	3	30	33	89/68	84/66	80/65	28	72	69	67
Laguna Beach	33 3	117 3	35	43	43	83/68	80/68	77/67	18	72	69	68
Livermore	37 4	122 0	545	24	27	100/69	97/68	93/67	24	71	70	68
Lompoc, Vandenburg AFB	34 4	120 3	552	35	38	75/61	70/61	67/60	20	63	61	60
Long Beach AP	33 5	118 1	34	41	43	83/68	80/68	77/67	22	70	69	68
Los Angeles AP (S)	34 0	118 2	99	41	43	83/68	80/68	77/67	15	70	69	68
Los Angeles CO (S)	34 0	118 1	312	37	40	93/70	89/70	86/69	20	72	71	70
Merced-Castle AFB	37 2	120 3	178	29	31	102/70	99/69	96/68	36	72	71	70
Modesto	37 4	121 0	91	28	30	101/69	98/68	95/67	36	71	70	69
Monterey	36 4	121 5	38	35	38	75/63	71/61	68/61	20	64	62	61
Napa	38 2	122 2	16	30	32	100/69	96/68	92/67	30	71	69	68
Needles AP	34 5	114 4	913	30	33	112/71	110/71	108/70	27	75	75	74
Oakland AP	37 4	122 1	3	34	36	85/64	80/63	75/62	19	66	64	63
Oceanside	33 1	117 2	30	41	43	83/68	80/68	77/67	13	70	69	68
Ontario	34 0	117 0	995	31	33	102/70	99/69	96/67	36	74	72	71

City	Lat.	Long.	Elev.	Winter 99%	Winter 97.5%	Mean Daily Range	Summer 1% DB/MWB	Summer 2.5% DB/MWB	Summer 5% DB/MWB	WB 1%	WB 2.5%	WB 5%
Oxnard	34 1	119 1	43	34	36	19	83/66	80/64	77/63	70	68	67
Palmdale AP	34 4	118 1	2517	18	22	35	103/65	101/65	98/64	69	67	66
Palm Springs	33 5	116 4	411	32	35	35	112/71	110/70	108/70	76	74	73
Pasadena	34 1	118 1	864	32	35	29	98/69	98/69	92/67	73	71	73
Petaluma	38 1	122 4	27	26	29	31	94/68	90/66	87/65	72	70	68
Pomona CO	34 0	117 5	871	30	30	30	102/70	99/69	95/68	74	72	71
Redding AP	40 3	122 1	495	29	31	31	105/68	102/67	100/66	71	69	68
Redlands	34 0	117 1	1318	31	33	33	102/70	99/69	96/68	74	72	71
Richmond	38 0	122 1	55	34	36	17	85/64	80/63	75/62	66	64	63
Riverside-March AFB (S)	33 5	117 1	1511	29	32	32	100/68	98/68	95/67	72	71	70
Sacramento AP	38 3	121 3	17	30	32	36	101/70	98/69	94/69	72	71	71
Salinas AP	36 4	121 4	74	30	30	24	74/61	70/60	67/59	62	61	59
San Bernardino, Norton AFB	34 1	117 1	1125	31	33	38	102/70	99/69	96/68	74	72	71
San Diego AP	32 4	117 1	19	42	44	12	83/69	80/69	78/68	71	70	68
San Fernando	34 1	118 1	977	37	39	38	95/68	91/68	88/67	71	70	69
San Francisco AP	37 4	122 4	8	35	38	20	82/64	77/63	73/62	65	64	62
San Francisco CO	37 5	122 5	52	38	40	14	74/63	71/62	69/61	64	62	61
San Jose AP	37 2	121 3	70r	34	34	26	85/66	81/65	77/64	68	67	65
San Luis Obispo	35 2	120 2	315	33	35	26	92/69	88/70	84/69	73	71	70
Santa Ana AP	33 4	117 5	115r	37	39	28	89/69	85/68	82/68	71	70	69
Santa Barbara MAP	34 3	119 5	10	34	36	24	81/67	77/66	75/65	68	67	66
Santa Cruz	37 0	122 0	125	35	38	28	75/63	71/61	68/61	64	62	61
Santa Maria AP (S)	34 5	120 2	238	31	31	23	81/64	76/63	73/62	65	64	63
Santa Monica CO	34 0	118 5	57	41	43	16	83/68	83/68	77/67	69	69	68
Santa Paula	34 1	119 0	263	33	35	36	90/68	86/67	84/66	71	69	68
Santa Rosa	38 3	122 5	167	27	29	34	99/68	95/67	91/66	70	68	67
Stockton AP	37 5	121 2	28	28	30	37	100/69	97/68	94/67	71	71	68
Ukiah	39 1	123 1	620	27	29	40	99/69	95/68	91/67	72	68	67
Visalia	36 2	119 2	354	28	30	38	102/70	100/69	97/68	72	71	71
Yreka	41 4	122 4	2625	13	17	38	95/65	92/64	89/63	67	65	64
Yuba City	39 1	121 4	70	29	31	36	104/68	101/67	99/66	71	69	68
COLORADO												
Alamosa AP	37 2	105 5	7536	-21	-16	35	84/57	82/57	80/57	62	61	60
Boulder	40 0	105 2	5385	-2	8	27	93/59	91/59	89/59	64	63	62

(continued)

TABLE 15.1 Climatic Conditions for the United States (*continued*)

Col. 1	Col. 2 Latitude °		Col. 3 Longitude °		Col. 4 Elevation ft	Winter, °F Col. 5 Design Dry-Bulb 99%	97.5%	Summer, °F Col. 6 Design Dry-Bulb and Mean Coincident Wet-Bulb 1%	2.5%	5%	Col. 7 Mean Daily Range	Col. 8 Design Wet-Bulb 1%	2.5%	5%
State and Station	°		°		Ft	99%	97.5%	1%	2.5%	5%	Range	1%	2.5%	5%
Colorado														
Springs AP	38	5	104	5	6173	−3	2	91/58	88/57	86/57	30	63	62	61
Denver AP	39	5	104	5	5283	−5	−1	93/59	91/59	89/59	28	64	63	62
Durango	37	1	107	5	6550	−1	4	89/59	87/59	85/59	30	64	63	62
Fort Collins	40	4	105	0	5001	−10	−4	93/59	91/59	89/59	28	64	63	62
Grand Junction AP (S)	39	1	108	3	4849	2	7	96/59	94/59	92/59	29	64	63	62
Greeley	40	3	104	4	4648	−11	−5	96/60	94/60	92/60	29	65	64	63
La Junta AP	38	0	103	3	4188	−3	3	100/68	98/68	95/67	31	72	70	69
Leadville	39	2	106	2	10177	−8	−4	84/51	81/51	78/50	30	56	55	54
Pueblo AP	38	2	104	4	4639	−7	0	97/61	95/61	92/61	31	67	66	65
Sterling	40	4	103	1	3939	−7	−2	95/62	93/62	90/62	30	67	66	65
Trinidad AP	37	2	104	2	5746	−2	3	93/61	91/61	89/61	32	66	65	64
CONNECTICUT														
Bridgeport AP	41	1	73	1	7	6	9	86/73	84/71	81/70	18	75	74	73
Hartford,														
Brainard Field	41	5	72	4	15	3	7	91/74	88/73	85/72	22	77	75	74
New Haven AP	41	2	73	0	6	3	7	88/75	84/73	82/72	17	76	75	74
New London	41	2	72	1	60	5	9	88/73	85/72	83/71	16	76	75	74
Norwalk	41	1	73	0	37	6	9	86/73	84/71	81/70	19	75	74	73
Norwich	41	3	72	0	20	3	7	89/75	86/73	83/72	18	76	75	74
Waterbury	41	3	73	0	605	−4	2	88/73	85/71	82/70	21	75	74	72
Windsor Locks,														
Bradley Field (S)	42	0	72	4	169	0	4	91/74	88/72	85/71	22	76	75	73
DELAWARE														
Dover AFB	39	0	75	3	38	11	15	92/75	90/75	87/74	18	79	77	76
Wilmington AP	39	4	75	3	78	10	14	92/74	89/74	87/73	20	77	76	75
DISTRICT OF COLUMBIA														
Andrews AFB	38	5	76	5	279	10	14	92/75	90/74	87/73	18	78	76	75
Washington														
National AP	38	5	77	0	14	14	17	93/75	91/74	89/74	18	78	77	76

FLORIDA

Belle Glade	26	4	80	4	16	41	44	92/76	91/76	89/76	79	78	78
Cape Kennedy AP	28	3	80	3	16	35	38	90/78	88/78	87/78	79	79	79
Daytona Beach AP	29	1	81	0	31	32	35	92/78	90/77	88/77	80	79	78
Fort Lauderdale	26	1	80	1	13	42	46	92/78	91/78	90/78	80	79	79
Fort Myers AP	26	4	81	5	13	41	44	93/78	92/78	91/77	80	79	79
Fort Pierce	27	3	80	2	10	38	42	91/78	90/78	89/78	80	79	78
Gainesville AP (S)	29	4	82	2	155	28	31	95/77	93/77	92/77	80	79	78
Jacksonville AP	30	3	81	4	24	29	32	96/77	94/77	92/76	79	79	78
Key West AP	24	3	81	5	6	55	57	90/78	90/78	89/78	79	79	79
Lakeland CO (S)	28	0	82	0	214	39	41	93/76	93/76	89/76	79	79	78
Miami AP (S)	25	5	80	2	7	44	47	91/77	90/77	89/77	79	79	79
Miami Beach CO	25	5	80	1	9	45	48	90/77	89/77	88/77	80	79	79
Ocala	29	1	82	1	86	31	34	95/77	93/77	92/76	80	79	78
Orlando AP	28	3	81	2	106r	35	38	94/76	93/76	91/76	79	79	78
Panama City, Tyndall AFB	30	0	85	4	22	33	33	92/78	90/77	89/77	81	80	79
Pensacola CO	30	3	87	1	13	29	29	94/77	93/77	91/77	80	80	79
St. Augustine	29	5	81	2	15	31	35	92/78	89/78	87/78	80	79	78
St. Petersburg	28	0	82	4	35	36	40	92/77	91/77	90/76	79	79	78
Sanford	28	5	81	2	14	35	38	94/76	93/76	91/76	79	79	79
Sarasota	27	2	82	3	30	39	42	93/77	92/77	90/76	79	79	78
Tallahassee AP (S)	30	2	84	2	58	27	30	94/77	94/77	90/76	79	79	78
Tampa AP (S)	28	0	82	3	19	36	40	92/77	91/77	90/76	79	79	78
West Palm Beach AP	26	4	80	1	15	41	45	92/78	91/78	90/78	80	79	79

GEORGIA

Albany, Turner AFB	31	3	84	1	224	25	29	97/77	95/76	93/76	80	79	78
Americus	32	0	84	2	476	21	25	97/77	94/76	92/75	79	78	77
Athens	34	0	83	2	700	18	22	94/74	92/74	90/73	78	77	76
Atlanta AP (S)	33	4	84	3	1005	17	22	94/74	92/74	90/73	77	76	75
Augusta AP	33	2	82	0	143	20	23	97/77	95/77	93/76	80	79	79
Brunswick	31	1	81	3	14	29	32	92/78	89/78	87/78	80	79	79
Columbus, Lawson AFB	32	3	85	0	242	24	24	95/76	93/76	91/75	78	78	78
Dalton	34	5	85	0	720	22	22	94/76	93/76	91/76	77	77	79

(continued)

TABLE 15.1 Climatic Conditions for the United States (continued)

State and Station	Lat.°	Lat.′	Long.°	Long.′	Elev. Ft	Winter 99%	Winter 97.5%	Summer DB/MCWB 1%	2.5%	5%	Mean Daily Range	Design WB 1%	2.5%	5%
Dublin	32	3	83	0	215	21	25	96/77	93/76	91/75	20	79	78	77
Gainesville	34	2	83	5	1254	16	21	93/74	91/74	89/73	21	77	76	75
Griffin (S)	33	1	84	2	980	18	22	93/76	90/75	88/74	21	78	77	76
La Grange	33	0	85	0	715	19	23	93/76	91/75	89/74	21	78	77	76
Macon AP	32	4	83	4	356	21	25	96/77	93/76	91/75	22	79	78	77
Marietta, Dobbins AFB	34	0	84	3	1016	17	21	94/74	92/74	90/74	21	78	77	77
Moultrie	31	1	83	4	340	27	30	95/77	92/76	91/76	20	80	79	78
Rome AP	34	2	85	1	637	17	22	94/76	93/76	91/76	23	79	78	77
Savannah-Travis AP	32	1	81	1	52	24	27	93/77	91/77	92/76	20	80	79	78
Valdosta-Moody AFB	31	0	83	2	239	28	31	96/77	94/77	91/77	20	80	79	78
Waycross	31	2	82	2	140	26	29	96/77	94/77	91/76	20	80	79	78
HAWAII														
Hilo AP (S)	19	4	155	1	31	61	62	84/73	83/72	82/72	15	75	74	74
Honolulu AP	21	2	158	0	7	62	63	87/73	86/73	85/72	12	76	75	75
Kaneohe Bay MCAS	21	2	157	5	18	65	66	85/75	84/74	83/74	12	76	75	74
Wahiawa	21	3	158	0	900	58	59	86/73	85/72	84/72	14	75	74	73
IDAHO														
Boise AP(S)	43	3	116	3	2842	3	10	96/65	94/64	91/64	31	68	66	65
Burley	42	5	113	5	4180	-3	2	99/62	95/61	92/60	35	64	63	61
Coeur d'Alene AP	47	5	116	5	2973	-8	-1	89/62	86/61	83/60	31	65	63	61
Idaho Falls AP	43	3	112	0	4730r	-11	-6	89/61	87/61	84/59	38	65	63	66
Lewiston AP	46	3	117	0	1413	-6	6	96/65	94/64	90/63	32	67	66	64
Moscow	46	4	117	0	2660	-7	0	90/63	87/62	84/61	32	65	64	62
Mountain Home AFB	43	0	115	5	2992	6	12	99/64	97/63	94/62	36	66	65	63
Pocatello AP	43	0	112	3	4444	-8	-1	94/61	91/60	89/59	35	64	63	61
Twin Falls AP (S)	42	3	114	3	4148	-3	2	99/62	95/61	92/60	34	64	63	61
ILLINOIS														
Aurora	41	5	88	2	744	-6	-1	93/76	91/76	88/75	20	79	78	76
Belleville, Scott AFB	38	3	89	5	447	1	6	94/76	92/76	89/75	21	79	78	76

	Lat °	Lat '	Long °	Long '	Elev	99%	97.5%	1%	2.5%	5%	Range	WB 1%	WB 2.5%	WB 5%
Bloomington	40	3	89	0	775	-6	-2	92/75	90/74	88/73	21	78	76	75
Carbondale	37	5	89	1	380	2	7	95/75	93/77	90/76	21	80	79	77
Champaign/Urbana	40	0	88	5	743	-3	2	95/75	92/74	88/72	20	78	77	75
Chicago, Midway AP	41	0	87	5	610	-5	0	94/74	91/73	89/74	20	77	76	74
Chicago, O'Hare AP	42	5	87	4	658	-8	-4	91/74	89/71	86/72	20	77	76	74
Chicago CO	41	0	87	4	594	-3	1	94/75	91/74	88/73	21	79	77	75
Danville	40	1	87	5	558	-4	1	93/75	90/74	88/73	21	78	77	75
Decatur	39	5	88	1	670	-3	2	93/75	91/74	88/73	21	78	77	75
Dixon	41	5	89	5	696	-7	-2	93/75	90/74	88/73	23	78	77	75
Elgin	42	2	88	2	820	-7	-2	91/75	89/73	86/73	24	78	76	75
Freeport	41	0	89	0	780	-9	-4	93/75	91/75	88/74	24	79	78	76
Galesburg	41	2	90	0	771	-7	-1	93/75	91/75	89/74	21	78	77	75
Greenville	39	0	89	0	563	-1	4	94/76	91/75	89/74	21	79	78	76
Joliet	41	3	88	0	588	-5	0	93/75	90/74	88/73	20	78	77	75
Kankakee	41	1	87	5	625	-4	1	93/75	90/74	88/73	21	78	77	75
La Salle/Peru	41	2	89	0	520	-7	-2	93/75	91/75	89/75	22	78	77	76
Macomb	40	3	90	4	702	-5	0	95/76	92/76	89/75	22	79	78	76
Moline AP	41	3	90	3	582	-9	-4	93/75	91/75	88/74	23	78	77	75
Mt Vernon	38	2	88	5	500	-4	5	95/76	92/75	89/74	21	79	78	76
Peoria AP	40	4	89	4	652	-8	-4	91/75	89/74	87/73	22	79	78	76
Quincy AP	40	0	91	1	762	-2	3	96/76	93/76	89/74	22	80	78	77
Rantoul, Chanute AFB	40	2	88	1	740	-4	1	94/75	91/74	89/73	21	78	77	75
Rockford	42	1	89	0	724	-9	-4	91/74	89/73	87/73	24	77	76	75
Springfield AP	39	5	89	4	587	-3	2	94/75	92/74	89/74	21	79	77	76
Waukegan	42	2	87	5	680	-6	-3	92/76	89/74	87/73	21	78	76	75
INDIANA														
Anderson	40	0	85	4	847	0	6	95/76	92/75	90/74	22	79	78	76
Bedford	38	5	86	3	670	0	5	95/76	92/75	89/74	22	79	78	76
Bloomington	39	1	86	3	820	0	5	95/76	92/75	89/74	22	79	78	76
Columbus, Bakalar AFB	39	2	85	5	661	3	7	95/76	92/75	90/74	22	79	78	76
Crawfordsville	40	0	86	5	752	3	7	95/76	94/75	91/75	22	79	78	76
Evansville AP	38	0	87	3	381	4	9	96/76	93/75	91/75	22	79	78	76
Fort Wayne AP	41	0	85	1	791	1	1	92/73	89/72	87/72	22	77	75	74
Goshen AP	41	3	85	5	823	1	1	91/73	89/72	86/72	23	77	75	74
Hobart	41	3	87	2	600	1	1	91/73	88/73	85/72	23	77	75	74
Huntington	40	5	85	3	802	1	1	92/73	89/72	87/72	23	77	75	74
Indianapolis AP (S)	39	4	86	2	793	-2	2	92/74	90/74	87/73	22	78	76	75

(continued)

TABLE 15.1   Climatic Conditions for the United States (*continued*)

Col. 1	Col. 2		Col. 3		Col. 4	Winter, °F Col. 5		Summer, °F Col. 6 Design Dry-Bulb Mean Coincident Wet-Bulb			Col. 7	Col. 8		
State and Station	Lati- tude		Longi- tude		Eleva- tion	Design Dry-Bulb					Mean Daily	Design Wet-Bulb		
	°	'	°	'	Ft	99%	97.5%	1%	2.5%	5%	Range	1%	2.5%	5%
Jeffersonville	38	2	85	5	455	5	10	95/74	93/74	91/73	23	79	77	76
Kokomo	40	2	86	5	790	−4	0	91/74	90/73	88/73	22	77	75	75
Lafayette	40	4	86	5	600	−3	3	94/74	91/73	88/73	22	78	76	75
La Porte	41	4	86	4	810	−3	3	93/74	90/74	87/73	22	78	76	75
Marion	40	3	85	4	791	−4	0	91/74	90/73	88/73	23	77	75	74
Muncie	40	1	85	2	955	−3	2	92/74	90/73	87/73	23	78	76	75
Peru, Grissom AFB	40	4	86	1	804	−6	−1	90/74	88/73	86/73	22	77	75	74
Richmond AP	39	5	84	5	1138	−2	2	92/74	90/74	87/73	22	78	76	75
Shelbyville	39	3	85	5	765	−1	3	93/74	91/74	88/73	22	78	76	75
South Bend AP	41	4	86	2	773	−3	1	91/73	89/73	86/72	22	77	75	74
Terre Haute AP	39	3	87	2	601	−2	4	95/75	92/74	89/73	22	79	77	76
Valparaiso	41	2	87	0	801	−3	1	93/74	90/74	87/73	22	78	76	75
Vincennes	38	4	87	3	420	1	6	95/75	93/74	90/73	22	79	77	76
IOWA														
Ames (S)	42	0	93	4	1004	−11	−6	93/75	90/74	87/73	23	78	76	75
Burlington AP	40	5	91	1	694	−7	−3	94/74	91/75	88/73	22	78	77	75
Cedar Rapids AP	41	5	91	4	863	−10	−5	91/76	88/75	86/74	23	78	77	76
Clinton	41	5	90	1	595	−8	−3	92/75	90/75	87/74	23	78	77	75
Council Bluffs	41	2	95	5	1210	−8	−3	94/76	91/75	88/74	22	78	77	76
Des Moines AP	41	3	93	4	948r	−10	−5	94/75	91/74	88/73	23	78	77	75
Dubuque	42	2	90	3	1065	−12	−7	90/74	88/74	86/72	22	77	75	74
Fort Dodge	42	3	94	1	1111	−12	−7	91/74	88/74	86/72	23	77	75	74
Iowa City	41	4	91	3	645	−11	−6	92/76	89/76	87/74	22	80	78	76
Keokuk	40	2	91	1	526	−5	0	95/75	92/75	89/74	22	79	77	76
Marshalltown	42	0	92	5	898	−12	−7	92/76	90/75	88/74	23	78	77	75
Mason City AP	43	1	93	2	1194	−15	−11	90/74	88/74	85/72	24	77	75	74
Newton	41	4	93	0	946	−10	−5	94/75	91/74	88/73	23	78	77	75
Ottumwa AP	41	1	92	2	842	−8	−4	94/75	91/74	88/73	22	78	77	75

Sioux City AP	42	2	96	2	1095	-11	-7	95/74	92/74	89/73	24	78	77	75
Waterloo	42	3	92	2	868	-15	-10	91/76	89/75	86/74	23	78	77	75
KANSAS														
Atchison	39	3	95	3	945	-2	2	96/77	93/76	91/76	23	81	79	77
Chanute AP	37	4	95	5	977	3	7	100/74	97/74	94/74	23	78	77	76
Dodge City AP (S)	37	5	100	0	2594	0	5	100/69	97/69	94/69	25	74	73	71
El Dorado	37	5	96	5	1282	3	5	101/72	98/73	96/73	24	77	77	76
Emporia	38	5	96	5	1209	-1	4	100/74	97/74	94/73	25	78	77	76
Garden City AP	38	0	101	1	2882	-1	1	99/69	99/66	94/69	28	74	73	71
Goodland AP	39	2	101	1	3645	-5	0	99/66	96/65	93/66	31	71	70	68
Great Bend	38	2	98	5	1940	0	4	101/73	98/73	95/73	28	78	76	75
Hutchinson AP	38	0	97	5	1524	4	8	102/72	99/72	97/72	28	77	78	77
Liberal	37	0	100	5	2838	2	7	99/68	96/68	94/68	28	77	75	74
Manhattan, Fort Riley (S)	39	0	96	5	1076	-1	3	99/75	97/75	92/74	24	76	76	76
Parsons	37	2	95	3	908	5	9	100/74	97/74	94/74	23	79	77	77
Russell AP	38	5	98	5	1864	0	4	101/73	98/73	95/73	29	78	76	75
Salina	38	5	97	4	1271	0	5	103/74	100/74	97/73	26	78	78	76
Topeka AP	39	0	95	5	877	0	4	99/75	96/75	93/74	24	79	78	76
Wichita AP	37	4	97	3	1321	3	7	101/72	98/73	96/73	23	77	76	75
KENTUCKY														
Ashland	38	0	82	5	551	5	10	94/76	91/74	89/73	22	78	77	75
Bowling Green AP	36	0	86	3	535	4	9	94/77	91/76	89/74	21	79	77	76
Corbin AP	37	0	84	1	1175	4	9	94/73	92/73	89/72	23	77	76	75
Covington AP	37	0	84	4	869	1	6	92/73	90/72	88/72	22	77	76	75
Hopkinsville, Ft. Campbell	36	4	87	3	540	4	10	94/77	92/75	90/72	21	79	77	77
Lexington AP (S)	38	0	84	4	979	3	8	94/73	91/73	88/72	22	77	76	77
Louisville AP	38	1	85	4	474	5	10	95/74	93/74	90/74	23	79	77	78
Madisonville	37	2	87	3	439	5	10	96/76	93/75	90/75	22	79	78	78
Owensboro	37	2	87	1	420	5	10	97/76	94/75	91/75	23	79	78	78
Paducah AP	37	0	88	5	398	7	12	98/76	95/75	92/75	20	80	78	78
LOUISIANA														
Alexandria AP	31	2	92	2	92	23	27	95/77	94/77	92/77	20	80	79	78
Baton Rouge AP	30	3	91	1	64	25	29	95/77	93/77	92/77	19	80	80	79
Bogalusa	30	5	89	5	103	24	28	95/77	93/77	92/77	19	80	80	79
Houma	29	5	90	4	13	31	35	95/78	93/78	92/77	15	81	80	79

(continued)

TABLE 15.1 Climatic Conditions for the United States (continued)

Col. 1	Col. 2		Col. 3		Col. 4	Col. 5		Col. 6			Col. 7	Col. 8		
	Latitude [b]		Longitude [b]		Elevation [f]	Winter [d], F Design Dry-Bulb		Summer [e], F Design Dry-Bulb and Mean Coincident Wet-Bulb			Mean Daily Range	Design Wet-Bulb		
State and Station	°		°		Ft	99%	97.5%	1%	2.5%	5%		1%	2.5%	5%
Lafayette AP	30	1	92	0	38	26	30	95/78	94/78	92/78	18	81	80	79
Lake Charles AP (S)	30	1	93	1	14	27	31	95/77	93/77	92/77	17	80	79	79
Minden	32	4	93	2	250	20	25	99/77	96/76	94/76	20	80	79	78
Monroe AP	32	5	92	0	78	20	25	99/77	96/76	94/76	20	79	79	78
Natchitoches	31	5	93	0	120	22	26	97/77	95/77	93/77	20	80	80	78
New Orleans AP	30	0	90	2	3	29	33	93/78	92/78	90/77	16	81	80	79
Shreveport AP(S)	32	3	93	5	252	20	25	99/77	96/76	94/76	20	79	79	78
MAINE														
Augusta AP	44	2	69	5	350	−7	−3	88/73	85/70	82/68	22	74	72	70
Bangor, Dow AFB	44	5	68	5	162	−11	−6	86/70	83/68	80/67	22	73	71	69
Caribou AP (S)	46	5	68	0	624	−18	−13	84/69	81/67	78/66	21	71	69	67
Lewiston	44	0	70	1	182	−7	−2	88/73	85/70	82/68	22	74	72	70
Millinocket AP	45	4	68	4	405	−13	−9	87/69	83/68	80/66	22	72	72	68
Portland (S)	43	4	70	2	61	−6	−1	87/72	84/71	81/69	22	74	72	70
Waterville	44	3	69	4	89	−8	−4	87/72	84/69	81/68	22	74	72	70
MARYLAND														
Baltimore AP	39	1	76	4	146	10	14	94/75	91/75	89/74	21	78	77	76
Baltimore CO	39	2	76	5	14	13	17	92/75	89/74	87/75	17	80	78	76
Cumberland	39	4	78	5	945	6	10	92/75	89/74	87/74	22	77	76	75
Frederick AP	39	3	77	3	294	8	12	94/76	91/75	89/74	22	78	76	75
Hagerstown	39	4	77	4	660	8	12	94/75	91/74	89/74	22	78	76	75
Salisbury (S)	38	2	75	3	52	12	16	93/75	91/75	88/74	18	79	77	76
MASSACHUSETTS														
Boston AP (S)	42	2	71	0	15	6	9	91/73	88/71	85/70	16	75	74	72
Clinton	42	4	71	4	398	−2	2	90/72	87/71	84/69	17	75	73	72
Fall River	41	4	71	1	190	5	9	87/72	84/71	81/69	18	74	73	72
Framingham	42	2	71	3	170	5	6	89/72	86/71	83/69	17	74	73	71
Gloucester	42	3	70	4	10	2	5	89/73	86/71	83/70	15	75	74	73
Greenfield	42	3	72	4	205	−7	−2	88/72	85/71	82/69	23	74	73	71
Lawrence	42	4	71	1	57	−6	0	90/73	87/72	84/70	22	76	74	73

Station	Lat °	Lat ′	Long °	Long ′	Elev	99%	97½%	1%	2½%	5%	DR			
Lowell	42	3	71	2	90	-4	1	91/73	88/72	85/70	21	76	74	73
New Bedford	41	3	71	0	70	5	9	87/72	82/71	80/69	19	74	73	72
Pittsfield AP	42	4	73	1	1170	-8	-3	85/71	84/70	81/68	23	73	72	70
Springfield,														
Westover AFB	42	1	72	3	247	-5	0	90/72	87/71	84/69	19	75	73	72
Taunton	41	5	71	1	20	5	9	89/73	86/72	83/70	18	75	74	73
Worcester AP	42	2	71	5	986	0	4	87/71	84/70	81/68	18	73	72	70
MICHIGAN														
Adrian	41	5	84	0	754	1	5	91/73	88/72	85/71	23	76	75	73
Alpena AP	45	2	83	3	689	-11	-6	89/70	85/70	83/69	27	73	72	70
Battle Creek AP	42	2	85	2	939	1	5	92/74	88/72	85/70	23	76	75	73
Benton Harbor AP	42	1	86	3	649	1	6	91/72	88/72	85/70	20	75	74	73
Detroit	42	2	83	0	633	3	6	91/73	88/72	86/71	20	76	74	73
Escanaba	45	4	87	0	594	-11	-7	87/70	83/69	80/68	17	73	72	69
Flint AP	42	0	83	4	766	-4	1	90/73	87/72	85/70	25	75	74	72
Grand Rapids AP	42	5	85	3	681	1	5	91/72	88/72	85/70	24	75	74	72
Holland	42	5	86	1	612	2	6	88/72	86/71	83/70	22	76	74	72
Jackson AP	42	2	84	2	1003	1	5	92/74	88/72	85/70	23	76	74	73
Kalamazoo	42	1	85	3	930	1	5	92/74	88/72	85/70	23	76	74	73
Lansing AP	42	5	84	4	852	1	5	92/72	87/72	84/70	24	75	74	72
Marquette CO	46	3	87	3	677	-12	-8	84/70	81/69	77/66	18	72	70	68
Mt Pleasant	43	4	84	4	796	0	4	91/73	87/72	85/71	24	76	74	72
Muskegon AP	43	1	86	1	627	2	6	86/73	84/70	82/70	21	75	73	72
Pontiac	42	4	83	1	974	0	4	90/73	87/72	85/71	21	76	74	73
Port Huron	43	0	82	2	586	0	4	90/73	87/72	83/71	21	76	74	72
Saginaw AP	43	3	84	1	662	0	4	91/73	87/72	84/71	23	76	74	72
Sault Ste. Marie AP (S)	46	3	84	2	721	-12	-8	84/72	81/69	77/66	23	72	70	68
Traverse City AP	44	4	85	4	618	-3	1	89/72	86/71	83/69	22	75	73	71
Ypsilanti	42	1	83	3	777	1	5	92/72	89/71	86/70	22	75	74	72
MINNESOTA														
Albert Lea	43	4	93	2	1235	-17	-12	90/74	87/72	84/71	24	77	75	73
Alexandria AP	45	5	95	4	1421	-22	-16	91/72	88/72	85/70	24	76	74	72
Bemidji AP	47	3	94	5	1392	-31	-26	88/69	85/69	81/67	24	71	71	69
Brainerd	46	2	94	1	1214	-20	-16	90/73	88/71	84/69	24	75	73	71
Duluth AP	46	5	92	1	1426	-21	-16	85/70	82/68	79/66	22	72	70	68

(continued)

TABLE 15.1 Climatic Conditions for the United States (continued)

| Col. 1 | Col. 2 | | Col. 3 | | Col. 4 | Winter, °F Col. 5 | | Summer, °F Col. 6 | | | Col. 7 | Col. 8 | | |
State and Station	Latitude °	'	Longitude °	'	Elevation Ft	Design Dry-Bulb 99%	97.5%	Design Dry-Bulb and Mean Coincident Wet-Bulb 1%	2.5%	5%	Mean Daily Range	Design Wet-Bulb 1%	2.5%	5%
Fairbault	44	2	93	3	1190	−17	−12	91/74	88/72	85/71	24	77	75	73
Fergus Falls	46	1	96	0	1210	−21	−17	91/72	88/72	85/70	24	76	74	72
International Falls AP	48	3	93	2	1179	−29	−25	85/68	83/68	80/66	26	71	70	68
Mankato	44	1	94	0	785	−17	−12	91/72	88/71	85/70	24	77	75	73
Minneapolis/ St Paul AP	44	5	93	1	822	−16	−12	92/75	89/73	86/71	22	77	75	73
Rochester AP	44	0	92	3	1297	−17	−12	90/74	87/72	84/71	24	77	75	73
St Cloud AP (S)	45	4	94	1	1034	−15	−11	91/74	88/72	85/70	24	76	74	72
Virginia	47	3	92	3	1435	−25	−21	85/69	83/68	80/66	23	71	70	68
Willmar	45	1	95	0	1133	−15	−11	91/74	88/72	85/71	24	76	74	72
Winona	44	1	91	4	652	−14	−10	91/75	88/73	85/72	24	77	75	74
MISSISSIPPI														
Biloxi,														
Keesler AFB	30	2	89	0	25	28	31	94/79	92/79	90/78	16	82	81	80
Clarksdale	34	1	90	3	178	14	19	96/77	94/77	92/76	21	80	79	78
Columbus AFB	33	4	88	3	224	15	20	95/77	93/77	91/76	22	80	79	78
Greenville AFB	33	3	91	1	139	15	20	95/77	93/77	91/76	21	80	79	78
Greenwood	33	3	90	1	128	15	20	95/77	93/77	91/76	21	80	79	78
Hattiesburg	31	2	89	2	200	21	24	96/78	94/77	92/77	21	81	80	79
Jackson AP	32	2	90	1	330	21	25	97/76	95/76	93/76	21	79	78	78
Laurel	31	4	89	0	264	24	27	96/78	94/77	92/77	21	81	80	79
McComb AP	31	1	90	3	458	21	26	96/77	94/76	92/76	18	80	79	78
Meridian AP	32	2	88	5	294	19	23	95/76	93/76	91/76	22	80	79	78
Natchez	31	4	91	1	168	23	27	96/78	94/78	92/77	21	81	80	79
Tupelo	34	2	88	4	289	10	14	96/77	94/77	92/76	22	80	79	78
Vicksburg CO	32	2	91	0	234	16	22	97/78	95/78	93/77	21	81	80	79
MISSOURI														
Cape Girardeau	37	1	89	3	330	8	13	98/76	95/75	92/75	21	79	78	77

	Lat. °	′	Long. °	′	Elev.	Winter		Summer design dry/wet bulb			Daily range	Design wet bulb		
Columbia AP (S)	39	0	92	2	778	−1	4	97/74	94/74	91/73	22	78	77	76
Farmington AP	37	5	90	5	928	−3	3	96/76	93/75	90/74	22	78	77	75
Hannibal	39	1	91	1	489	−2	8	96/76	93/76	90/76	22	80	78	77
Jefferson City	38	4	92	1	640	2	7	98/75	95/74	92/74	23	78	77	76
Joplin AP	37	1	94	1	982	6	10	100/73	97/73	94/73	24	78	77	76
Kansas City AP	39	1	94	4	742	2	6	99/75	96/74	93/74	20	78	77	76
Kirksville AP	40	1	92	0	966	−5	0	96/74	93/74	90/73	24	78	77	76
Mexico	39	1	92		775	−1	4	97/74	94/74	91/73	22	78	77	76
Moberly	39	3	92		850	−2	3	97/74	94/74	91/73	23	78	77	76
Poplar Bluff	36	5	90	0	322	11	16	98/78	95/76	92/76	22	81	79	78
Rolla	38	0	91	2	1202	3	9	94/77	91/75	89/74	22	78	79	78
St Joseph AP	39	5	95	0	809	−3	2	96/77	95/76	91/76	23	81	79	77
St Louis AP	38	5	90	2	535	2	6	97/75	94/75	91/74	21	78	77	76
St Louis CO	38	4	90	2	465	3	8	98/75	94/75	91/74	18	78	77	76
Sedalia, Whiteman AFB	38	4	93	3	838	−1	4	95/76	92/76	90/75	22	79	78	76
Sikeston	36	5	89	3	318	9	15	98/77	96/76	92/75	21	80	78	77
Springfield AP	37	1	93	2	1265	3	9	96/73	93/74	91/74	23	78	77	75
MONTANA														
Billings AP	45	5	108	3	3567	−15	−10	94/64	91/64	88/63	31	67	66	64
Bozeman	45	5	111	0	4856	−20	−14	90/61	87/60	84/59	32	63	62	60
Butte AP	46	0	112	3	5526r	−24	−17	86/58	83/56	80/56	35	60	58	57
Cut Bank AP	48	4	112	2	3838r	−25	−20	88/61	85/61	82/60	35	64	62	61
Glasgow AP (S)	48	1	106	4	2277	−22	−18	92/64	89/63	85/62	29	68	66	64
Glendive	47	1	104	4	2076	−18	−13	95/66	92/64	89/62	29	69	67	65
Great Falls AP (S)	47	3	111	2	3664r	−21	−15	91/60	88/60	85/59	28	64	62	60
Havre	48	3	109	4	2488	−18	−11	94/65	90/64	87/63	33	68	66	65
Helena AP	46	4	112	0	3893	−21	−16	90/60	88/60	85/59	33	64	62	61
Kalispell AP	48	2	114	2	2965	−14	−7	91/62	87/61	84/60	34	65	63	62
Lewiston AP	47	0	109	3	4132	−22	−16	90/62	87/61	83/60	30	65	63	62
Livingston AP	45	4	110	3	4653	−20	−14	90/61	87/60	84/59	32	63	63	60
Miles City AP	46	3	105	5	2629	−20	−15	98/66	95/66	92/65	30	70	68	67
Missoula AP	46	5	114	1	3200	−13	−6	92/62	88/61	85/60	36	65	63	62
NEBRASKA														
Beatrice	40	2	96	5	1235	−5	−2	99/75	95/74	92/74	24	78	77	76

(continued)

TABLE 15.1   Climatic Conditions for the United States (continued)

	Col. 2 Latitude		Col. 3 Longitude		Col. 4 Eleva-tion	Col. 5 Design Dry-Bulb		Col. 6 Design Dry-Bulb and Mean Coincident Wet-Bulb			Col. 7 Mean Daily Range	Col. 8 Design Wet-Bulb		
State and Station	°	′	°	′	Ft	99%	97.5%	1%	2.5%	5%		1%	2.5%	5%
Chadron AP	42	5	103	0	3300	−8	−3	97/66	94/65	91/65	30	71	69	68
Columbus	41	3	97	2	1442	−6	−2	98/74	95/73	92/73	25	77	76	75
Fremont	41	3	96	3	1203	−6	−2	98/75	95/74	92/74	22	78	77	76
Grand Island AP	41	0	98	2	1841	−8	−3	97/72	94/71	91/71	28	75	74	73
Hastings	40	4	98	3	1932	−7	−3	97/72	94/71	91/71	27	75	74	73
Kearney	40	4	99	1	2146	−9	−4	96/71	93/70	90/70	28	74	73	72
Lincoln CO (S)	40	5	96	5	1150	−5	−2	99/75	95/74	92/74	24	78	77	76
McCook	40	1	100	4	2565	−6	−4	98/69	95/69	91/69	28	74	72	71
Norfolk	42	0	97	3	1532	−8	−4	97/74	93/74	90/73	30	78	75	75
North Platte AP (S)	41	1	100	4	2779	−8	−4	97/69	94/69	90/69	28	74	72	71
Omaha AP	41	2	95	5	978	−8	−3	94/76	91/75	88/74	22	78	77	75
Scottsbluff AP	41	5	103	3	3950	−8	−3	95/65	92/65	90/64	31	70	68	67
Sidney AP	41	1	103	0	4292	−8	−3	95/65	92/65	90/64	31	70	68	67
NEVADA														
Carson City	39	1	119	5	4675	4	9	94/60	91/59	89/58	42	63	61	60
Elko AP	40	5	115	5	5075	−8	−2	94/59	92/59	90/58	42	63	62	62
Ely AP (S)	39	1	114	5	6257	−10	−4	89/57	87/56	85/55	39	60	59	58
Las Vegas AP (S)	36	1	115	1	2162	25	28	108/66	106/65	104/65	30	71	70	69
Lovelock AP	40	0	118	3	3900	8	12	98/63	96/63	93/62	42	66	65	64
Reno AP (S)	39	3	119	5	4404	5	10	95/61	92/60	90/59	45	64	63	61
Reno CO	39	3	119	5	4490	6	11	96/61	93/60	91/59	45	64	62	61
Tonopah AP	38	0	117	1	5426	5	10	94/60	92/59	90/58	40	64	62	61
Winnemucca AP	40	5	117	5	4299	−1	3	96/60	94/60	92/60	42	64	62	61
NEW HAMPSHIRE														
Berlin	44	3	71	1	1110	−14	−9	87/71	84/69	81/68	22	73	71	70
Claremont	43	2	72	2	420	−9	−4	89/72	86/70	83/69	24	74	73	71
Concord AP	43	1	71	3	339	−8	−3	90/72	87/70	84/69	26	74	73	71
Keene	43	0	72	1	490	−12	−7	90/72	87/70	85/70	25	74	73	71
Laconia	43	3	71	3	505	−10	−5	89/72	86/70	83/69	25	74	73	71
Manchester, Grenier AFB	43	0	71	3	253	−8	−3	91/72	88/71	85/70	24	75	74	72
Portsmouth, Pease AFB	43	1	70	5	127	−2	2	89/73	85/71	83/70	22	75	74	72

State and Station	Lat. °	′	Long. °	′	Elev. ft	Winter 99%	97½%	Summer 1%	2½%	5%	Range	WB 1%	2½%	5%
NEW JERSEY														
Atlantic City CO	39	3	74	3	11	10	13	92/74	89/74	86/72	18	78	77	75
Long Branch	40	2	74	0	20	10	13	93/74	90/73	87/72	18	78	77	75
Newark AP	40	4	74	1	11	10	14	94/74	91/73	88/72	18	77	76	75
New Brunswick	40	3	74	3	86	6	10	92/74	89/73	86/72	19	77	76	75
Paterson	40	5	74	1	100	6	11	94/74	91/73	88/72	21	76	76	74
Phillipsburg	40	4	75	1	180	1	6	92/73	89/72	85/71	21	76	75	75
Trenton CO	40	1	74	5	144	11	14	92/73	89/74	85/73	19	78	75	75
Vineland	39	3	75	0	95	8	11	91/75	89/74	86/73	19	78	76	75
NEW MEXICO														
Alamogordo, Holloman AFB	32	5	106	0	4070	14	19	98/64	96/64	94/64	30	69	68	67
Albuquerque AP (S)	35	0	106	4	5310	12	16	96/61	94/61	92/61	27	66	65	64
Artesia	32	5	104	2	3375	13	19	103/67	100/67	97/67	28	72	71	70
Carlsbad AP	32	2	104	2	3234	13	19	103/67	100/67	97/67	28	72	71	70
Clovis AP	34	2	103	1	4279	8	13	95/65	93/65	91/65	28	69	68	67
Farmington AP	36	4	108	0	5495	0	6	95/63	93/62	90/59	30	66	65	64
Gallup	35	5	108	5	6465	−1	4	90/59	88/58	86/58	32	64	62	61
Grants	35	1	107	5	6520	−1	5	89/59	87/59	86/58	32	64	62	61
Hobbs AP	32	4	103	0	3664	13	18	101/66	99/66	96/66	29	71	70	69
Las Cruces	32	2	106	2	3900	15	20	99/64	96/64	94/64	30	69	67	67
Los Alamos	35	5	106	3	7410	−1	9	89/60	87/60	85/60	32	62	61	60
Raton AP	36	5	104	3	6379	5	4	91/60	89/60	87/60	34	65	64	63
Roswell, Walker AFB	33	2	104	4	3643	13	18	100/66	98/66	96/66	33	71	70	70
Santa Fe CO	35	4	105	0	7045	10	4	90/61	88/61	86/61	28	63	62	61
Silver City AP	32	4	108	1	5373	13	17	94/60	92/60	90/60	30	66	64	63
Socorro AP	34	0	106	5	4617	17	13	97/62	95/62	93/62	30	67	66	65
Tucumcari AP	35	1	103	4	4053	8	13	99/66	96/66	93/65	28	70	69	68
NEW YORK														
Albany AP (S)	42	4	73	5	277	−6	−1	91/73	88/72	85/70	23	75	73	72
Albany CO	42	4	73	5	19	−4	1	91/73	88/72	85/70	20	75	73	72
Auburn	42	5	76	3	715	−3	2	90/73	87/71	84/70	22	74	73	72
Batavia	43	0	78	1	900	−1	5	90/72	87/71	84/70	22	73	72	70
Binghamton AP	42	1	75	5	1590	−2	2	86/71	83/69	81/68	20	72	71	70
Buffalo AP	43	0	78	4	705r	2	6	88/71	85/70	83/69	21	73	72	71
Cortland	42	4	76	1	1129	0	5	88/71	85/71	83/69	23	73	72	71
Elmira AP	42	1	76	5	860	−4	1	89/71	86/71	85/71	24	74	73	72
Geneva (S)	42	5	77	0	590	−3	1	90/73	87/71	85/71	22	75	74	72
Glens Falls	43	2	73	4	321	−11	−5	88/72	85/71	82/69	23	74	73	71
Gloversville	43	1	74	2	790	−8	−2	89/72	86/71	83/69	23	75	74	72

(continued)

TABLE 15.1   Climatic Conditions for the United States (continued)

Col. 1	Col. 2		Col. 3		Col. 4	Col. 5		Col. 6			Col. 7	Col. 8		
	Latitude[b]		Longitude[c]		Elevation[d]	Winter,[d] °F		Summer,[e] °F						
						Design Dry-Bulb		Design Dry-Bulb and Mean Coincident Wet-Bulb			Mean Daily	Design Wet-Bulb		
State and Station	°	'	°	'	Ft	99%	97.5%	1%	2.5%	5%	Range	1%	2.5%	5%
Hornell	42	2	77	4	1325	-4	0	88/71	85/70	82/69	24	74	73	72
Ithaca (S)	42	3	76	3	950	-5	0	88/71	85/71	82/70	24	74	73	71
Jamestown	42	1	79	2	1390	-1	3	88/70	86/70	83/69	20	74	72	71
Kingston	42	0	74	0	279	-3	2	91/73	88/72	85/70	22	76	74	73
Lockport	43	1	78	4	520	4	7	89/74	86/72	84/71	21	76	74	73
Massena AP	45	1	75	0	202r	-13	-8	86/70	83/69	80/68	20	73	72	70
Newburg-Stewart AFB	41	3	74	1	460	-1	4	90/73	88/72	85/71	21	76	74	73
NYC-Central Park (S)	40	5	74	0	132	11	15	92/74	89/73	87/72	17	76	75	74
NYC-Kennedy AP	40	4	73	5	16	12	15	90/73	87/72	84/71	16	76	75	74
NYC-La Guardia AP	40	5	73	5	19	11	15	92/74	89/73	87/72	16	76	75	74
Niagara Falls AP	43	1	79	0	596	4	7	89/74	86/72	84/71	20	76	74	73
Olean	42	1	78	3	1420	-2	2	87/71	84/71	81/70	23	74	73	71
Oneonta	42	3	75	0	1150	-7	-1	86/71	83/69	80/68	24	73	72	70
Oswego CO	43	3	76	3	300	1	7	86/73	83/71	80/70	20	75	73	72
Plattsburg AFB	44	4	73	3	165	-13	-8	86/70	84/69	80/68	22	75	73	72
Poughkeepsie	41	4	73	5	103	0	6	92/74	89/74	86/72	21	77	75	74
Rochester AP	43	1	77	4	543	1	5	91/73	88/71	85/70	22	75	73	72
Rome-Griffiss AFB	43	1	75	2	515	-11	-5	88/71	85/70	83/69	22	75	74	73
Schenectady (S)	42	5	73	5	217	-4	1	90/73	87/72	84/70	22	75	74	73
Suffolk County AFB	40	5	72	4	57	7	10	86/72	83/71	80/70	16	76	75	73
Syracuse AP	43	1	76	1	424	-3	2	90/73	87/71	84/70	20	75	74	73
Utica	43	1	75	2	714	-12	-6	88/73	85/71	82/70	22	75	74	72
Watertown	44	0	76	0	497	-11	-6	86/73	83/71	81/70	20	75	73	72
NORTH CAROLINA														
Asheville AP	35	3	82	3	2170	10	14	89/73	87/72	85/71	21	75	74	72
Charlotte AP	35	0	81	0	735	18	22	95/74	93/74	91/74	20	77	76	76
Durham	36	0	78	5	406	16	20	94/75	92/75	90/75	20	78	77	76
Elizabeth City AP	36	2	76	1	10	12	19	93/78	91/77	89/76	18	80	78	78
Fayetteville, Pope AFB	35	1	79	0	95	17	20	95/76	92/76	90/75	20	79	78	77

Station	Lat. °	Long. °	Elev. ft	Winter 99%	Winter 97½%	Summer 1%	Summer 2½%	Summer 5%	Daily Range	WB a	WB b	WB c
Goldsboro, Seymour-Johnson AFB	35	78	88	18	21	94/77	91/76	89/75	18	77	78	79
Greensboro AP (S)	36	80	887	14	18	93/77	91/76	89/73	21	75	76	77
Greenville	35	77	25	18	21	93/77	91/76	89/75	19	77	78	79
Henderson	36	78	510	12	15	95/77	92/76	90/76	20	77	78	79
Hickory	35	81	1165	14	18	92/78	90/72	88/72	21	73	74	80
Jacksonville	34	77	24	20	24	92/78	90/78	88/77	18	78	79	79
Lumberton	34	79	132	18	21	95/76	92/76	90/75	20	77	78	80
New Bern AP	35	77	17	20	24	92/78	90/78	88/77	18	78	79	80
Raleigh/Durham AP (S)	35	78	433	16	20	94/75	92/75	90/75	20	76	77	78
Rocky Mount	36	77	81	16	19	94/77	91/76	89/77	19	77	78	78
Wilmington AP	34	78	30	23	26	93/79	91/78	89/79	18	79	80	81
Winston-Salem AP	36	80	967	16	20	94/74	91/73	89/73	20	74	75	76
NORTH DAKOTA												
Bismarck AP (S)	46	100	1647	−23	−19	95/68	91/68	88/67	27	70	71	73
Devil's Lake	48	98	1471	−21	−17	91/69	88/68	85/66	25	69	71	71
Dickinson AP	46	102	2595	−17	−13	94/66	90/66	87/65	25	68	69	71
Fargo AP	46	96	900	−18	−14	92/73	89/71	86/69	25	72	74	76
Grand Forks AP	48	97	832	−26	−22	91/70	89/69	87/68	26	70	72	74
Jamestown AP	47	98	1492	−22	−18	94/70	90/69	87/68	25	71	74	72
Minot AP	48	101	1713	−24	−20	92/68	89/67	86/65	25	68	70	72
Williston	48	103	1877	−25	−21	91/68	88/65	85/65	25	68	70	72
OHIO												
Akron-Canton AP	41	81	1210	1	6	89/72	86/71	84/70	21	72	73	75
Ashtabula	42	80	690	4	9	88/73	85/72	83/71	18	74	74	75
Athens	39	82	700	0	6	95/75	92/74	90/73	22	74	76	78
Bowling Green	41	83	675	−2	2	92/73	89/73	86/71	23	73	75	76
Cambridge	40	81	800	4	7	93/75	90/73	87/73	23	75	76	78
Chillicothe	39	83	638	0	6	95/75	92/74	90/73	22	74	76	78
Cincinnati CO	41	84	761	1	6	92/73	90/72	88/72	21	74	75	77
Cleveland AP (S)	40	81	777	0	5	91/73	88/72	86/71	21	73	74	76
Columbus AP (S)	40	83	812	−1	5	92/73	90/73	87/72	20	74	75	77
Dayton AP	39	84	997	−1	5	91/73	89/72	86/71	24	73	75	77
Defiance	41	84	700	2	4	94/74	91/73	88/72	24	74	76	77
Findlay AP	41	83	797	0	3	92/74	90/73	88/72	24	74	76	77
Fremont	41	83	600	−3	1	90/73	88/72	85/71	24	73	75	76
Hamilton	39	84	650	0	5	92/73	90/72	87/71	23	75	75	76
Lancaster	39	82	920	0	5	93/74	91/73	88/72	23	74	75	77

(continued)

TABLE 15.1   Climatic Conditions for the United States (continued)

Col. 1	Col. 2 Latitude		Col. 3 Longitude		Col. 4 Elevation Ft	Col. 5 Winter Design Dry-Bulb 99%	97.5%	Col. 6 Summer Design Dry-Bulb and Mean Coincident Wet-Bulb 1%	2.5%	5%	Col. 7 Mean Daily Range	Col. 8 Design Wet-Bulb 1%	2.5%	5%
State and Station	°	'	°	'										
Lima	40	4	84	0	860	-0	5	94/74	91/73	88/72	24	77	76	74
Mansfield AP	40	4	82	3	1297	-0	5	90/73	87/72	85/72	22	76	74	73
Marion	40	4	83	1		0	5	93/74	91/73	88/72	23	77	76	74
Middletown	39	3	84	3	635	5	5	92/73	90/73	87/71	22	76	75	73
Newark	40	1	82	2	825	-1	1	94/73	92/73	89/72	23	76	75	74
Norwalk	41	1	82	4	720	-3	1	90/73	88/73	85/71	22	76	75	73
Portsmouth	38	5	83	0	530	5	10	95/76	92/74	89/73	22	78	77	75
Sandusky CO	41	3	82	4	606	1	6	93/73	91/72	88/71	21	78	77	75
Springfield	40	0	83	5	1020	-1	3	91/74	89/73	87/72	21	77	76	74
Steubenville	40	2	80	4	992	1	5	89/72	86/71	84/70	22	74	73	72
Toledo AP	41	4	83	3	676r	-3	1	90/73	88/73	85/71	25	74	73	73
Warren	41	2	80	5	900	0	5	89/71	87/71	85/70	23	75	73	71
Wooster	41	5	82	0	1030	-1	6	89/71	86/71	84/70	23	75	74	72
Youngstown AP	41	2	80	4	1178	-1	4	88/71	86/71	84/70	23	74	73	71
Zanesville AP	40	0	81	5	881	1	7	93/75	90/74	87/73	23	78	76	75
OKLAHOMA														
Ada	34	5	96	4	1015	10	14	100/74	97/74	95/74	23	77	76	75
Altus AFB	34	4	99	1	1390	11	16	102/73	100/74	98/73	25	77	75	75
Ardmore	34	2	97	0	880	13	17	100/74	98/74	95/74	23	77	76	76
Bartlesville	36	5	96	0	715	6	10	101/73	98/74	95/74	23	78	77	76
Chickasha	35	0	98	0	1085	10	14	101/74	98/74	95/74	24	78	77	76
Enid-Vance AFB	36	2	98	0	1287	9	13	103/74	100/74	97/74	24	79	77	76
Lawton AP	34	3	98	2	1108	12	16	101/74	99/74	96/74	24	78	78	76
Mc Alester	34	5	95	5	760	14	19	99/74	96/74	93/74	23	78	77	75
Muskogee AP	35	4	95	2	610	10	15	101/74	98/75	95/75	23	79	78	77
Norman	35	1	97	3	1109	9	13	99/74	96/74	94/74	24	79	77	75
Oklahoma City AP (S)	35	2	97	4	1280	9	13	100/74	97/74	95/73	23	78	77	76
Ponca City	36	4	97	1	996	5	9	101/74	97/74	94/74	24	77	76	76
Seminole	36	2	96	4	865	11	15	99/74	96/74	94/73	23	78	77	75
Stillwater (S)	36	1	97	0	884	8	13	100/74	96/74	93/74	24	78	76	77
Tulsa AP	36	1	95	5	650	8	13	101/74	96/74	95/75	22	79	78	77
Woodward	36	3	99	3	1900	6	10	100/73	97/73	94/73	26	77	76	75

Station	Lat°	′	Long°	′	Elev	Winter 99%	97.5%	Summer 1%	2.5%	5%	Range	WB 1%	2.5%	5%
Albany	44	4	123	5	224	18	22	92/67	89/66	86/65	31	69	67	66
Astoria AP (S)	46	5	123	5	8	25	29	72/65	71/62	68/60	16	65	63	62
Baker AP	44	5	117	5	3368	-1	6	92/65	90/62	89/61	30	65	63	61
Bend	44	1	121	5	3599	-3	4	89/60	87/60	84/59	33	64	62	61
Corvallis (S)	44	3	123	2	221	18	22	92/67	89/66	86/65	31	69	67	66
Eugene AP	44	1	123	1	364	17	22	92/67	89/66	86/65	31	69	67	68
Grants Pass	42	3	123	2	925	20	24	99/69	96/68	93/67	33	71	69	68
Klamath Falls AP	42	1	121	5	4091	4	9	90/61	87/60	84/59	36	63	61	60
Medford AP (S)	42	2	122	5	1298	19	23	98/68	94/67	91/66	35	70	68	67
Pendleton AP (S)	45	4	118	5	1492	-2	5	97/65	93/64	90/62	29	66	65	63
Portland AP	45	4	122	4	21	17	23	89/68	85/67	81/65	23	69	67	66
Portland CO	45	3	122	4	57	18	24	90/68	86/67	82/65	23	69	67	66
Roseburg AP	43	1	123	2	505	18	23	93/67	90/66	87/65	30	69	69	66
Salem AP	45	0	123	0	195	18	21	92/68	88/66	84/65	31	69	68	66
The Dalles	45	4	121	1	102	13	19	93/69	89/68	85/66	28	70	68	67
PENNSYLVANIA														
Allentown AP	40	4	75	3	376	4	9	92/73	88/72	86/72	22	76	74	73
Altoona CO	40	2	78	2	1468	0	5	90/72	87/71	84/70	23	74	74	72
Butler	40	4	80	1	1100	1	6	91/73	88/72	85/73	22	75	76	73
Chambersburg	40	0	77	4	640	4	8	93/75	90/74	87/73	23	75	76	73
Erie AP	42	1	80	1	732	4	9	88/73	85/72	83/71	18	75	74	72
Harrisburg AP	40	1	76	5	335	9	11	94/75	91/74	88/73	21	77	76	75
Johnstown	40	2	78	5	1214	-3	2	86/70	83/70	80/68	23	72	71	70
Meadville	40	4	80	1	255	0	4	93/75	90/74	87/73	21	77	74	75
New Castle	41	0	80	2	1065	2	7	88/71	85/70	82/70	21	73	72	75
Philadelphia AP	39	5	75	1	825	10	14	93/75	91/73	89/71	21	77	74	75
Pittsburgh AP	40	3	80	1	1137	1	7	92/73	90/74	87/72	22	77	76	75
Pittsburgh CO	40	2	80	0	749r	3	7	89/72	86/71	84/70	19	74	73	74
Reading CO	40	2	76	0	226	9	13	92/73	89/72	86/72	19	74	73	74
Scranton/Wilkes-Barre	41	2	75	4	940	1	5	90/72	87/71	84/70	19	74	73	72
State College (S)	40	5	77	5	1175	5	9	90/72	87/71	84/70	21	74	73	72
Sunbury	40	5	76	5	480	3	7	92/73	89/72	86/72	22	76	75	73
Uniontown	39	5	79	4	1280	1	9	92/74	89/74	86/73	22	76	73	73
Warren	41	5	79	1	1280	-5	1	89/71	86/71	83/72	20	77	76	74
West Chester	40	0	75	4	440	9	13	92/73	89/74	86/72	20	77	74	76
Williamsport AP	41	1	77	0	527	2	7	89/74	86/71	86/73	22	77	74	74
York	40	0	76	4	390	9	12	94/73	91/74	88/73	21	77	75	75
RHODE ISLAND														
Newport (S)	41	3	71	2	20	5	9	88/73	85/72	82/70	16	76	75	73
Providence AP	41	4	71	3	55	5	9	89/73	86/72	83/70	19	75	74	73

(continued)

TABLE 15.1   Climatic Conditions for the United States (continued)

Col. 1	Col. 2 Latitude[b]		Col. 3 Longitude[b]		Col. 4 Elevation[c]	Col. 5 Winter,[d] °F Design Dry-Bulb		Col. 6 Summer,[e] °F Design Dry-Bulb and Mean Coincident Wet-Bulb			Col. 7 Mean Daily Range	Col. 8 Design Wet-Bulb		
State and Station	°	'	°	'	Ft	99%	97.5%	1%	2.5%	5%		1%	2.5%	5%
SOUTH CAROLINA														
Anderson	34	3	82	4	764	19	23	94/74	92/74	90/74	21	77	76	75
Charleston AFB (S)	32	5	80	0	41	24	27	93/78	91/78	89/77	18	81	80	79
Charleston CO	32	5	80	0	9	25	28	94/78	92/78	90/77	13	81	80	80
Columbia AP	34	0	81	0	217	20	24	97/76	95/75	90/75	22	79	78	77
Florence AP	34	1	79	4	146	21	25	94/77	92/77	89/74	21	80	79	78
Georgetown	33	2	79	2	14	23	26	92/79	90/78	88/77	18	81	80	79
Greenville AP	34	5	82	1	957	18	22	91/74	89/74	89/74	21	78	76	75
Greenwood	34	1	82	1	671	19	23	95/75	93/74	91/74	21	78	77	76
Orangeburg	33	3	80	5	244	20	24	97/75	95/75	93/75	20	79	78	77
Rock Hill	35	0	81	0	470	19	23	96/75	94/74	92/74	20	79	78	76
Spartanburg AP	35	0	82	0	816	18	22	93/74	91/74	89/74	20	77	76	75
Sumter-Shaw AFB	34	0	80	3	291	22	25	95/77	92/76	90/75	21	79	78	77
SOUTH DAKOTA														
Aberdeen AP	45	3	98		1296	−19	−15	94/73	91/72	88/70	27	77	75	73
Brookings	44	2	96	5	1642	−17	−13	95/73	92/72	89/71	25	77	75	73
Huron AP	44	2	98	1	1282	−18	−14	96/72	93/72	90/71	28	77	75	73
Mitchell	43	4	98	0	1346	−15	−10	96/72	93/71	90/70	28	76	75	73
Pierre AP	44	2	100	2	1718r	−15	−10	99/71	96/71	92/69	29	75	74	72
Rapid City AP (S)	44	0	103	0	3165	−11	−7	95/66	92/65	89/65	28	71	69	67
Sioux Falls AP	43	3	96	4	1420	−15	−11	94/73	91/72	88/71	24	76	75	73
Watertown AP	45	0	97	1	1746	−20	−15	93/72	90/71	88/71	26	75	73	73
Yankton	43	0	97	2	1280	−13	−7	94/73	91/72	90/75	25	77	76	74
TENNESSEE														
Athens	35	3	84		940	13	18	95/74	92/73	90/73	22	77	76	75
Bristol-Tri City AP	36	3	82	2	1519	9	14	91/72	89/72	87/71	22	75	75	73
Chattanooga AP	35	0	85	1	670	13	18	96/75	94/74	91/74	22	78	77	76
Clarksville	36	4	87	2	470	8	12	95/76	93/74	90/74	21	78	77	76
Columbia	35	3	87	0	690	10	15	97/75	94/77	91/76	21	79	78	78
Dyersburg	36	0	89	2	334	11	15	96/78	94/77	91/76	21	81	80	78
Greenville	35	3	82	5	1320	11	16	92/73	90/72	88/72	22	76	75	74
Jackson AP	35	4	88	5	413	13	18	95/75	92/75	90/73	21	79	78	77
Knoxville AP	35	5	84	0	980	13	19	94/74	92/73	90/73	21	79	78	77
Memphis AP	35	0	90	0	263	13	18	95/76	93/76	90/75	21	80	79	78
Murfreesboro	35	5	86		608	9	18	95/77	93/76	93/76		77	75	78

Column headings are continued from the preceding page. Each station row lists, left to right: latitude (° and additional figure), longitude (° and additional figure), elevation (ft), winter design 99% and 97.5% dry-bulb, summer design dry-bulb/mean coincident wet-bulb at 1%, 2.5% and 5%, mean daily range, and design wet-bulb at 1%, 2.5% and 5%.

Station	Lat		Long		Elev	99%	97½%	1%	2½%	5%	DR	1%	2½%	5%
Nashville AP (S)	36	1	86	4	577	9	14	97/75	94/74	91/74	21	78	77	76
Tullahoma	35	2	86	1	1075	8	13	96/74	93/73	91/73	22	77	76	75
TEXAS														
Abilene AP	32	3	99	4	1759	15	20	101/71	99/71	97/71	22	75	74	74
Alice AP	27	4	98	0	180	31	34	100/78	98/77	95/77	20	82	81	79
Amarillo AP	35	1	101	4	3607	6	11	98/67	95/67	93/67	26	71	70	70
Austin AP	30	2	97	4	597	24	28	100/74	98/74	97/74	22	78	78	77
Bay City	29	0	96	0	18	29	33	96/77	94/77	92/77	16	80	79	79
Beaumont	30	0	94	0	18	27	31	95/79	93/78	91/78	19	81	80	80
Beeville	28	2	97	2	225	30	33	99/78	97/77	95/77	18	82	81	79
Big Spring AP (S)	32	2	101	3	2537	16	20	100/74	97/69	92/77	18	74	73	72
Brownsville AP (S)	25	5	97	3	16	35	39	94/77	93/77	94/77	18	80	79	79
Bryan AP	30	4	96	2	275	29	33	98/78	96/76	94/76	20	79	78	78
Corpus Christi AP	27	5	97	3	43	32	36	95/78	94/78	92/78	19	80	80	79
Corsicana	32	0	96	3	425	22	26	100/75	98/75	96/75	21	79	78	78
Dallas AP	32	5	96	5	481	22	27	102/75	100/75	97/75	20	78	78	77
Del Río, Laughlin AFB	29	2	100	0	1072	28	31	100/73	98/73	97/73	24	79	78	77
Denton	33	1	97	1	655	17	22	101/74	99/74	97/74	22	78	77	76
Eagle Pass	28	5	100	3	743	32	32	101/73	99/73	99/73	24	79	78	77
El Paso AP (S)	31	5	106	5	3918	20	24	100/64	98/64	96/64	27	69	68	68
Fort Worth AP (S)	32	5	97	0	544r	17	22	101/74	99/74	97/74	22	78	77	76
Galveston AP	29	2	94	5	5	31	36	90/79	89/79	88/78	10	81	80	80
Greenville	33	0	96	1	575	19	22	101/74	99/74	99/74	21	78	77	76
Harlingen	26	1	97	4	37	35	39	96/77	94/77	93/77	19	80	79	79
Houston AP	29	4	95	2	50	28	32	96/77	94/77	92/77	18	80	79	79
Houston CO	29	5	95	2	158r	28	33	96/77	95/77	93/77	18	80	79	79
Huntsville	30	4	95	3	494	22	27	100/75	98/75	96/75	20	80	78	77
Killeen-Gray AFB	31	1	97	5	1021	22	25	99/73	97/73	95/73	22	77	76	75
Lamesa	32	4	102	0	2965	13	17	99/69	96/69	94/69	26	73	72	71
Laredo AFB	27	2	99	3	343	32	36	102/73	101/73	99/76	23	78	78	77
Longview	32	3	94	3	345	24	24	99/76	97/76	94/69	20	80	79	78
Lubbock AP	33	2	101	4	3243	10	15	99/69	96/69	94/69	26	73	72	71
Lufkin AP	31	1	94	2	286	25	29	97/77	95/77	93/77	21	80	79	78
McAllen	26	2	98	1	122	35	39	100/77	98/69	96/69	25	80	79	78
Midland AP (S)	32	3	102	1	2815r	16	21	100/69	99/74	98/74	26	73	72	71
Mineral Wells AP	32	2	98	0	934	17	22	101/74	99/74	98/76	22	79	79	78
Palestine CO	31	1	95	5	580	23	27	100/76	98/76	96/76	20	79	78	78
Pampa	35	3	101	1	3230	7	12	99/67	96/67	94/67	26	71	70	70
Pecos	31	2	103	3	2580	12	16	100/69	99/68	96/68	27	73	72	70
Plainview	34	1	101	4	3400	7	13	98/68	96/68	94/68	26	72	72	70

(continued)

TABLE 15.1   Climatic Conditions for the United States (continued)

Col. 1	Col. 2		Col. 3		Col. 4	Col. 5		Col. 6			Col. 7	Col. 8		
	Latitude[b]		Longitude[c]		Elevation[c]	Winter,[d] °F		Summer,[e] °F						
						Design Dry-Bulb		Design Dry-Bulb and Mean Coincident Wet-Bulb			Mean Daily Range	Design Wet-Bulb		
State and Station	°	′	°	′	Ft	99%	97.5%	1%	2.5%	5%		1%	2.5%	5%
Port Arthur AP	30	0	94	0	16	27	31	95/79	93/78	91/78	19	80	80	80
San Angelo Goodfellow AFB	31	2	100	2	1878	18	22	101/71	99/71	97/70	24	75	74	73
San Antonio AP (S)	29	3	98	3	792	25	30	99/72	97/73	96/73	19	77	76	76
Sherman Perrin AFB	33	4	96	4	763	15	20	100/75	98/75	95/74	22	78	77	76
Snyder	32	3	101	2	2325	13	18	100/70	98/70	96/70	26	74	73	72
Temple	31	1	97	2	675	22	27	100/74	99/74	97/74	22	78	77	77
Tyler AP	32	2	95	2	527	19	24	99/76	97/76	95/76	21	80	79	78
Vernon	34	1	99	1	1225	13	17	102/73	100/73	97/73	24	77	77	75
Victoria AP	28	5	97	0	104	29	32	98/78	96/77	94/77	18	82	81	79
Waco AP	31	4	97	1	500	21	26	101/75	99/75	97/75	22	78	78	77
Wichita Falls AP	34	0	98	3	994	14	18	103/73	101/73	98/73	24	77	76	75
UTAH														
Cedar City AP	37	4	113	1	5613	−2	5	93/60	91/60	89/59	32	65	63	62
Logan	41	4	111	5	4775	−3	2	93/62	91/61	88/60	33	65	64	63
Moab	38	5	109	3	3965	6	11	100/60	98/60	96/60	30	65	64	63
Ogden AP	41	1	112	0	4455	1	5	93/63	91/61	88/61	33	65	65	64
Price	39	4	110	5	5580	−2	5	93/60	91/60	89/60	33	65	64	63
Provo	40	1	111	4	4470	1	6	98/62	96/62	94/61	32	66	65	64
Richfield	38	5	112	0	5300	−2	3	93/60	91/60	89/59	34	65	65	63
St George CO	37	1	113	3	2899	21	14	103/65	101/65	99/64	33	70	68	67
Salt Lake City AP (S)	40	5	112	0	4220	3	8	97/62	95/62	92/61	32	66	65	64
Vernal AP	40	3	109	3	5280	−5	0	91/60	89/60	86/59	32	64	63	62
VERMONT														
Barre	44	1	72	3	1120	−16	−11	84/71	81/69	78/68	23	71	71	70
Burlington AP (S)	44	3	73	1	331	−12	−7	88/72	85/70	82/69	23	74	72	71
Rutland	43	3	73	0	620	−13	−8	87/72	84/70	81/69	23	74	72	71
VIRGINIA														
Charlottesville	38	0	78	3	870	14	18	94/74	91/74	88/73	23	77	76	75
Danville AP	36	3	79	2	590	14	16	94/74	92/73	90/73	21	78	77	76
Fredericksburg	38	2	77	3	50	10	14	96/76	93/75	90/74	21	78	77	76
Harrisonburg	38	3	78	5	1340	12	16	93/72	91/72	88/71	23	75	74	73
Lynchburg AP	37	2	79	1	947	12	16	93/74	90/74	88/73	21	77	76	75

Station	Lat.°	Long.°	Elev. ft	Winter 99%	Winter 97.5%	Summer 1%	Summer 2.5%	Summer 5%	Daily Range	WB 1%	WB 2.5%	WB 5%
Norfolk AP	36	76	20	20	22	95/76	92/76	90/75	20	79	78	77
Petersburg	37	77	194	14	17	95/76	92/76	90/75	20	79	78	77
Richmond AP	37	77	162	14	17	95/76	92/76	90/75	21	79	78	77
Roanoke AP	37	80	1174r	12	16	93/72	91/72	88/71	23	75	74	73
Staunton	38	79	1480	12	16	93/72	91/72	88/71	23	75	74	73
Winchester	39	78	750	6	10	93/75	90/74	88/74	21	77	76	75
WASHINGTON												
Aberdeen	47	123	12	24	28	80/65	77/62	73/61		65	63	62
Bellingham AP	48	122	150	21	25	81/67	77/65	74/63		68	65	63
Bremerton	47	122	162	21	25	82/65	78/64	75/62		66	64	63
Ellensburg AP	47	120	1729	2	6	94/65	91/64	87/62		66	65	63
Everett– Paine AFB	47	122	598	21	25	80/65	76/64	73/62		67	64	63
Kennewick	46	119	392	5	11	99/68	96/67	92/66		70	68	67
Longview	46	123	12	19	24	88/68	85/67	81/65		69	67	66
Moses Lake, Larson AFB	47	119	1183	1	7	97/66	94/65	90/63		67	66	64
Olympia AP	47	122	190	16	22	87/66	83/65	79/64		67	66	64
Port Angeles	48	123	99	24	27	72/62	69/61	67/60		64	62	61
Seattle– Boeing Field	47	122	14	26	32	84/68	81/66	77/65		69	67	65
Seattle CO (S)	47	122	14	27	33	85/68	82/66	78/65		69	67	65
Seattle– Tacoma AP (S)	47	122	386	21	26	84/65	80/64	76/62		66	64	63
Spokane AP (S)	47	117	2357	−6	2	93/64	90/63	87/62		65	64	62
Tacoma– Mc Chord AFB	47	122	350	19	24	86/66	82/65	79/63		68	66	64
Walla Walla AP	46	118	1185	7	11	97/67	94/66	90/65		69	67	66
Wenatchee	47	120	634	0	7	99/67	96/66	92/64		68	67	65
Yakima AP	46	120	1061	−2	5	96/65	93/65	89/63		68	66	65
WEST VIRGINIA												
Beckley	37	81	2330	−2	5	83/71	81/69	79/69	22	73	71	70
Bluefield AP	37	81	2850	−2	5	83/71	81/69	79/69	22	73	71	70
Charleston AP	38	81	939	7	11	92/74	90/73	87/72	20	76	75	74
Clarksburg	39	80	977	6	11	92/74	90/73	87/72	21	76	72	74
Elkins AP	38	79	1970	1	6	86/72	84/70	82/70	22	74	72	71
Huntington CO	38	82	565r	5	10	94/76	91/74	89/73	21	78	76	75
Martinsburg AP	39	78	537	4	9	93/75	90/73	87/73	21	77	76	74
Morgantown AP	39	80	1245	7	11	90/74	87/73	85/73	22	76	75	74
Parkersburg CO	39	81	615r	1	7	93/75	90/74	88/73	21	77	76	75
Wheeling	40	80	659	−1	5	89/72	86/71	84/70	21	74	73	72

(continued)

TABLE 15.1 Climatic Conditions for the United States (continued)

Col. 1	Col. 2 Latitude °		Col. 3 Longitude °		Col. 4 Elevation Ft	Winter, °F Col. 5 Design Dry-Bulb		Summer, °F Col. 6 Design Dry-Bulb and Mean Coincident Wet-Bulb			Col. 7 Mean Daily Range	Col. 8 Design Wet-Bulb		
State and Station						99%	97.5%	1%	2.5%	5%		1%	2.5%	5%
WISCONSIN														
Appleton	44	2	88	2	742	−14	−9	89/74	86/72	83/71	23	76	74	72
Ashland	46	3	90	5	650	−21	−16	85/70	82/68	79/66	23	72	70	68
Beloit	42	3	89	0	780	−7	−3	92/75	90/75	88/74	24	78	75	73
Eau Claire AP	44	5	91	3	888	−15	−11	92/75	89/73	86/71	23	77	75	73
Fond du Lac	43	5	88	3	760	−12	−8	89/74	86/72	84/71	23	77	74	72
Green Bay AP	44	3	88	1	683	−13	−9	88/74	85/72	83/71	23	76	74	72
La Crosse AP	43	5	91	2	652	−13	−9	91/75	88/73	85/72	22	77	75	74
Madison AP (S)	43	1	89	2	858	−11	−7	91/74	88/73	85/71	22	77	75	73
Manitowoc	44	0	87	4	660	−11	−7	89/74	86/72	83/71	21	76	74	72
Marinette	45	0	87	5	605	−11	−11	87/73	84/71	82/70	20	75	73	71
Milwaukee AP	43	0	87	5	672	−8	−4	90/74	87/73	84/71	21	76	74	73
Racine	42	4	87	5	640	−6	−2	91/75	88/73	85/72	21	77	75	74
Sheboygan	43	5	87	4	648	−10	−6	92/75	88/73	83/72	20	77	75	73
Stevens Point	43	0	89	5	1079	−11	−7	92/75	89/73	86/71	22	77	75	73
Waukesha	43	0	88	0	860	−9	−5	90/74	87/73	84/71	22	76	75	74
Wausau AP	44	6	89	1	1196	−12	−8	91/74	88/72	85/70	23	77	74	72
WYOMING														
Casper AP	42	5	106	2	5319	−11	−5	92/58	90/57	87/57	31	63	61	60
Cheyenne AP	41	1	104	5	6126	−9	−1	89/58	86/58	84/57	30	63	62	60
Cody AP	44	3	109	0	5090	−19	−13	89/60	86/60	83/59	32	63	63	61
Evanston	41	2	111	0	6860	−9	−3	86/55	84/55	82/55	32	59	58	57
Lander AP (S)	42	5	108	4	5563	−16	−11	91/61	88/61	85/60	32	64	63	61
Laramie AP (S)	41	2	105	4	7266	−14	−6	84/56	81/56	79/55	28	61	60	59
Newcastle	43	5	104	1	4480	−17	−12	91/64	89/64	86/63	30	69	68	66
Rawlins	41	5	107	1	6741	−15	−9	87/63	85/63	83/62	40	62	61	60
Rock Springs AP	41	4	109	0	6741	−9	−3	86/57	84/55	81/54	32	59	58	57
Sheridan AP	44	5	107	0	3942	−14	−8	94/62	91/62	88/61	32	66	65	63
Torrington	42	0	104	1	4098	−14	−8	94/62	91/62	88/61	30	66	65	63

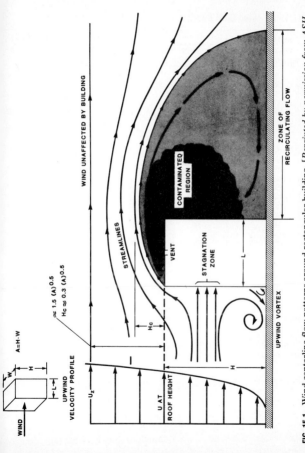

FIG. 15.1 Wind centerline flow patterns around rectangular building. [*Reprinted by permission from ASH-RAE Handbook—1985 Fundamentals (Inch-Pound Edition)*.]

TABLE 15.2   Wind Chill Chart[a]

Ambient Winds,[b] mph (km/h)	Ambient Temperature, F (°C)							
	35 (1.7)	30	25 (-3.9)	20	15 (-9.4)	10	5 (-15)	0
0	35	30	25	20	15	10	5	0
(0)	(1.7)		(-3.9)		(-9.4)		(-15)	
5	33	27	21	16	12	7	1	-6
(8)	(0.6)		(-6.1)		(-11.1)		(-17.2)	
10	21	16	9	2	-2	-9	-15	-22
(16)	(-6.1)		(-13)		(-19)		(-26)	
15	16	11	1	-6	-11	-18	-25	-33
(24)	(-8.9)		(-17)		(-24)		(-32)[c]	
20	12	3	-4	-9	-17	-24	-32	-40
(32)	(-11.1)		(-20)		(-27)		(-36)	
25	7	0	-7	-15	-22	-29	-37	-45
(40)	(-13.9)		(-22)		(-30)		(-38)	
30	5	-2	-11	-18	-26	-33	-41	-49
(48)	(-15)		(-24)		(-32)		(-41)	
35	3	-4	-13	-20	-27	-35	-43	-52
(56)	(-16)		(-25)		(-33)		(-42)	
40	1	-4	-15	-22	-29	-36	-45	-54
(64.4)	(-17)		(-26)		(-34)		(-43)	
45	1	-6	-17	-24	-31	-38	-46	-54
(72.4)	(-17.2)		(-27)		(-35)		(-43)	
50	0	-7	-17	-24	-31	-38	-47	-56
(80.5)	(-17.8)		(-27)		(-35)		(-44)	

[a]Equivalent in cooling power on exposed flesh under calm conditions. Equivalent temperatures (t_{eq}) are above those calculated at 10°F intervals; for additional levels, use $5/9(F - 32)$.

[b]Wind speeds greater than 40 mi/h (64.4 km/h) have little added chilling effect.

SOURCE: Reprinted by permission from *ASHRAE Handbook—1981 Fundamentals*. ASHRAE's source of data: Environmental Science Services Administration.

−5 (−20.6)	−10	−15 (−26)	−20	−25 (−32)	−30	−35 (−37)	−40	−45 (−43)
− 5 (−20.6)	−10	−15 (−26)	−20	−25 (−32)	− 30	− 35 (−37)	− 40	− 45 (−43)
−11 (−24)	−15	−20 (−29)	−26	−31 (−35)	− 35	− 41 (−40.6)	− 47	− 54 (−48)
−27 (−33)	−31	−38 (−39)	−45	−52 (−47)	− 58	− 64 (−53)	− 70	− 77 (−61)
−40 (−40)	−45	−51 (−46)	−60	−65 (−54)	− 70	− 78 (−61)	− 85	− 90 (−68)
−46 (−43)	−52	−60 (−51)	−68	−76 (−60)	− 81	− 88 (−67)	− 96	−103 (−75)
−52 (−47)	−58	−67 (−55)	−75	−83 (−64)	− 89	− 96 (−71)	−104	−112 (−80)
−56 (−49)	−63	−70 (−57)	−78	−87 (−66)	− 94	−101 (−74)	−109	−117 (−83)
−60 (−51)	−67	−72 (−58)	−83	−90 (−68)	− 98	−105 (−76)	−113	−123 (−86)
−62 (−52)	−69	−76 (−60)	−87	−94 (−70)	−101	−107 (−77)	−116	−128 (−89)
−63 (−53)	−70	−78 (−61)	−87	−94 (−70)	−101	−108 (−78)	−181	−128 (−89)
−63 (−53)	−70	−79 (−62)	−88	−96 (−71)	−103	−110 (−79)	−120	−128 (−89)

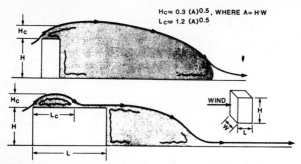

$H_C \approx 0.3\,(A)^{0.5}$, WHERE $A = H \cdot W$
$L_C \approx 1.2\,(A)^{0.5}$

FIG. 15.2 Effect of building length on air flow reattachment. [*Reprinted by permission from ASHRAE Handbook—1985 Fundamentals (Inch-Pound Edition).*]

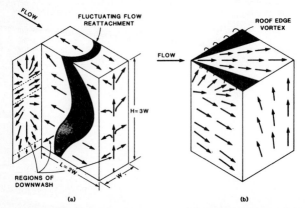

FIG. 15.3 Surface wind flow patterns on high-rise buildings. [*Reprinted by permission from ASHRAE Handbook—1985 Fundamentals (Inch-Pound Edition).*]

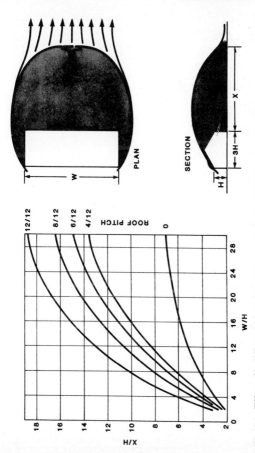

FIG. 15.4 Effect of building width and roof pitch on zone of recirculating air flow. [*Reprinted by permission from ASHRAE Handbook—1985 Fundamentals (Inch-Pound Edition)*].

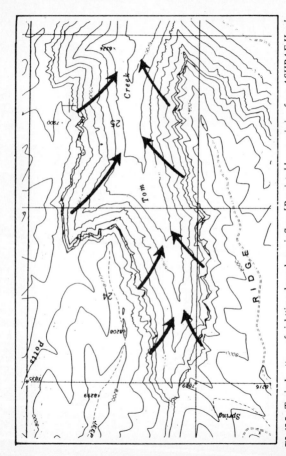

FIG. 15.5 Typical pattern of nighttime downslope air flow. [*Reprinted by permission from ASHRAE Handbook—1985 Fundamentals (Inch-Pound Edition).*]

TABLE 15.3 Beaufort Scale of Wind Velocity

Beaufort scale No.	Description	Indicators of velocity	Velocity, mi/h
0	Calm	Calm air; smoke rises vertically	<1
1	Light air	Direction of wind shown by smoke drift but not by wind vanes	1–3
2	Slight breeze	Wind felt on face; leaves rustle; ordinary vane moved by wind	4–7
3	Gentle breeze	Leaves and small twigs in constant motion; wind extends light flag	8–12
4	Moderate breeze	Raises dust and loose paper; small branches are moved	13–18
5	Fresh breeze	Small trees in leaf sway; crested wavelets form on inland waters	19–24
6	Strong breeze	Large branches in motion; whistling heard in telegraph wires; umbrellas used with difficulty	25–31
7	High wind	Whole trees in motion; inconvenience when walking against wind	32–38
8	Gale	Breaks twigs off trees; wind generally impedes progress	39–46
9	Strong gale	Slight structural damage occurs to signs; branches broken	47–54
10	Whole gale	Trees uprooted or broken; considerable structural damage occurs	55–63
11	Storm	Very rarely experienced; accompanied by widespread damage	64–75
12	Hurricane		Over 75

**Zones Include Areas with Design Temperatures about
as follows: Zone I, −20 F and lower; Zone II,
0 to −20 F; and Zone III, above 0 F. Note that
cross hatched areas are outside of Zones I and III.**

FIG. 15.6   Condensation zones in the United States. [*Reprinted by permission from ASHRAE Handbook—1985 Fundamentals (Inch-Pound Edition).*]

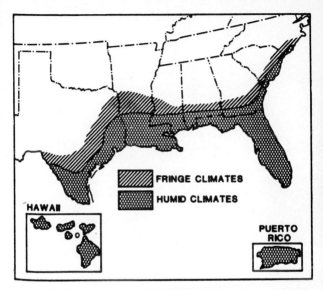

FIG. 15.7 Humid climates in the continental United States, Hawaii, and Puerto Rico. [*Reprinted by permission from ASHRAE Handbook—1985 Fundamentals (Inch-Pound Edition).*]

TABLE 15.4   Checklist for Condensation and Mold Growth

| Symptom | Possible Causes or Influence | |
	Heating System	Ventilation
Dampness/mold growth in cold weather only	Low or intermittent heat input leading to inadequate background heating	Heating and ventilation not matched; usually too little ventilation due to absence of natural paths of ventilation
Dampness/mold growth in rooms with large areas of exposed walls, roofs, or floors	Low or intermittent heat input leading to inadequate background heating	Heating and ventilation not matched
Dampness/mold growth in bedrooms	Unheated bedrooms; lack of background heat in dwelling generally	Poor ventilation not matched with heating; moist air entering from kitchens and bathrooms
Dampness/mold growth in or behind cupboards or in upper corners of rooms		Stagnant air pockets; lack of air movement behind furniture; lack of air movement in cupboards, etc.
Dampness/mold growth at times of rain or snow, especially with recurrence in summer during periods of rain		
Moisture staining, tidemarks on ceilings, walls, etc.	Radiator, boiler, pipe leaks	

Building	Householder	Heating System				Ventilation				Building						Action by Householder			
		Maintain	Supplement, Livingroom	Supplement, Bedroom	Supplement, Hall, Ldg.	Exhaust Vent, Kitchen	Exhaust Vent, Bathroom	Window Vent (Slot Type)	Cupboard Ventilators	Insulate Roof	Insulate Walls	Insulate Floor	Double Glazing	Alter Fittings	Mold Treatment	Adequate Heat Input	Adequate Ventilation	Reduce Use of LPG, etc.	Mold Treatment
Inappropriate construction for intermittent heating, i.e., slow response, heavy weight	Heat input too low, particularly in bedrooms; intermittent heating; drying clothes indoors; use of LPG or wood	○					○	○		○	○	○				●	●	●	○
Exposed or projecting areas of building that are poorly insulated (incl. stairwells); downstair bedrooms; large windows, esp. full-height windows	Heat input too low, particularly in bedrooms; uninsulated water pipes; intermittent heating; use of LPG or wood	○	●	●	●		○	●		●	●	●	●			○	○	○	○
Exposed or projecting bedrooms, poorly insulated, esp. north-facing unheated bedrooms downstairs; bedrooms over unheated spaces, e.g. walkways, garages	Heat input too low, particularly at night; leaving kitchen and bathroom doors open; use of LPG or wood			●	○		○	●	○		○	○				○	●	○	○
Built-in cupboards or wardrobes on outside walls	Bad positioning of furniture; overfilling of cupboards; putting clothes away damp						○	○	○					●	○		○	○	○
Dampness from water penetration; building cracks; defective flashing or pointing; blocked gutters; cracked gutters/rainwater pipes; exposure to driving rain		Repair, overhaul, replace as necessary																	
Dampness from plumbing leaks; cracked water pipes, tanks, etc.		Repair, overhaul, replace as necessary																	

SOURCE: Reprinted by permission from *ASHRAE Handbook—1985 Fundamentals* (Inch-Pound Edition).

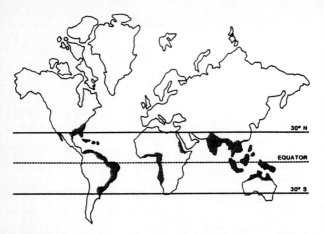

FIG. 15.8 Humid climates of the world. Humid climates usually occur in latitudes 30°S to 30°N, in low coastal regions; for any specific location, local weather records should be consulted. [*Reprinted by permission from ASHRAE Handbook—1985 Fundamentals (Inch-Pound Edition).*]

15.1 WEATHER EXTREMES IN CANADA AND THE UNITED STATES[1]

1. U.S. lowest annual mean temperature [−13°C (9°F)] and coolest summer average [2°C (36°F)]: Barrow, AK

2. U.S. lowest temperature [−62°C (−80°F)]: Prospect Creek, Endicott Mts., AK

3. Alaska's greatest snowfall in 24 h [157.5 cm (62 in)], in one storm [445.5 cm (175 in)], and in one season [2475 cm (974.5 in)]: all at Thompson Pass, AK

4. Canada's greatest snowfall in a climatological day [118 cm (46 in)]: Lakelse Lake, British Columbia

5. North America's greatest average yearly precipitation [650 cm (256 in)]: Henderson Lake, British Columbia, Canada

6. Canada's greatest 24-h rainfall [49 cm (19 in)]: Ucluelet Brynnor Mines, British Columbia

7. U.S. West Coast's foggiest place (average 2552 h/yr): Cape Disappointment, WA

8. North America's greatest snowfall in one season [2850 cm (1122 in)]: Rainier Paradise Ranger Station, WA

9. North America's greatest snowfall in one storm [480 cm (189 in)]: Mt. Shasta Ski Bowl, CA

10. North America's greatest depth of snow on the ground [1145.5 cm (451 in)]: Tamarack, CA

11. Western Hemisphere's highest temperature [57°C (134°F)] and highest summer average [37°C (98°F)]; U.S. highest annual mean temperature [26°C (78°F)] and lowest average

[1]Adapted from material appearing in Pauline Riordan and Paul G. Bourget, *World Weather Extremes,* compiled by the Geographic Sciences Laboratory, U.S. Army Engineer Topographic Laboratories, Ft. Belvoir, VA, December 1985. Used with permission.

yearly precipitation [4.1 cm (1.63 in)]: all at Death Valley, CA

12. U.S. longest dry period (767 days): Bagdad, CA

13. U.S. coldest winter average temperature [−26.5°C (−16°F)]: Barter Island, AK

14. North America's lowest temperature, excluding Greenland [−63°C (−81°F)]: Snag, Yukon Territory, Canada

15. North America's highest sea-level air pressure [106.76 kPa (31.53 in)]: Mayo, Yukon Territory, Canada

16. Canada's greatest snowfall in one season [2446.5 cm (964 in)]: Revelstoke, Mt. Copeland, British Columbia

17. U.S. largest 24-h temperature fall [56°C (100°F)]: Browning, MT

18. U.S. lowest temperature, excluding Alaska [−56.5°C (−70°F)]: Rogers Pass, MT

19. North America's greatest 24-h snowfall [192.5 cm (76 in)]: Silver Lake, CO

20. Canada's heaviest hailstone [290 g (10.23 oz)]: Cedoux, Saskatchewan

21. Canada's highest temperature [45°C (113°F)]: Midale and Yellow Grass, Saskatchewan

22. U.S. largest 2-min temperature rise [27°C (49°F)]: Spearfish, SD

23. Three temperature rises and two falls of 22°C (40°F) or over during a 3-h, 10-min period: Rapid City, SD

24. World's greatest 42-min rainfall [30.5 cm (12 in)]: Holt, MO

25. U.S. largest hailstone circumference [44.5 cm (17.5 in)]: Coffeyville, KS

26. A 24-h rainfall of 109 cm (43 in): Alvin, TX

27. North America's lowest mean temperature for a month, ex-

cluding Greenland [−48°C (−54°C)], and Canada's lowest annual mean temperature [−19°C (−3°F)]: both at Eureka, Northwest Territories

28. Canada's highest average annual wind speed [36 km/h (22 mi/h)]: Cape Warwick, Resolution Island, Northwest Territories

29. Canada's highest maximum observed hourly wind speed [201 km/h (125 mi/h)]: Cape Hopes Advance, Quebec

30. U.S. East Coast's foggiest place (average 1580 h/yr): Moose Peak Lighthouse, Mistake Island, ME

31. World's highest surface wind peak gust [372 km/h (231 mi/h)] and 5-min wind speed [303 km/h (188 mi/h)]; U.S. highest average annual wind speed [56 km/h (35 mi/h)]: Mt. Washington, NH; Mt. Washington also had a mean wind speed for 24 h of 206 km/h (128 mi/h) and a mean wind speed for a month of 112 km/h (70 mi/h)

32. World's's greatest 1-min rainfall [3.1 cm (1.23 in)]: Unionville, MD

33. North America's lowest sea-level air pressure [89.23 kPa (26.35 in)]: Matecumbe Key, FL

34. World's greatest average yearly precipitation [1168 cm (460 in)]: Mt. Waialeale, Kauai, HI

35. U.S. warmest winter average temperature [23°C (73°F)]: Honolulu, HI

36. U.S. greatest rainfall in 12 months [1878 cm (739 in)]: Kukui, Maui, HI

15.2 WORLD WEATHER EXTREMES[2]

1. North America's lowest temperature (excluding Greenland) [−63°C (−81°F)]: Snag, Yukon Territory, Canada

2. North America's greatest snowfall in one season [2850 cm (1122 in)]: Rainier Paradise Ranger Station, WA

3. North America's greatest snowfall in one storm [480 cm (189 in)]: Mt. Shasta Ski Bowl, CA

4. North America's greatest depth of snow on the ground [1145.5 cm (451 in)]: Tamarack, CA

5. Western Hemisphere's highest temperature [57°C (134°F)] and hottest summer average [37°C (98°F)]: both at Death Valley, CA

6. North America's lowest average yearly precipitation [3 cm (1.2 in)]: Bataques, Mexico

7. World's greatest average yearly precipitation [1168 cm (460 in)]: Mt. Waialeale, Kauai, HI

8. North America's lowest mean temperature for a month (excluding Greenland) [−48°C (−54°F)]: Eureka, Northwest Territories, Canada

9. North America's greatest average yearly precipitation [650 cm (256 in)]: Henderson Lake, British Columbia, Canada

10. U.S. largest 24-h temperature fall [56°C (100°F)]: Browning, MT

11. U.S. largest 2-min temperature rise [27°C (49°F)]: Spearfish, SD

12. North America's greatest 24-h snowfall [192.5 cm (76 in)]: Silver Lake, CO

[2]Adapted from material appearing in Pauline Riordan and Paul G. Bourget, *World Weather Extremes,* compiled by the Geographic Sciences Laboratory, U.S. Army Engineer Topographic Laboratories, Ft. Belvoir, VA, December 1985. Used with permission.

13. World's highest surface wind peak gust [372 km/h (231 mi/h)] and 5-min wind speed [303 km/h (188 mi/h)]: Mt. Washington, NH; Mt. Washington also had a mean wind speed for 24 h of 206 km/h (128 mi/h) and a mean wind speed for a month of 112 km/h (70 mi/h)

14. World's greatest 1-min rainfall [3.1 cm (1.23 in)]: Unionville, MD

15. World's greatest 42-min rainfall [30.5 cm (12 in)]: Holt, MO

16. U.S. largest hailstone circumference [44.5 cm (17.5 in)]: Coffeyville, KS

17. 24-h rainfall of 109 cm (43 in) (possibly the world's greatest on flat terrain): Alvin, TX

18. Thule, Greenland, had a peak wind gust of 333 km/h (207 mi/h)

19. Greenland's lowest temperature [−66°C (−87°F)]: Northice

20. South America's greatest average yearly precipitation [899 cm (354 in)]: Quibdo, Colombia

21. World's lowest average yearly precipitation [0.08 cm (0.03 in)] and no rain for longer than 14 consecutive years: Arica, Chile

22. South America's highest temperature [49°C (120°F)]: Rivadavia, Argentina

23. South America's lowest temperature [−33°C (−27°F)]: Sarmiento, Argentina

24. Bahia Felix, Chile, averages 325 days/yr with rain

25. Bessans, France, had a snowfall of 172 cm (68 in) in 19 h

26. Europe's highest temperature [50°C (122°F)]: Seville, Spain

27. Europe's greatest average yearly precipitation [465 cm (183 in)]: Crkvice, Yugoslavia

28. World's greatest 20-min rainfall [20.5 cm (8.10 in)]: Curtea-de-Arges, Romania

29. Europe's lowest temperature [−55°C (−67°F)]: Ust'Shchugor, U.S.S.R.

30. Europe's lowest average yearly precipitation [16 cm (6.4 in)]: Astrakhan, U.S.S.R.

31. World's highest sea-level air pressure [108.38 kPa (32.01 in)]: Agata, U.S.S.R.

32. Africa's lowest temperature [−24°C (−11°F)]: Ifrane, Morocco

33. World's highest temperature [58°C (136°F)]: El Azizia, Libya

34. Africa's greatest average yearly precipitation [1029 cm (405 in)]: Debundscha, Cameroon; Debundscha, Cameroon, also has a 191-cm (75-in) average variability of annual precipitation

35. Africa's lowest average yearly precipitation [<0.25 cm (<0.1 in)]: Wadi Halfa, Sudan

36. Dallol, Ethiopia, has an annual mean temperature of 35°C (94°F) (possibly the world's highest)

37. Assab, Ethiopia, has 29°C (84°F) average afternoon dewpoint in June

38. Kampala, Uganda, averages 242 days/yr with thunderstorms

39. World's greatest 12-h rainfall [135 cm (53 in)]: Belouve, La Réunion Island

40. World's greatest 24-h rainfall [188 cm (74 in)] and 5-day rainfall [386 cm (152 in)]: Cilaos, La Réunion Island

41. Themed, Israel, has a 94 percent variability of annual precipitation

42. Asia's highest temperature [54°C (129°F)]: Tirat Tsvi, Israel

43. Persian Gulf had a 36°C (96°F) sea-surface temperature

44. Asia's lowest average yearly precipitation [4.6 cm) (1.8 in)]: Aden, South Yemen

45. Verkhoyansk, U.S.S.R., has a difference of 102°C (183°F) between lowest and highest recorded temperatures

46. Northern hemisphere's lowest temperature [−68°C (−90°F)]: Verkhoyansk and Oimekon, U.S.S.R.

47. Eastern Sayan Region of U.S.S.R. has 81°C (146°F) mean annual temperature range

48. Lhasa, Tibet, had a 100 percent relative variability of annual precipitation

49. World's greatest 12-month rainfall [2647 cm (1042 in)] and 1-month rainfall [930 cm (366 in)]; Asia's greatest average yearly precipitation [1143 cm (450 in)]: Cherrapunji, India

50. Miyakojima Island, Ryukyu Islands, had a peak gust of 306 km/h (190 mi/h)

51. Northern hemisphere's greatest 24-h rainfall [125 cm (49 in)]: Paishih, Taiwan

52. World's lowest sea-level air pressure [87.00 kPa (25.69 in)] (estimated by dropsonde in eye of typhoon tip): in area of 17°N 138°E

53. Bogor, Indonesia, averaged 322 days/yr with thunderstorms

54. Marble Bar, Western Australia, had temperatures of 38°C (100°F) or above on 162 consecutive days

55. Australia's highest temperature [53°C (128°F)]: Cloncurry, Queensland

56. Australia's lowest average yearly precipitation [10 cm (4.05 in)]: Mulka, South Australia

57. Australia's greatest average yearly precipitation [455 cm (179 in)]: Tully, Queensland

58. Australia's greatest 24-h rainfall [91 cm (36 in)]: Crohamhurst, Queensland

59. Australia's lowest temperature [−22°C (−8°F)]: Charlotte Pass, New South Wales

60. South Pole's highest temperature: $-14°C$ ($7.5°F$)

61. South Pole has 463 W/m² (955 ly) average daily insolation in December

62. Plateau Station, Antarctica, had a mean temperature for a month of $-73°C$ ($-100°F$) and an annual mean temperature of $-57°C$ ($-70°F$)

63. World's lowest temperature [$-89°C$ ($-129°F$)]: Vostok, Antarctica

64. Port Martin, Antarctica, had a mean wind speed for 24 h of 174 km/h (108 mi/h) and a mean wind speed for a month of 105 km/h (65 mi/h)

65. Vanda Station, Antarctica, had a maximum temperature of $15°C$ ($59°F$) (possibly Antarctica's highest)

NOTES

NOTES

ABBREVIATIONS AND SYMBOLS

ABBREVIATIONS AND SYMBOLS

TABLE 16.1   Abbreviations of Organizations and Agencies

The names of technical societies, trade organizations, and governmental agencies are abbreviated as follows in HVAC documents:

AABC	Associated Air Balance Association
ABMA	American Boiler Manufacturers Association
AGA	American Gas Association
ADC	Air Diffusion Council
AMCA	Air Moving and Conditioning Association
ANSI	American National Standards Institute
ASHRAE	American Society of Heating, Refrigerating, and Air-Conditioning Engineers
ASME	American Society of Mechanical Engineers
ASTM	American Society for Testing and Materials
AWWA	American Water Works Association
FIA	Factory Insurance Association
FM	Factory Mutual Insurance Company
IBR	Institute of Boiler and Radiator Manufacturers
MCAA	Mechanical Contractors Association of America
NFPA	National Fire Protection Association
NBS	National Bureau of Standards
NEC	National Electrical Code NFPA pamphlet No. 70
NEMA	National Electrical Manufacturers Association
SBI	Steel Boiler Institute
SMACNA	Sheet Metal and Air Conditioning Contractors National Association
UL	Underwriters' Laboratories Incorporated

SOURCE: A. M. Khashab, *Heating, Ventilating, and Air Conditioning Systems Estimating Manual*, 2d ed., McGraw-Hill, New York, © 1984. Used with permission of the publisher.

TABLE 16.2   General Abbreviations

HVAC terms and units are abbreviated as follows. SI and ANSI abbreviations are shown in parentheses when they differ from the common HVAC abbreviations

ac	Alternating current
ACU	Air-conditioning unit
AHU	Air-handling unit
amp (A)	Ampere
ATC	Automatic temperature control
atm	Atmospheric
auto	Automatic
avg	Average
AWG	American wire gauge
BE	Beveled end
BF	Board foot
bhp	Brake horsepower or boiler horsepower
bldg	Building
bsmt	Basement
Btu	British thermal unit
Btuh (Btu/h)	British thermal unit per hour
°C	Degree centigrade (celsius)
CC	Cooling coil
C-C	Center to center
cc (cm³)	Cubic centimeter
cf (ft³)	Cubic foot
cfm (ft³/min)	Cubic feet per minute
CH	Chiller
CI	Cast iron
cm	Centimeter

TABLE 16.2 General Abbreviations (*continued*)

const	Construction
CS	Commercial standard
CT	Cooling tower
cu ft (ft³)	Cubic foot
cu in (in³)	Cubic inch
cu m (m³)	Cubic meter
cu yd (yd³)	Cubic yard
CV	Constant volume
CY (yd³)	Cubic yard
db (dB)	Decibel
DBT	Dry-bulb temperature
dc	Direct current
DD	Dual duct
deg (°C or °F)	Degree
dia	Diameter
dia-in	Diameter inch
dia-mm	Diameter millimeter
disch	Discharge
dist	Distribution
dn	Down
DPT	Dew-point temperature
DWV	Drainage, waste, vent
D-X	Direct-expansion
EDR	Equivalent direct radiation
eff	Efficiency

SOURCE: A. M. Khashab, *Heating, Ventilating, and Air Conditioning Systems Estimating Manual*, 2d ed., McGraw-Hill, New York, © 1984. Used with permission of the publisher.

TABLE 16.3   Mechanical Drawing Symbols

NOTE: THESE ARE STANDARD SYMBOLS AND MAY NOT ALL APPEAR ON THE PROJECT DRAWINGS; HOWEVER, WHEREVER THE SYMBOL ON THE PROJECT DRAWINGS OCCURS, THE ITEM SHALL BE PROVIDED AND INSTALLED.

Symbol	Description	Abbrev.	Meaning
——S——	STEAM PIPE	MBH	THOUSAND BTU PER HOUR
——C——	CONDENSATE RETURN PIPE	GPM	GALLONS PER MINUTE
——HWS——	HOT WATER SUPPLY PIPE	CFM	CUBIC FEET PER MINUTE
——HWR——	HOT WATER RETURN PIPE	○	ROUND
——CWS——	CHILLED WATER SUPPLY PIPE	□	SQUARE
——CWR——	CHILLED WATER RETURN PIPE	SA	SUPPLY AIR
——HCS——	COMB HOT - CHILLED WATER SUPPLY	RA	RETURN AIR
——HCR——	COMB HOT - CHILLED WATER RETURN	OA	OUTSIDE AIR
——CS——	CONDENSER WATER SUPPLY PIPE	EA	EXHAUST AIR
——CR——	CONDENSER WATER RETURN PIPE	HSWR	HIGH SIDEWALL REGISTER
——D——	DRAIN PIPE FROM COOLING COIL	HSWG	HIGH SIDEWALL GRILLE
——FOS——	FUEL OIL SUPPLY PIPE	LSWR	LOW SIDEWALL REGISTER
——FOR——	FUEL OIL RETURN PIPE	LSWG	LOW SIDEWALL GRILLE
——R——	REFRIGERANT PIPE	CSR	CEILING SUPPLY REGISTER
⊙	PIPE RISING	CR	CEILING REGISTER
⊙	PIPE TURNING DOWN	CG	CEILING GRILLE
		FR	FLOOR REGISTER
	UNION	FG	FLOOR GRILLE
	REDUCER - CONCENTRIC	CD	CEILING DIFFUSER
	REDUCER - ECCENTRIC	TV	TURNING VANES
	STRAINER	AE	AIR EXTRACTOR
	GATE VALVE	SD	SPLITTER DAMPER
	GLOBE VALVE	MD	MANUAL DAMPER
	VALVE IN RISER		

	Symbol	Description
		(cut off) IN TANK
	PRESSURE RELIEF VALVE	
	SQUARE HEAD COCK	
	BALANCING VALVE	
	3-WAY CONTROL VALVE	
	2-WAY CONTROL VALVE	
	PITCH PIPE MINIMUM 1"/40'	
	ANCHOR LOCATION	
	FLEXIBLE PIPE CONNECTION	
	IN-LINE PUMP	
	BOTTOM TAKE-OFF	
	TOP TAKE-OFF	
	PRESSURE GAUGE	
	THERMOMETER	
	HOT WATER RISER	
	CHILLED WATER RISER	
	FAN COIL UNIT	
	EQUIPMENT AS INDICATED	
	AIR INTO REGISTER	
	AIR OUT OF REGISTER	
	AIR FLOW THRU UNDERCUT OR LOUVERED DOOR	
	TURNING VANES	
	AIR EXTRACTOR	

AHU	AIR HANDLING UNIT
BU	BLOWER UNIT
FCU	FAN COIL UNIT
HWC	HOT WATER CONVECTOR
UV	UNIT VENTILATOR
WH	WALL HEATER
UH	UNIT HEATER
WF	WALL FIN RADIATION
PRV	POWER ROOF VENTILATOR
UVS	UTILITY VENT SET
PF	PROPELLER FAN
⊖T	THERMOSTAT
⊖N	NIGHT THERMOSTAT
⊖M	THERMOSTAT - HEATING ONLY
⊖C	THERMOSTAT - COOLING ONLY
⊖	THERMOSTAT - REMOTE BULB
6'-8"	MOUNTING HEIGHT ABOVE FINISHED FLOOR
NIC	NOT IN CONTRACT
	SUPPLY AIR DUCT SECTION
	RETURN OR EXHAUST DUCT SECTION
	FLEXIBLE DUCT CONNECTION

SOURCE: John E. Traister, *Practical Drafting for the HVAC Trades*, Prentice-Hall, Englewood Cliffs, NJ. Used with permission of the author.

SYMBOL MEANING	SYMBOL	SYMBOL MEANING	SYMBOL
POINT OF CHANGE IN DUCT CONSTRUCTION (BY STATIC PRESSURE CLASS)		SUPPLY GRILLE (SG)	20 · 12 SG / 700 CFM
DUCT (1ST FIGURE, SIDE SHOWN 2ND FIGURE, SIDE NOT SHOWN)	20 · 12	RETURN (RG) OR EXHAUST (EG) GRILLE (NOTE AT FLR OR GLG)	20 · 12 RG / 700 CFM
ACOUSTICAL LINING DUCT DIMENSIONS FOR NET FREE AREA		SUPPLY REGISTER (SR) (A GRILLE · INTEGRAL VOL CONTROL)	20 · 12 SR / 700 CFM
DIRECTION OF FLOW		EXHAUST OR RETURN AIR INLET CEILING (INDICATE TYPE)	20 · 12 GR / 700 CFM
DUCT SECTION (SUPPLY)	S 30 · 12	SUPPLY OUTLET, CEILING, ROUND (TYPE AS SPECIFIED) INDICATE FLOW DIRECTION	20 / 700 CFM
DUCT SECTION (EXHAUST OR RETURN)	E OR R / 20 · 12	SUPPLY OUTLET, CEILING, SQUARE (TYPE AS SPECIFIED) INDICATE FLOW DIRECTION	12 · 12 / 700 CFM
INCLINED RISE (R) OR DROP (D) ARROW IN DIRECTION OF AIR FLOW	R	TERMINAL UNIT (GIVE TYPE AND OR SCHEDULE)	T U
TRANSITIONS GIVE SIZES NOTE F O T FLAT ON TOP OR F O B FLAT ON BOTTOM IF APPLICABLE		COMBINATION DIFFUSER AND LIGHT FIXTURE	
STANDARD BRANCH FOR SUPPLY & RETURN (NO SPLITTER)	S R	DOOR GRILLE	DG / 12 · 6
SPLITTER DAMPER		SOUND TRAP	ST
VOLUME DAMPER MANUAL OPERATION	VD	FAN & MOTOR WITH BELT GUARD & FLEXIBLE CONNECTIONS	
AUTOMATIC DAMPERS MOTOR OPERATED	SEC MOD	VENTILATING UNIT (TYPE AS SPECIFIED)	
ACCESS DOOR (AD) ACCESS PANEL (AP)	OR AD	UNIT HEATER (DOWNBLAST)	
FIRE DAMPER SHOW ◄VERTICAL POS SHOW ◆HORIZ POS	FD AD	UNIT HEATER (HORIZONTAL)	
SMOKE DAMPER	AD SD	UNIT HEATER (CENTRIFUGAL FAN) PLAN	
CEILING DAMPER OR ALTERNATE PROTECTION FOR FIRE RATED CLG		THERMOSTAT	T
TURNING VANES		POWER OR GRAVITY ROOF VENTILATOR-EXHAUST (ERV)	
FLEXIBLE DUCT FLEXIBLE CONNECTION		POWER OR GRAVITY ROOF VENTILATOR-INTAKE (SRV)	
GOOSENECK HOOD (COWL)		POWER OR GRAVITY ROOF VENTILATOR-LOUVERED	
BACK DRAFT DAMPER	BDD	LOUVERS & SCREEN	36 · 24L

FIG. 16.1 Symbols for ventilation and air conditioning. (*From HVAC Duct Construction Standards—Metal and Flexible, 1st ed.,* © 1985 by the Sheet-Metal and Air Conditioning Contractor's National Association. Used with permission of the copyright holder.)

TABLE 16.4 Graphical Symbols for Drawing Heating, Air Conditioning, Refrigeration, and Plumbing Piping System Plans

Heating

High Pressure Steam	——HPS——
Medium Pressure Steam	——MPS——
Low Pressure Steam	——LPS——
High Pressure Return	——HPR——
Medium Pressure Return	——MPR——
Low Pressure Return	——LPR——
Boiler Blow Down	——BBD——
Condensate Pump Discharge	——CP——
Vacuum Pump Discharge	——VPD——
Makeup Water	——MU——
Air Relief Line (Vent)	——V——
Fuel Oil Flow	——FOF——
Fuel Oil Return	——FOR——
Fuel Oil Tank Vent	——FOV——
Low Temperature Hot Water Supply	——HWS——
Medium Temperature Hot Water Supply	——MTWS——
High Temperature Hot Water Supply	——HTWS——
Low Temperature Hot Water Return	——HWR——
Medium Temperature Hot Water Return	——MTWR——
High Temperature Hot Water Return	——HTWR——
Compressed Air	——A——
Vacuum (Air)	——VAC——
Existing Piping	——(NAME)——
Pipe to be Removed	—*—(NAME)—*—

Air Conditioning and Refrigeration

Refrigerant Discharge	——RD——
Refrigerant Suction	——RS——
Brine Supply	——B——
Brine Return	——BR——
Condenser Water Supply	——C——
Condenser Water Return	——CR——
Chilled Water Supply	——CHWS——
Chilled Water Return	——CHWR——
Fill Line	——FILL——
Humidification Line	——H——
Drain	——D——

Plumbing

Soil, Waste, or Leader (Above Grade)	————
Soil, Waste, or Leader (Below Grade)	————
Vent	————
Cold Water	————
Hot Water	————
Hot Water Return	————
Gas	——G——G——
Acid Waste	——ACID——
Drinking Water Flow	——DW——
Drinking Water Return	——DWR——
Vacuum (Air)	——VAC——
Compressed Air	——A——
Chemical Supply Pipes	——(NAME+)——

SOURCE: Reproduced by permission from *ASHRAE Handbook—1981 Fundamentals.*

TABLE 16.5   Graphical Symbols for Drawing Fittings

The following fittings are shown with screwed connections. The symbol for the body of a fitting is the same for all types of connections, unless otherwise specified. The types of connections are often specified for a range of pipe sizes, but are shown with the fitting symbol where required. For example, an elbow would be:

Flanged   Screwed   Belt & Spigot

Welded[2]   Soldered   Solvent Cement

Fitting	Symbol
Bushing	
Cap	
Connection, Bottom	
Connection, Top	
Coupling (Joint)	
Cross	
Elbow, 90°	
Elbow, 45°	
Elbow, turned up	
Elbow, turned down	
Elbow, reducing, show sizes	
Elbow, base	
Elbow, long radius	

TABLE 16.5   Graphical Symbols for Drawing Fittings (*continued*)

Elbow, double branch	
Elbow, side outlet, outlet up	
Elbow, side outlet, outlet down	
Lateral	
Reducer, concentric	
Reducer, eccentric straight invert	
Reducer, eccentric straight crown	
Tee	
Tee, outlet up	
Tee, outlet down	
Tee, reducing, show sizes	
Tee, side outlet, outlet up	
Tee, side outlet, outlet down	
Tee, single sweep	
Union	

SOURCE: Reprinted by permission from *ASHRAE Handbook—1981 Fundamentals*.

TABLE 16.6   Graphical Symbols for Drawing Piping Specialties

Air Eliminator	
Air Separator	
Alignment Guide	
Anchor, Intermediate	PA
Anchor, Main	PA
Ball Joint	
Expansion Joint	EJ-1
Expansion Loop	
Flexible Connector	
Floor Drain	FD
Flowmeter, Orifice	OFM-1
Flowmeter, Venturi	VFM-1
Flow Switch	FS
Funnel Drain, Open	
Hanger, Rod	H
Hanger, Spring	H
Heat Exchanger, Liquid	
Heat Transfer Surface, indicate type	RAD-1
Pitch of Pipe, Rise (R) Drop (D)	P→R

TABLE 16.6   (continued)

Pressure Gage and Cock	
Pressure Switch	PS
Pump, indicate use	CW-1
Pump Suction Diffuser	PSD
Spool Piece, Flanged	
Strainer	
Strainer, Blow Off	
Strainer, Duplex	
Tank, indicate use	FO
Thermometer	
Thermometer Well, only	
Thermostat, Electric	
Thermostat, Pneumatic	
Thermostat, Self-Contained	
Traps, Steam, indicate type	F&T
Trap, Water	
Unit Heater, indicate type	UH

SOURCE: Reprinted by permission from *ASHRAE Handbook—1981 Fundamentals.*

TABLE 16.7   Graphical Symbols for Drawing Air-Moving, Refrigeration, Control, Auxiliary Equipment, Energy Recovery Equipment, and Power Source Devices and Components

Air Moving Devices and Components

Fans (indicate use)

Axial Flow		Turning Vanes	
Centrifugal		Detectors, Fire and/or Smoke	
Propeller		*Dampers*	
		Adjustable Blank Off	
Roof Ventilator, Intake	SRV-I	Back Draft Damper	
Roof Ventilator, Exhaust	ERV-I	Control, Electric	
Roof Ventilator, Louvered		Control, Pneumatic	
Ductwork		Fire Damper and Sleeve, provide access door	
Direction of Flow		Manual Volume	
Duct Size, first figure is side shown		Manual Splitter	
Duct Section, Positive Pressure, first figure is top		Smoke Damper, provide access door	
Duct Section, Negative Pressure			

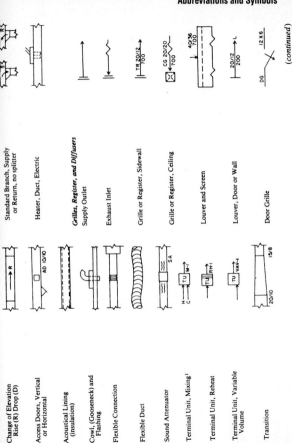

Change of Elevation Rise (R) Drop (D)	
Access Doors, Vertical or Horizontal	
	Standard Branch, Supply or Return, no splitter
	Heater, Duct, Electric
Acoustical Lining (insulation)	*Grilles, Register, and Diffusers*
	Supply Outlet
Cowl, (Gooseneck) and Flashing	Exhaust Inlet
Flexible Connection	
Flexible Duct	Grille or Register, Sidewall
	Grille or Register, Ceiling
Sound Attenuator	
	Louver and Screen
Terminal Unit, Mixing[1]	
Terminal Unit, Reheat	Louver, Door or Wall
Terminal Unit, Variable Volume	Door Grille
Transition	

(continued)

SOURCE: Reprinted by permission from *ASHRAE Handbook—1981 Fundamentals.*

TABLE 16.7 Graphical Symbols for Drawing Air-Moving, Refrigeration, Control, Auxiliary Equipment, Energy Recovery Equipment, and Power Source Devices and Components (*continued*)

Undercut Door		*Evaporators* Finned Coil	
Ceiling Diffuser, Rectangular		Forced Convection	
Ceiling Diffuser, Round		Immersion Cooling Unit	
Diffuser, Linear		Plate Coil	
Diffuser and Light Fixture Combination		Pipe Coil	
Transfer Grille Assembly		*Liquid Chillers* (Chillers only) Direct Expansion	
		Flooded	
Refrigeration		Tank, Closed	
Compressor Centrifugal		Tank, Open	
Reciprocating		*Chilling Units* Absorption	
Rotary			

Centrifugal

Reciprocating

Rotary Screw

Controls

Refrigerant Controls

Capillary Tube

Expansion Valve, Hand

Expansion Valve, Automatic

Expansion Valve, Thermostatic

Float Valve, High Side

Float Valve, Low Side

Thermal Bulb

Rotary Screw

Condensers

Air Cooled

Evaporative

Water Cooled,
specify type

Condensing Units

Air Cooled

Water Cooled

*Condenser-Evaporator
(Cascade System)*

Cooling Towers

Cooling Tower

Spray Pond

(continued)

TABLE 16.7   Graphical Symbols for Drawing Air-Moving, Refrigeration, Control, Auxiliary Equipment, Energy Recovery Equipment, and Power Source Devices and Components *(continued)*

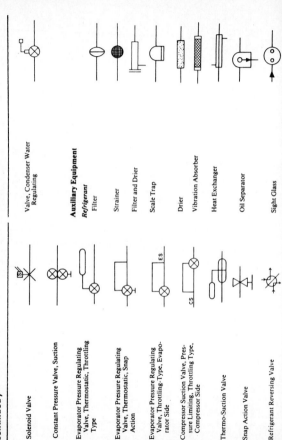

Solenoid Valve	Valve, Condenser Water Regulating
Constant Pressure Valve, Suction	**Auxiliary Equipment**
	Refrigerant
Evaporator Pressure Regulating Valve, Thermostatic, Throttling Type	Filter
	Strainer
Evaporator Pressure Regulating Valve, Thermostatic, Snap Action	Filter and Drier
	Scale Trap
Evaporator Pressure Regulating Valve, Throttling Type, Evaporator Side	Drier
	Vibration Absorber
Compressor Suction Valve, Pressure Limiting, Throttling Type, Compressor Side	Heat Exchanger
Thermo-Suction Valve	Oil Separator
Snap Action Valve	Sight Glass
Refrigerant Reversing Valve	

Temperature or Temperature-Actuated Electrical or Flow Controls

Thermostat, Self-Contained

Thermostat, Remote Bulb

Pressure of Pressure-Actuated Electrical or Flow Controls

Pressure Switch

Pressure Switch, Dual (High-Low)

Pressure Switch, Differential Oil Pressure

Automatic Reducing Valve

Automatic Bypass Valve

Valve, Pressure Reducing

Fusible Plug

Rupture Disc

Receiver, High Pressure, Horizontal

Receiver, High Pressure, Vertical

Receiver, Low Pressure

Intercooler

Intercooler/Desuperheater

Energy Recovery Equipment

Condenser, Double Bundle

(continued)

TABLE 16.7   Graphical Symbols for Drawing Air-Moving, Refrigeration, Control, Auxiliary Equipment, Energy Recovery Equipment, and Power Source Devices and Components *(continued)*

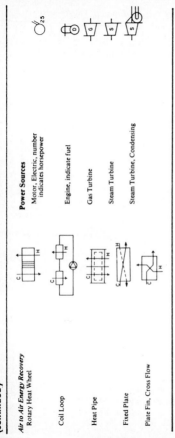

Air to Air Energy Recovery

Rotary Heat Wheel

Coil Loop

Heat Pipe

Fixed Plate

Plate Fin, Cross Flow

Power Sources

Motor, Electric, number indicates horsepower

Engine, indicate fuel

Gas Turbine

Steam Turbine

Steam Turbine, Condensing

TABLE 16.8 Graphical Symbols for Drawing Electrical Equipment

Symbols for electrical equipment shown on mechanical drawings are usually geometric figures with an appropriate name or abbreviation, with details described in the specifications. The following are some common examples.

Motor Control	☐ **MC**
Disconnect Switch	☐ **DS**
Disconnect Switch, Fused	☐ **DSF**
Time Clock	☐ **TC**
Automatic Filter Panel	☐ **AFP**
Lighting Panel	◼ **LP**
Power Panel	▨ **PP**

SOURCE: Reprinted by permission from *ASHRAE Handbook—1981 Fundamentals.*

BASIC WELD SYMBOLS

BACK	FILLET	PLUG OR SLOT	SQUARE	GROOVE OR BUTT					
				V	BEVEL	U	J	FLARE V	FLARE BEVEL
◡	△	□	‖	∨	⩗	⊻	⊬	⋁	⫫

SUPPLEMENTARY WELD SYMBOLS

BACKING	SPACER	WELD ALL AROUND	FIELD WELD	CONTOUR		For other basic and supplementary weld symbols, see AWS A2.4-79
				FLUSH	CONVEX	
▭M	▭M▭	◯	⦧	—	⌒	

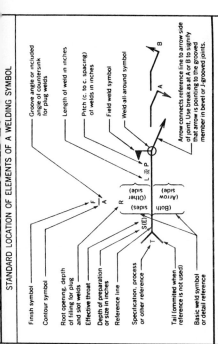

STANDARD LOCATION OF ELEMENTS OF A WELDING SYMBOL

Finish symbol

Contour symbol

Root opening, depth of filling for plug and slot welds

Effective throat

Depth of preparation or size in inches

Reference line

Specification, process or other reference

Tail (omitted when reference is not used)

Basic weld symbol or detail reference

Groove angle or included angle of countersink for plug welds

Length of weld in inches

Pitch (c. to c. spacing) of welds in inches

Field weld symbol

Weld-all-around symbol

Arrow connects reference line to arrow side of joint. Use break as at A or B to signify that arrow is pointing to the grooved member in bevel or J-grooved joints.

(Other side)

(Arrow side)

(Both sides)

Note:

Size, weld symbol, length of weld and spacing must read in that order from left to right along the reference line. Neither orientation of reference line nor location of the arrow alter this rule.

The perpendicular leg of $\triangle$, V, $\triangleright$, $\triangleright$ weld symbols must be at left.

Arrow and Other Side welds are of the same size unless otherwise shown. Dimensions of fillet welds must be shown on both the Arrow Side and the Other Side Symbol.

The point of the field weld symbol must point toward the tail.

Symbols apply between abrupt changes in direction of welding unless governed by the "all around" symbol or otherwise dimensioned.

These symbols do not explicitly provide for the case that frequently occurs in structural work, where duplicate material (such as stiffeners) occurs on the far side of a web or gusset plate. The fabricating industry has adopted this convention: that when the billing of the detail material discloses the existence of a member on the far side as well as on the near side, the welding shown for the near side shall be duplicated on the far side.

FIG. 16.2 Welded joints standard symbols. (*Reproduced courtesy American Institute of Steel Construction.*)

NOTES

INDEX

ABOUT THE AUTHOR

Robert O. Parmley, P.E., CMfgE, is President and Consulting Engineer of Morgan & Parmley, Ltd., Professional Consulting Engineers, Ladysmith, Wisconsin. Mr. Parmley is also a member of the National Society of Professional Engineers, the American Society of Mechanical Engineers, the Construction Specifications Institute, the American Institute for Design Drafting, and the Society of Manufacturing Engineers, and is listed in the AAES *Who's Who in Engineering*. He holds a BSME and a MSCE from Columbia Pacific University and is a registered professional engineer in Wisconsin, California, and Canada. Mr. Parmley is also a certified manufacturing engineer under SME's national certification program. In a career covering more than 25 years, Mr. Parmley has worked on the design and construction supervision of a wide variety of structures, systems, and machines—from dams and bridges to pollution control systems and municipal projects. The author of over 40 technical articles, he is also the Editor in Chief of the *Standard Handbook of Fastening and Joining, Field Engineer's Manual,* and *Mechanical Components Handbook,* all published by McGraw-Hill. Mr. Parmley lives in Ladysmith, Wisconsin.